Routledge Revivals

The Geometry of Environment

Originally published in 1971 *The Geometry of Environment* is a fusion of art and mathematics introducing stimulating ideas from modern geometry, using illustrations from architecture and design. The revolution in the teaching of mathematics and the advent of the computer in design challenge traditional ways of appreciating the space about us, and expand the 'structural' understanding of our surroundings through such concepts as transformations, symmetry groups, sets and graphs. This book aims to show the relevance of 'new maths' and encourages exploration of the widening intellectual horizons of environmental design and architecture.

The Geometry of Environment

An Introduction to Spatial Organization in Design

by Lionel March and Philip Steadman

Routledge Revivals
Routledge
Taylor & Francis Group

First published in 1971
by RIBA Publications Limited

This edition first published in 2020 by Routledge
2 Park Square, Milton Park, Abingdon, Oxon, OX14 4RN
and by Routledge
711 Third Avenue, New York, NY 10017

Routledge is an imprint of the Taylor & Francis Group, an informa business

Publisher's Note
The publisher has gone to great lengths to ensure the quality of this reprint but points out that some imperfections in the original copies may be apparent.

Disclaimer
The publisher has made every effort to trace copyright holders and welcomes correspondence from those they have been unable to contact.

A Library of Congress record exists under LCCN: 71887283

ISBN 13: 978-0-367-36023-8 (hbk)
ISBN 13: 978-0-429-34334-6 (ebk)

MIX
Paper from responsible sources
FSC™ C013985
www.fsc.org

Printed in the United Kingdom
by Henry Ling Limited

The geometry of environment

Lionel March and Philip Steadman

The geometry of environment

An introduction to spatial organization in design

RIBA Publications Limited

First published 1971

Designed and published by RIBA Publications Limited
66 Portland Place London W1N 4AD

Contents

Foreword

by Sir Leslie Martin, Professor of Architecture, University of Cambridge

This is a book about the new mathematics and architecture. One of its stated objectives is to suggest to the younger reader with a mathematical background that architecture can be an interesting and possibly an absorbing subject. The point is that if mathematics is thought of as a 'logical pattern of entities and relationships' then these may perhaps be seen to be reflected in the physical and spatial arrangement of buildings. Indeed precisely this same phrase 'a logical pattern of entities and relationships' built around activities might well be used as a generalized definition of architecture. And if we think of the subject in this way then there is no doubt about the value of this book.

But it goes beyond that. To recognize and to admit this relationship is to deepen and stimulate a whole area of thought about architecture. We become aware of another way of looking at a design problem through which we can consider more effectively and rigorously the ranges of choice that are open to us. The study which may start with a building is found to be one which extends right through the whole environmental field.

This is indeed an area of theoretical study to which this book makes a fundamental contribution. It provides a base from which to build a more systematic design theory and indicates a direction which this might follow with advantage.

Preface

In the past, geometry and architecture have been seen to have much in common. Indeed, it seems evident that recurring practical problems encountered by early builders such as those of Egypt and Greece led to geometrical discoveries. Today, however, there is not just one kind of geometry but many, and the architect is unlikely to find many of them of direct use to him. He will still, like his engineering colleagues, use descriptive geometry to enable him to draw plans, elevations and projections of his scheme. It is possible that space-frame and geodesic dome structures, although limited in their application, will bring some architects in touch with the geometrical properties of space-filling polygons and polyhedra which pack together. Other architects may use a variety of geometrical devices, mostly concerned with congruence and similarity, to sharpen the sense of aesthetic order in their work. This book does not attempt to cover these uses of geometry in architecture although we occasionally refer to them. We believe that they are dealt with adequately elsewhere.

Our specific concern is to introduce the student of architecture, whatever his age or experience, to some of the concepts of new mathematics which seem to us to have potential value in describing and helping us to understand some of the geometrical relationships which arise when we organize shape within buildings. We could have expanded the field of environment to cover urban and regional systems, but a number of recent books illustrate extremely well, if sometimes implicitly, the direction such a study of geometry and the geographical environment might take. Not only would our contribution here have been redundant, but it would have required us inevitably to discuss probability theory and geometrical probability in particular. Such a discussion would have made the book unwieldy and, we suspect, off-putting for the non-mathematical reader.

Our aim is twofold: one, to help bridge the gap between the new mathematics and the older generation; and two, to suggest to the youthful reader, perhaps with a science and mathematics background, that architecture is an exciting subject – it is neither wholly looking at old churches, nor laboriously calculating stresses in beams and loads in columns. We would expect that the book would be of value in sixth-forms as an introduction to modern ideas of architectural form and spatial organization; in schools of architecture, planning and environmental studies as a course suggesting new mathematical methodologies in design; and in professional practice as a stimulus to thought and inquiry. We hope that this introduction will suggest ways in which modern geometry can contribute to the progress of architectural design, especially at a time when computer-aids are developing so rapidly.

In attempting to achieve our purpose for the architectural student we have deliberately avoided a rigorous approach. At the same time we have tried to convey a *feeling* for mathematical structure – a logical pattern of entities and relationships. Deliberately, we have not avoided mathematical notation, but wherever possible mathematical statements are accompanied by architectural illustrations. In our own experience this helps enormously, and as soon as confidence is gained in reading mathematical notation it is so much more rewarding to follow up specialist works by mathematicians themselves. On the other hand, for readers who are not architects we have limited our illustrations, in the main, to the works of a few well-known international architects. Because these works represent distinguished contributions to architectural design, the reader should be able to find out more about them from books which are readily available in libraries and the larger bookshops.

Perhaps the chief difference between the traditional treatment of geometry in architecture and the one presented here, is that, previously, geometry was employed to *measure* properties of space such as area, volume, angle, whereas the new mathematical theories of sets, groups and graphs – to name but a few – enable us to describe *structural relationships* which cannot be expressed in metrical forms, for example, 'adjacent to', 'in the neighbourhood of', 'contained by'. Claude Lévi-Strauss has drawn attention to a similar trend in the social sciences where the growth of structural studies is seen to be 'the direct outcome of modern developments in mathematics which have given increasing importance to the qualitative point of view in contradistinction to the quantitative point of view of traditional mathematics.' The advent of computer methods has meant that architectural elements and relationships need to be given mathematical representation of some kind. This book does not concern itself with computing as such, but it does introduce some of the algebraic and geometrical structures of the new mathematics which appear to be similar to – or isomorphic with – physical and spatial aspects of buildings.

The first notable application of the new mathematics in architectural design occurs in Christopher Alexander's outstanding *Notes on the Synthesis of Form*. Since its publication in 1964, Alexander and his colleagues have made further important contributions to design methods and to what, perhaps prematurely, became known as relational theory. To Alexander must go the credit of breaking down the barriers of prejudice and habit, but of those who followed him past the barriers not all have had the advantages of his dual training in mathematics and architecture and, worse, not all have maintained the scrupulous self-critical attitude of their leader. For example, in the aftermath, there grew up an extreme point of view which seemed to be claiming that the objective structural analysis of the functional requirements of a social organization would, *ipso facto*, generate the design of the building or environment to accommodate it. That is to say, if we knew enough about the elaborate relations existing between pupils, pupils and staff, members of staff and so on, we could design a school. Alexander himself never took this view. In his more recent papers he has stressed the importance of the geometrical constraints which a particular design solution impose on a human situation. Nevertheless, it is true that his work tends to emphasize functional factors rather than formal. This emphasis is surely right in a field where so much guesswork and untutored intuition have ruled, for lack of a systematic approach. In the confusion designers find it necessary to grasp at the more easily comprehended formal attributes to achieve anything worthwhile at all. Our hope is that this introduction to the geometry of spatial organization will complement Alexander's work to some degree. Certainly we would recommend *Notes on the Synthesis of Form* to our readers in its own right, but particularly to counter our total disregard in this book for any but the most simple functional requirements of a design programme. Without doubt, geometry may be used to make fun shapes, but in architectural design that must never be at the expense of satisfying people's psychological, social, cultural or economic desires.

This book is in fourteen chapters. The first discusses the fundamental idea of mapping. Typically, an architect's drawing can be said to be a map of the real building: transformation mappings such as isometry, similarity, affinity, perspectivity, topology are introduced. The next two chapters are concerned with symmetry as it is understood in modern mathematics and, after a historical account of its development in mineralogy, crystallography, and plant and animal morphology, the notions of automorphic groups, of rotations, of reflections, and translations are discussed. The fourth chapter introduces vectors and matrices and establishes the mathematical foundations for some of the subsequent topics.

The next three chapters, 5, 6, and 7, centre on problems related to the description of shape. They describe some of the difficulties encountered in trying to give mathematical expression to shape. The chapters progress from a discussion of modular and rectangular forms used to introduce set-theoretic concepts, to non-modular, rectangular spaces and, finally, to irregular polygonal shapes. Set theory, vectors and matrices are employed, together with the concept of convexity which is basic to linear programming. Number theory and modular coordination are the themes of Chapter 8, which takes the form of a review of number combinations and their permutations in the general context of Diophantine equations. Chapter 9 attempts to remove much of the mystique, traditional among some art historians and architects, from the subject of proportional systems, and to give some of these systems a set-theoretic formulation.

The next two chapters, 10 and 11, introduce graph theory. In particular, they develop the idea of mapping rooms or spaces onto the vertices of a graph in problems where some rooms are constrained to be next to others, a relationship which is identified by the presence or absence of an edge to the graph. The value of the graph lies in the capacity it has for showing up the essential structure of a set of relationships (edges) between a number of elements (vertices). An interesting analogy with Kirchhoff's laws for electrical flow is used to determine the plan arrangement of a house. Finally, the last three chapters, 12, 13, and 14, take a critical look at some computer techniques currently under development for generating building plans and allocating activities, and for evaluating circulation patterns. These chapters illustrate some uses of classification trees and matrix representations, and demonstrate the immense combinatorial problems some automatic design methods imply. The final chapters also serve to introduce some of the algorithmic procedures employed in computing, for example, shortest paths through networks.

We include specific references in each chapter, but more general reading will be found at the end of the book together with a list of the mathematical symbols which we use.

We would like to thank the Professional Literature Committee of the Royal Institute of British Architects for identifying this particular 'gap' in the literature, and for inviting us to attempt to fill it. Our colleagues at the School of Architecture and at the centre for Land Use and Built Form Studies in Cambridge have all contributed to the work, sometimes explicitly and on such occasions specific references will be found at the appropriate point in the text, but more often unwittingly by their good humour and constant intellectual stimulation. Especially, we wish to acknowledge the help that our colleague, Philip Tabor, has given us in preparing Chapters 12, 13, and 14. Substantial sections of these chapters are based on his original research.

We should also remark on the curious sustenance derived from the *genius loci* of our University. We are conscious of both past influences and common endeavours. Although Christopher Alexander graduated from Cambridge some years ago, his presence is never far away, constantly challenging and provoking. His has been a tremendous influence. Then we must mention Bruce Martin, Colin Rowe and Colin St John Wilson who as lecturers both stimulated and infuriated us in a way which made sure that at the very least we thought for ourselves, but who also drew our attention, in their diverse styles, to the intellectual content of architectural discipline. Outside our own faculty, 'quantitative revolutions' in the humanities have been in the air in Cambridge for some time, in linguistics, economics, geography, history of populations, archaeology and anthropology. These activities and concurrent developments in the sciences were bound to impinge on architecture in the end. Above all, however, we are greatly indebted to Sir Leslie Martin, Professor of Architecture, who has given us unbounded encouragement and support in our work for several years now.

Many people helped in the preparation of the copy. However, there is one in particular, Catherine Cooke, without whose enthusiasm, intelligence and skill in drawing and designing more than half the illustrations the whole enterprise might well have foundered. Numerous typists aided us at various times, but in the end most of the work was done by Mrs Hogg and Mrs Skoyles to whom we are especially grateful.

Lionel March
Philip Steadman
Cambridge, 1970

1 Mappings and transformations

We are all familiar with the idea of a map. The architect's drawing may be considered to be a map. In the case of a measured drawing from an existing building the draughtsman plots selected points of the real building, such as roof lines and the corners of openings, and 'maps' these onto his drawing. There is in this instance a one-to-one correspondence between the points in reality and their representation on the drawing, and vice versa. In a mapping mathematically understood, this is not always so. Nor does a mapping necessarily preserve spatial characteristics such as length, area, angles, sense (left-handedness, right-handedness) and shape.

The Union Jack has found new popularity as an ornamental design. These uses are all mappings of the original. Any straight replication is a mapping and the geometry of the original is unchanged except for size. When the flag is stretched around a conical waste-paper bin it is distorted, as it is when it is squashed into a circle on a clock-face. In these examples, whilst a one-to-one correspondence is maintained, the only geometrical property which is preserved – apart from sense and that not always – is that of neighbourliness. Points which are near-neighbours in the original remain near-neighbours in the various mappings. We say that these mappings are topologically equivalent. When the flag is cut up to make a jig-saw puzzle there is again one-to-one correspondence: but unassembled, for the flag as a whole, the topological equivalence has gone, together with all the other geometrical attributes. Finally we have seen shirts and hats made up from Union Jacks. Usually one-to-one correspondence is not maintained since it is necessary, in order that the clothes fit, to add gussets, to cut out holes and to take in tucks and darts. Thus there may well be pieces of the original flag design used more than once whilst others are not used at all.

Mappings need not be visual representations as our examples have been so far. When an Englishman says his living-room is on the first floor, his American friend may imagine that it opens directly onto the garden. This is because the American first floor is the English ground floor, the American second floor the English first, and so on. Such a mapping is very simple and can be represented by means of the notation $a \to b$, meaning 'a maps onto b', thus

$$0 \to 1,\ 1 \to 2,\ 2 \to 3, \ldots,\ n \to n + 1$$

or more concisely

$$E \to A = E + 1$$

where E stands for the English description and A the American. The mapping is clearly one-to-one since

$$A \to E = A - 1.$$

Incidentally, the English word 'storey' maps onto the American spelling 'story', but this mapping is seen not to be one-to-one since given just the American word 'story' we would not know which of the two English words 'storey' or 'story' was meant.

A change of units gives us another example of mapping. Thus with metrication of imperial units we map measures in one system onto measures in another. For example,

$$I \to M = 304{\cdot}8I$$

maps imperial feet onto metric millimetres, but for practical purposes in the building industry it has been proposed that the mapping take the form

$$I \to M = 300I$$

where the traditional foot measure will in future be approximated by a 300-mm module.

When a draughtsman measures the window opening as being one metre wide he will draw a line one centimetre long to represent it if his chosen scale is 1:100. If the scale is $l : k$, the real length L is represented by the mapped length M measured in the same units according to the mapping

$$L \to M = \frac{l}{k}.L.$$

The mapped length may sometimes be in different units as it used to be when a designer spoke of the 'eighth scale'. Then he referred to a scale which mapped feet onto one-eighths of an inch, a scale of 1 : 96 when the same units are used, that is to say, 96 ft in reality are represented by 1 ft on the plan. It is obviously essential that this kind of mapping be one-to-one.

Similarly, we use percentages to reduce a set of figures to some common frame of reference. Thus

$$S \rightarrow P \text{ where } P_i = 100s_i / \Sigma s_i$$

and the ith figure s_i in the set of figures S maps onto the percentage P_i of the set P of percentages. P_i is given by the expression on the right where Σs_i is the sum total of the original figures for all i. So far we have been rather slack in our use of notation. We certainly do not intend to be rigorous in a strict mathematical manner any more than is necessary to convey a feeling for quantitative appreciation.

Let us define a *set* as a well-defined collection of objects known as the *elements* or members of the set. Sets are usually denoted by capital letters and their elements by lower-case letters. When we want to spell out a set in full we write the elements between braces in the following way:

$$A = \{a, m, o, r\}.$$

Order is not important in a set. Thus if

$$B = \{r, o, m, a\}$$

we may agree with Sophia Loren that $A = B$. It is often necessary to express the fact that an element belongs to a set and to do this we use a symbol $\in$ as shorthand for 'exists in', thus

$$a \in A.$$

But

$$b \notin A$$

uses the slash through a sign to state that element b does *not* exist in the set. A vertical bar, |, is used as shorthand for 'such that'. Going back to the percentage mapping we could have expressed the mapping more precisely as

$$M\colon \{s \in S\} = S \rightarrow P = \{p \in P \mid p = 100s/\Sigma s\}$$

or in plain English: 'The mapping, M, takes the set, S, comprising elements s onto the set, P, of percentages consisting of the elements p such that p equals one hundred times s divided by the total sum of all the elements s.'

In many cases percentages can be a useful scale of comparison, but they can be misleading. An example will make this clear. An English architect studying archaeological remains in Rome was worried by the prospect of the midday sun. Before leaving his hotel he asked the porter how hot it would be later on in the day. The porter said that the temperature would be 100 % higher at noon than at breakfast. The Englishman thought the Italian must be boasting and placed a bet against such an event. At noon, however, the temperature was indeed 100 % higher and the Englishman had to admit he had lost, although in his opinion it had only gone up 45 %. Why? Quite simply, the porter used the Celsius scale and the Englishman the Fahrenheit. The mapping of one onto the other may be represented

$$M\colon \{c \in C\} = C \rightarrow F = \{f \in F \mid f = 1{\cdot}8c + 32\}$$

and when the temperature rose from 15 °C in the morning to 30 °C at noon, the Italian could claim that it had risen by 100 %. For the Englishman, however, the temperature had gone from 59 °F to 86 °F – a mere 45 %. This demonstrates the workings of the political trick whereby one party claims great economic advancement and the other points to stagnation. The problem arises not so much because of the scale on which things are measured but of the base, or zero, from which measurements are taken. Politicians naturally tend to adopt that year as base which most enhances the result they are demonstrating.

Throughout this book we shall have occasion to refer to sets of numbers. We shall always be concerned with real numbers. The *real number system*, R, may be visualized as a one-to-one mapping of numbers onto points of a straight line. We mark an origin on the line corresponding to the number zero. We then choose any other point to the right to correspond to the number 1. This gives us a unit of measure. By marking further points at unit intervals to the right we generate the *natural numbers* $\{1, 2, 3, 4, \ldots\}$. We call this set N.

Figure 1.1

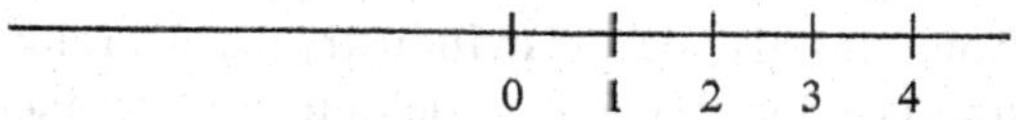

By marking off unit intervals to the left we map out the negative integers. The set of all negative and positive integers $\{\ldots -4, -3, -2, -1, 0, 1, 2, 3, 4, \ldots\}$ is usually referred to simply as the set of *integers*, Z.

Figure 1.2

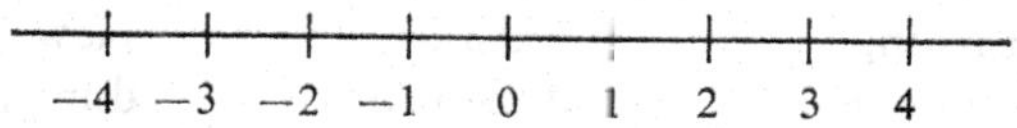

We will have occasion to refer to the *positive integers* (*including zero*) and for this we shall use the character Z_+. If we now subdivide these unit intervals in integral proportions we shall produce a mapping of the *rational numbers* for which it is customary to use the letter Q. We have $1/4 \in Q$, $-8/11 \in Q$, and, in general, $m/n \in Q$ if $m, n \in Z$.

Numbers such as $\sqrt{2}$ or π which cannot be expressed as the ratio of two whole numbers are said to be *irrational*, Q'. Every point of our line now has a corresponding number in the real number system, a system which includes the natural numbers, the integers, the rational numbers and the irrational. To summarize:

$$\begin{aligned}
N &= \{1, 2, 3, \ldots\} \\
Z &= \{0, \pm 1, \pm 2, \pm 3, \ldots\} \\
Z_+ &= \{0, 1, 2, 3, \ldots\} \\
Q &= \{m/n \mid m, n \in Z\} \\
Q' &= \{x \in R \mid x \notin Q\}.
\end{aligned}$$

The last statement says that the set of irrational numbers Q' consists of elements x in the real number system such that x is not a rational number in Q. Clearly $N, Z, Z_+, Q, Q' \in R$.

Frequently, in practical situations, it is necessary to 'round' figures to the nearest fraction or decimal part within a given degree of accuracy. For example,

To the nearest	0·001	0·1	0·25	1·0
$\pi = 3\cdot1415927...$	3·142	3·1	3·25	3·0

Rounding also occurs in industry when a range of *ad hoc* component sizes is reduced to a set of definite modular increments. Suppose, for example, a product is made in five lengths – 5660 mm, 5630 mm, 5600 mm, 5560 mm and 5530 mm – and it is considered that in future these sizes may be limited to modular dimensions based on 100 mm, then we map (round) the set of existing dimensions into new ones:

Existing components	Existing dimensions		New dimensions	New components
A	5660	→	5700	*F*
B	5630 }			
C	5600 }	→	5600	*G*
D	5560 }			
E	5530	→	5500	*H*

Another kind of mapping, which is also many-to-one, occurs in clock arithmetic. We are all familiar with the way in which we usually enumerate the time of the day 'ten o'clock, eleven o'clock, twelve noon, one o'clock, two o'clock' unless we are travelling, when schedules are listed on the 24-hour system as 'ten hundred hours, eleven hundred, twelve hundred, thirteen hundred, fourteen hundred.' We say that in the first instance the number of hours is measured *modulo 12*, and in the second *modulo 24*. In general,

$$a = b \text{ modulo } r$$

means that a, or the quantity b modulo r (more briefly b mod r), is the remainder after b has been divided by a positive non-zero integer, r, a whole number of times. We may express this in set notation as follows

$$M\colon \{b \in Z\} = B \rightarrow A = \{a \in Z_+ \mid 0 \leqslant a = b - ir < r, i \in Z\}.$$

For example, mod 3 of the numbers 0 to 10 maps onto $\{0, 1, 2\}$

Figure 1.3

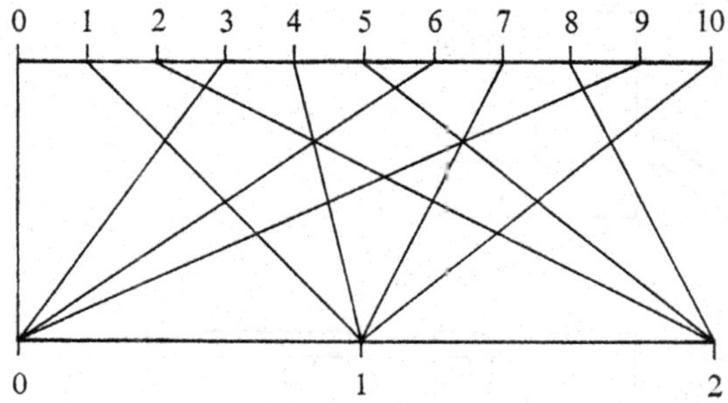

Another useful mapping occurs when we measure lengths between points. For example, the length between the point labelled 2 on the real line R and the point marked 3 is clearly $3 - 2 = 1$, but in general we obtain the result $a_1 - a_2$ for the points 'a_1' and 'a_2' and this could be a *negative* quantity if a_1 happened to be less than a_2. A negative length is meaningless, or rather it means too much. It means not only the *magnitude* of length involved, but also the *direction* in which the measurement is made, that is from right to left gives a negative quantity while left to right gives a positive. We therefore use two vertical bars, for example $|(a_1 - a_2)|$, to tell us to ignore signs and to state simply the positive *magnitude* or *absolute value* of the quantity between bars. Thus, if $i \in Z$ we could define Z_+ as

$$Z_+ = \{j \in Z \mid j = |i|, \text{for every } i \in Z\}$$

which represents the two-to-one mapping

$$i \to j = |i|.$$

We said earlier that the architect's drawings may be considered to be a map. In fact a variety of mappings are used in everyday practice by architects and designers. To start with the simplest: when a pencil drawing of an elevation is inked in, the mapping is the *identity* transformation of the elevation onto itself. This may seem trivial, but the identity mapping is of considerable importance in mathematical argument and we shall return to it in our discussion on symmetry and the description of shape.

When we take prints we produce *isometries* of the original; that is, having *iso-*, equal, *-metries*, lengths. These are the transformations that preserve lengths, but permit changes in position caused by rotation,

[1] For an introduction to Le Corbusier's work see Françoise Choay, *Le Corbusier*, New York, George Braziller Inc., 1960. The designs for Maisons Minimum are illustrated on pages 126–7 of Le Corbusier and Pierre Jeanneret, *Oeuvre Complète 1910–1929*, Zurich, Editions Girsberger, 1937.

[2] See *Oeuvre Complète 1910–1929*, pages 78–86.

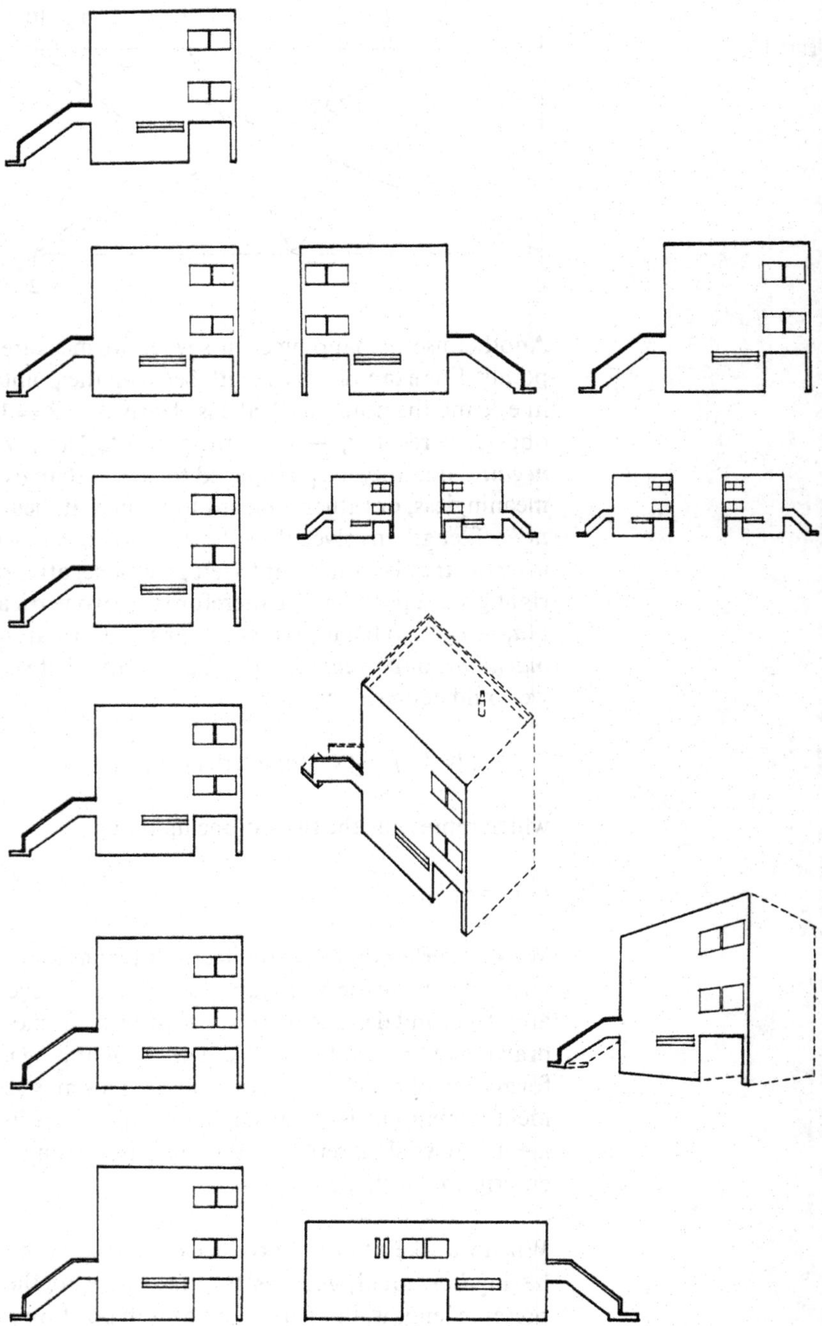

Figure 1.4
Mappings of Le Corbusier's elevation of a Maison Minimum:

a, identity

b, isometry

c, similarity

d, affinity

e, perspectivity

f, topology

Figure 1.5
Reflected house plans

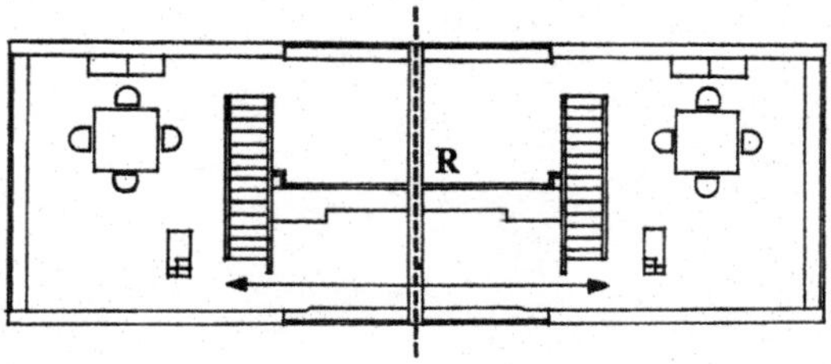

Figure 1.6
Translated house plans

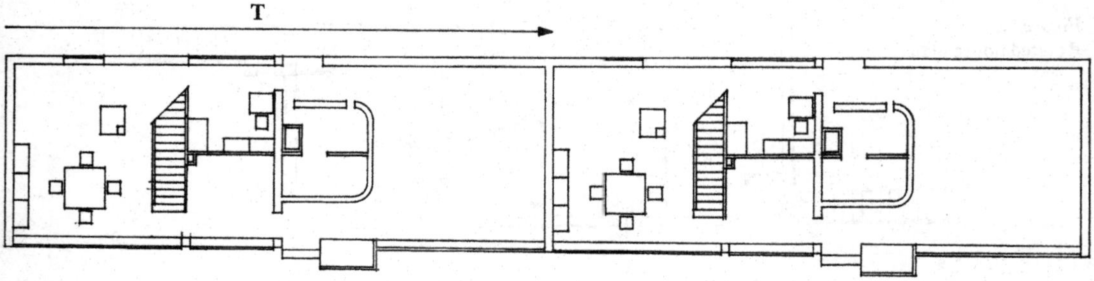

Figure 1.7
House plans showing reflection and translation

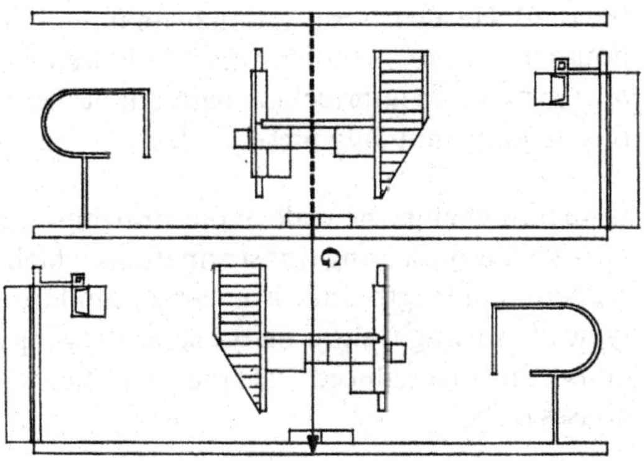

reflection or repetition of the original. If a negative is put back to front through a print-machine the lettering on the print will read backwards and the elevation will appear as if in a mirror. We say the elevation is reflected.

In Figure 1.4a we show the elevation of Le Corbusier's design for a minimum house (1925).[1] If the architect had wanted to arrange these houses along a street he might well have handed them alternately so that a pair of neighbouring houses were reflections of one another. Figure 1.4b shows such a set of isometries. We can also see isometries in the plans of semi-detached and terraced housing. At Pessac, near Bordeaux, Le Corbusier designed three types of house.[2] In one type he reflected the plan about the party wall to form two semi-detached dwellings. The two plans are identical apart from position and the fact that they are handed or *reflected* (Figure 1.5). In the second type the house is simply repeated along a line to form a terrace. The plans are not handed, and we say they are *translated* (Figure 1.6). In the third type he creates a terrace by *reflecting* (Figure 1.7) the plan about its midpoint

Figure 1.8
Rotated house plans

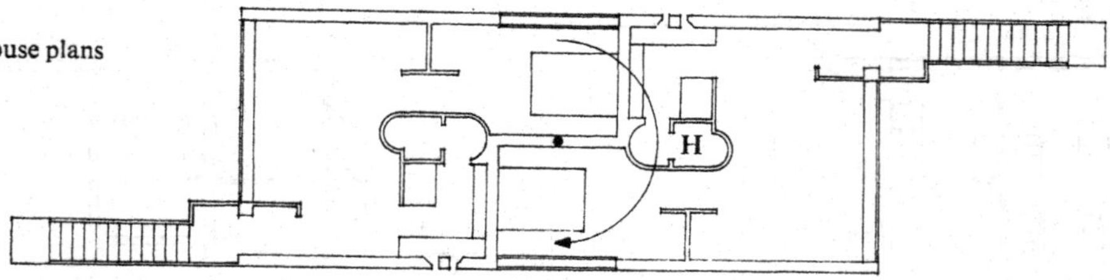

and *gliding* it one bay so that houses alternately 'front' and 'back' onto the road. He also uses rotation in an unbuilt design for a semi-detached house. Here Le Corbusier *rotates* (Figure 1.8) the plan through a *half-turn* to make an interlocking pair. All of these plans exhibit isometry: this mapping not only preserves lengths, but angles as well.

If we now change the scale of our drawings we introduce the *similarities*—angle-preserving transformations which permit enlargement or reduction of length. Like isometries, similarities may be handed. Quite typically we might show on the same drawing an elevation of a standard house, and at a reduced scale the street elevation of a series of such houses (Figure 1.4c).

When we project an elevation to give an angled view of it in such a way as to map parallel lines onto parallel lines we make what is called by mathematicians an *affine* transformation; that is, a transformation having an affinity with, or likeness to, the original. Architects frequently use *affine* transformations in which the lengths of vertical lines and those parallel to the two horizontal rectangular axes are preserved. Such a projection is often called 'isometric'. In mathematical terms this is not strictly true since the lengths of lines *not* parallel to the principal axes are not preserved. An 'isometric' drawing is not to be confused with the mathematical isometric transformation for which *all* lengths are preserved. Nevertheless, whilst an 'isometric' drawing is an example of an affine transformation it is not typical: in general an affine transformation does not preserve lengths, only parallelism. In Figure 1.4d we show an affine projection of the elevation for Le Corbusier's minimum house, in this case 'isometric'.

Figure 1.4e shows a perspective view of the same elevation. Angles and lengths are not preserved but straight lines remain straight. This is a special case of the class of projectivities to be found in modern geometry.

Figure 1.9

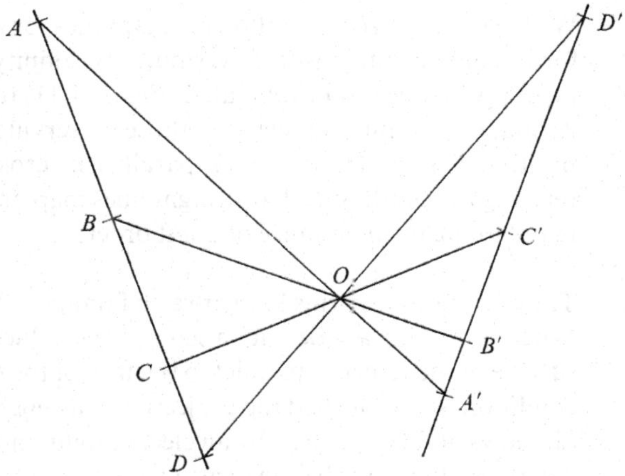

It is known as *perspectivity*. Essentially a perspectivity is a mapping (Figure 1.9) of typical points A, B, C, D... onto A', B', C', D'... so that the lines AA', BB', CC', DD'... meet at a point O called the *centre of the perspective*. It is clear that length is not preserved: $AB \neq A'B'$. Nor, unlike a similarity, is the ratio of lengths preserved: $AB/BC \neq A'B'/B'C'$. The Greeks knew, however, that 'ratios of ratios' are preserved:

$$\frac{AC/BC}{AD/BD} = \frac{A'C'/B'C'}{A'D'/B'D'}.$$

This ratio is usually abbreviated (AB,CD) and is known as the *cross-ratio*: it is fundamental in the development of projective geometry and the theory of linear transformations.

Figure 1.10

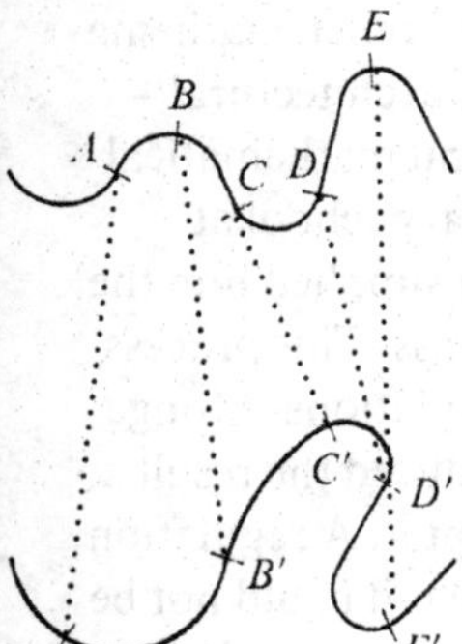

Finally in the elevation of Le Corbusier's two-storey version we have a mapping in which lengths, ratios and even ratios of ratios are not preserved. What is preserved is 'neighbourliness'. Points which are near to one another in the original elevation continue to remain near in the new elevation. Such a mapping, as we said in our early example of the Union Jack, is *topological*. Due to the preservation of 'neighbourliness', the order of a set of points along a line – not necessarily straight – is kept under a topological transformation (Figure 1.10). Thus, as we see in Figure 1.4f the semi-detached version of Le Corbusier's Maison Minimum is topologically equivalent to the detached house.

We have considered the following mappings, or more precisely, transformations: identity, isometry, similarity, affinity, perspectivity, topology. These can be tabulated (Figure 1.11) to show clearly how the various constraints are relaxed – those preserving the invariance of position, lengths, angles, ratios, parallelism, cross-ratios – finally relaxing the conditions that straight lines map onto straight lines whilst maintaining the invariance of point order.

Transformations of this kind are also familiar off the drawing-board. When projecting a slide onto a screen we produce a similarity only if the surface of the screen is parallel to the face of the slide. When this condition is not satisfied the projection is a general perspectivity. Shadows of furniture cast by an electric light onto the floor or wall are perspectivities. But shadows cast by the sun onto a flat surface are affinities. This is because the sun's rays are for most intents and purposes parallel and, consequently, parallel lines are mapped onto parallels. We see that an affine projection is a special case of a perspectivity when the centre of the perspective is at infinity.

The obstruction to daylight reaching a point within a room is found by mapping the solid buildings and objects of the external environment onto the window opening according to a perspectivity from that point. (We discuss this further at the end of Chapter 7.) This can be done because daylight, unlike sunlight, is considered to be non-directional. For sunlight, however, the obstruction is represented by an affine projection onto the window plane. Projections of this kind onto curved or irregular surfaces produce topological transformations.

Perspective drawing and shadow projection (sciagraphy) are practical applications of descriptive geometry invented by the French mathematician Gaspard Monge around 1765. It was a specific architectural – albeit military – problem that led Monge to his general mathematical formulation. One day, as a draughtsman in the military school at Mézières, he was required to work out from the data supplied him the *défilade*, the gunfire-free shadow, of a proposed fortress. This process could then be done only by lengthy arithmetical calculations. Monge substituted for this a geometrical method which produced the result so quickly that the commandant at first refused to accept it. A regulation time having been allocated, the commandant just *knew* it could not be done in less! But soon the value of the discovery was recognized.

Figure 1.11

invariant / mapping	position	length	angle and ratio	parallelism	cross-ratio	neighbourliness
identity 	●	●	●	●	●	●
isometry 		●	●	●	●	●
similarity 			●	●	●	●
affinity 				●	●	●
perspectivity 					●	●
topology 						●

[3] For a modern textbook see R. G. Robertson, *Descriptive Geometry*, London, Pitman, 1966.

[4] For a discussion of this see D. O. Hebb, *The Organisation of Behaviour, a Neuropsychological Theory*, London, Chapman and Hall, 1949.

The method is one that architects, engineers and designers use all the time, consisting simply of the orthogonal projections of a solid object onto a set of planes mutually perpendicular to each other to give plans and elevations. So familiar is this that it may come as a surprise to learn that the method was discovered as late as the mid-eighteenth century, and further that it remained a military secret for thirty years – officers instructed in the method were strictly forbidden to communicate it even to those engaged in other branches of the public service – until the Revolution, when Monge was at last free to lecture on the subject at the new *École Polytechnique*, his notes being subsequently published as *Géométrie Descriptive* in 1795.[3]

We have said that some mappings are topological. It is interesting that in human perception the patterns of cortical excitation aroused by looking at different parts of the same pattern have been shown to be topologically equivalent.[4] A square is mapped via the retina onto the cortex as a curvilinear quadrangle. Usually, in fact, there is not just one cortical pattern but two bilateral ones, depending on what point our attention is fixed upon.

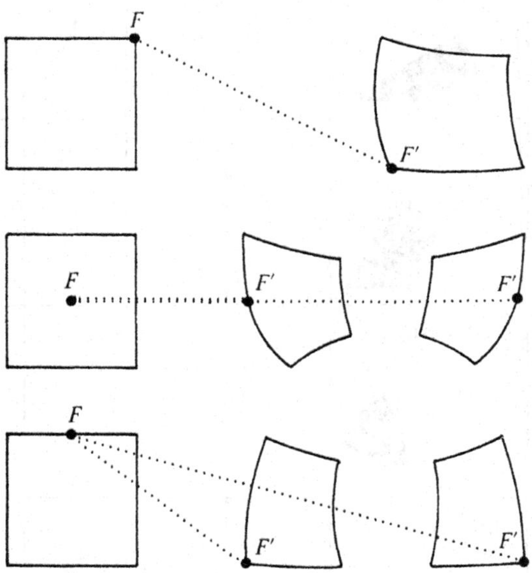

Figure 1.12
Conjectural diagram of the topological mappings in cortical projection of a square when the fixation point, F, is changed:

a, fixation on the upper right corner of the square, which thus falls in the lower left visual field and produces an excitation in the upper right cortex only

b, bilateral projection with fixation on the centre of the square

c, bilateral fixation on the midpoint of the top edges. After D. O. Hebb, 1949

With bilateral and topological projection onto the cortex in mind, we can appreciate how young children so easily confuse the lower-case letters:

b **d**
p **q.**

In many type faces, especially those used in children's readers, these letters are precise isometries of each other and children find it difficult to distinguish between them: a distinction to be made solely according to the letter's vertical orientation and handedness. It may also be seen how children learn first to recognize the difference between the letter pairs **b**, **d** and **p**, **q** since reflection in a horizontal plane does not occur in cortical projection, whereas in the vertical plane it gives rise, for example, to a bilateral projection onto the cortex of a **p** *and* a **q** for each single observed **p** *or* **q**. This point does not appear to have been recognized in the design of the initial teaching alphabet, although many other alphabets – from primitive hieratic Egyptian to the sophisticated Cyrillic notation, invented by a ninth-century monk called Cyril, and the basis of modern Russian – appear to avoid this particular problem.

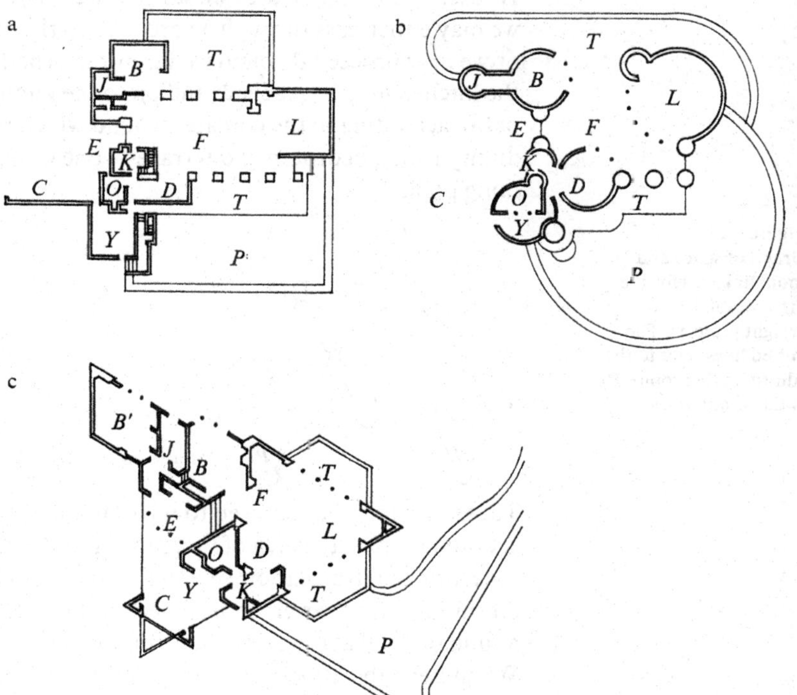

Figure 1.13
Three house projects by Frank Lloyd Wright:

a, *Life* 'House for a family of $5000–$6000 income', 1938

b, Ralph Jester House, Palos Verdes, California, 1938

c, Vigo Sundt House, near Madison, Wisconsin, 1941

B bedroom
B' Sundt bedroom
C car port
D dining-room
E entrance
F family room
J bathroom
K kitchen
L living-room
O office
P pool
T terrace
Y yard

Sometimes, objects which appear to be very dissimilar on first acquaintance may be seen, later, to share an underlying structural pattern. In Figure 1.13 we see three houses designed by the American architect

[5] A good guide to this prolific architect's work is *Frank Lloyd Wright; Writings and Buildings*, selected by Edgar Kaufmann and Ben Raeburn, New York, Horizon Press, 1960. See also Vincent Scully, *Frank Lloyd Wright*, New York, George Braziller Inc., 1960.

[6] *Introduction to Mathematical Philosophy*, London, Allen and Unwin, 1960, p. 3.

Frank Lloyd Wright.[5] In them he uses a range of 'grammars', by which he meant, above all, the controlling geometric unit which ordered the plan and pervaded the details. The unit is an equilateral triangle in one house, a four-foot square in another, and a circle in the third. Each building has extremely marked individuality, so much so that each looks completely different from the other. Two of the houses were designed in the same year, 1938, and the third, the Sundt House, two years later. Whilst they may look different, they are in fact topologically equivalent. If each functional space is mapped onto a point and if, when two spaces interconnect, a line is drawn between their representative points we produce a mapping known as a graph (Figure 1.14). Having done this for the three houses we find that they are topologically equivalent in plan, excepting the additional bedroom in the Sundt House. In all three houses we see that we arrive under a car port and we may either pass through a yard area to the office and kitchen, or we may proceed through to the entrance hall or loggia, from which the kitchen may be reached, and on to the family room or covered patio, according to the climate, around which radiate the living-room, dining-room, bedroom and terraces – one of which overlooks a swimming pool.

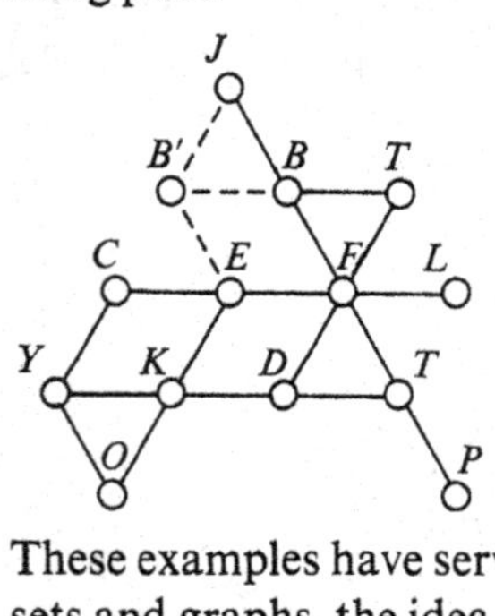

Figure 1.14
Graph of space and room linkages for the three Frank Lloyd Wright projects. The dotted lines refer to the additional bedroom, *B'*, in the Sundt house

These examples have served to introduce mathematical entities such as sets and graphs, the idea of mapping, and the concepts of rotation, reflection and translation. We shall return to these in later chapters. At this point, however, let us reflect on the idea of mapping. Mapping is a fundamental act in any process of abstraction or pattern recognition. When Bertrand Russell writes, 'It must have required many ages to discover that a brace of pheasants and a couple of days were both instances of the number 2', he is describing an exceedingly germane act of mapping which we all repeat in our childhood at some time or other.[6] As we have seen, mapping is not necessarily a means of visual representation. It is a way of structuring information.

[7] See James D. Watson, *The Double Helix*, London, Weidenfeld and Nicolson, 1968.

A given set of data usually only acquires significance when we map it onto a pattern of some kind, indeed it is easy to recognize when we make new observations of the world about us that we generally map them onto our old observations and experience. There is a remarkable passage in Captain Cook's *Voyages* where he describes how the Australian aborigines on the shore seemingly failed to 'see' his ship sailing in: we might assume that they had no previous experience upon which to map the new phenomenon. Mathematics is the subject, without equal, for the making of new, often unfamiliar synthetic patterns, and the history of modern science is full of examples of their use, such as the classic mapping of observations into a non-Euclidean frame of reference which led to the theory of relativity, or the mapping of experimental facts into the abstract matrix of the double helix in the discovery by Crick and Watson of the structure of the DNA molecule.[7] In the latter case, Pauling's abortive attempt to fit observations to a treble helix stands as a warning that data are not in themselves structured but require structuring, and that the mappings we choose may as much blur and confuse as they clarify and resolve.

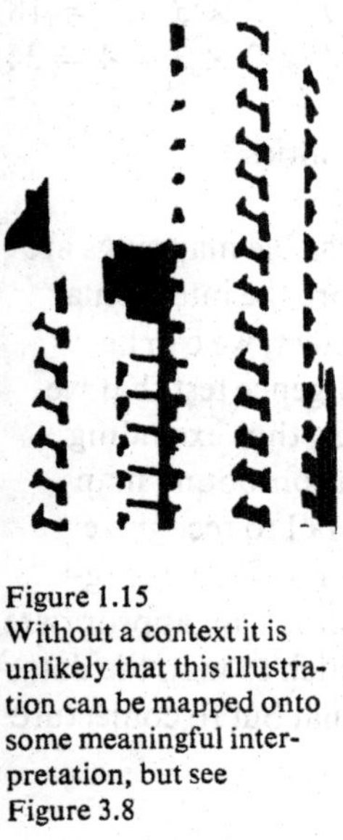

Figure 1.15
Without a context it is unlikely that this illustration can be mapped onto some meaningful interpretation, but see Figure 3.8

For the creative worker in the arts or sciences it would seem essential to cultivate as many mental sets as possible upon which to map the observed world, and in this the synthetic, mathematical generation of patterns is seen to be an aid. A set of data rarely has just one pattern to the exclusion of others. Often one pattern is so obvious that we tend to think of it as the only pattern of the set. This is a mistake. The only thing that is unique about the pattern is our manner of seeing it – our mapping of it onto some preconceived notion – for the data themselves have all the patterns of which they are capable (Figure 1.15).

It is clear that the context in which the data are presented is important in regard to the kind of mappings we think appropriate. In an intelligence test we would, in general, complete the series 2, 4, 6,... with 8, 10, 12. But outside, at a football match, we might well think of '2, 4, 6, 8, who do we appreciate', or playing cards '2, 4, 6, 8, 10, Queen', or rolling a dice on successive sides '2, 4, 6, 5, 3, 1', or cooking '2, 4, 6, off, 2, 4' as we turn the hot-plate control. In each context the answer is reasonable and the pattern valid. In the examination hall we must close our minds to all these possibilities and prove how intelligent we are at passing examinations by asking the pernicious question 'What answer does the examiner expect?'.

We *assume* he expects the answers to be taken from the set of natural numbers

$$\{1, 2, 3, 4, \ldots n, n+1, \ldots\}$$

and since *he* has selected the first three even numbers, *we* write the next three. This is a possible answer. It is certainly not a probable answer. Given just the first hundred natural numbers to choose from at random the probability is about one in a million that 8, 10, 12 would occur. The pattern is of our making, and there are patterns in 2, 4, 6 which might lead us to extrapolate quite differently

Figure 1.16

$2 = \mathit{2}$	$2 = 2$	$1 \times 1 \times 2 = 2$
$2 + \mathit{2} = 4$	$2 \times 2 = 4$	$2 \times 1 \times 2 = 4$
$\mathit{2} + 4 = \mathit{6}$	$\mathit{3} \times 2 = 6$	$\mathit{1} \times \mathit{1} \times 2 \times \mathit{3} = 6$
$4 + \mathit{6} = 10$	$\mathit{3} \times \mathit{3} = 9$	$\mathit{2} \times \mathit{1} \times 2 \times \mathit{3} = 12$
$\mathit{6} + 10 = \mathit{16}$	$\mathit{3} \times 4 = 12$	$\mathit{3} \times \mathit{1} \times 2 \times \mathit{3} = 18$
$10 + \mathit{16} = 26$	$4 \times 4 = 16$	$1 \times 1 \times 2 \times 3 \times 4 = 24$

These are just a few not so convention-bound possibilities.

When we consider the environment about us, the possible mappings are innumerable. Yet, if we assume the role, and therefore the intellectual ambience, of a planner, a traffic engineer or an architect, we can be almost as sure as we were about answering the intelligence test that we will confine our range of mappings to a particular set, thus excluding many others. The danger in this is that these convention-bound mappings may cease to fit the facts; or at least they may fail to recognize emergent patterns. Our argument is that a new pattern will be recognized only by an observer who has available, or develops, an appropriate range of mental sets, abstract or otherwise, upon which to map the data, and who actively seeks not to corroborate the habitual but to conjecture potentiality.

Two artists have testified to this process. Leonardo da Vinci describes a course of study for an artist and tells how to increase talent and stimulate various inventions:

> '... look into the stains of walls, or ashes of a fire, or clouds, or mud or like places, in which, if you consider them well, you may find really marvellous ideas. The mind of the painter is stimu-

[8] From I. A. Richter, *Notebooks of Leonardo da Vinci*, London, Oxford University Press, 1952, p. 182.

[9] Werner Haftmann. *The Mind and Work of Paul Klee*, London, Faber & Faber, 1954.

lated to new discoveries, the composition of battles of animals and men, various compositions of landscapes and monstrous things, such as devils and similar things, because by indistinct things the mind is stimulated to new inventions.'[8]

But the mind needs to be receptive, and a wide range of potential mappings must be made before – click! a 'good integrated form' emerges.

Werner Haftmann has described how Paul Klee, perhaps the most seminal of all modern artists, had heard colourful Oriental fairy-tales told to him and illustrated by his grandmother.[9] And how, in a restau-

Figure 1.17
Berne, 1910. Drawing by Paul Klee from Werner Haftmann, 1954

[10] Robert Hooke. *Micrographia, or Some Physiological Descriptions of Minute Bodies Made by Magnifying Glasses with Observations and Inquiries thereupon*, facsimile reproduction of the first edition published by the Royal Society in 1665, New York, Dover, 1961. See particularly pages 112–16.

rant owned by his uncle, there were tables with marble tops whose fine veinings he would stare at for hours on end, until the complex patterns induced elaborate dream-images. Such were the origins of Klee's first fantastic illustrative drawings, which were closely related to the fairy-tale world about which he had heard as a child. Klee's drawing of Berne makes explicit this kind of relationship, and allows us to participate in such 'discovery' for ourselves. A city seen in the grain of wood, or the veins of marble, or the circles and triangulations of the geometer's art, or the cellular structure of sponge: in each a mapping is involved, one set into another.

Certainly one of the most striking examples in the history of planning concerns the proposal for the rebuilding of London after the Great Fire of 1666. Sir Christopher Wren had been Sevilian Professor of Astronomy at Oxford and had just returned at the time of the Fire from an extended visit to France where he had been most impressed by French Renaissance architecture and planning. He was at the time, according to Newton, 'one of the three best geometers'. Dr Robert Hooke was Gresham Professor of Geometry, Secretary of the Royal Society, and the author of *Micrographia*, the first important work on the world as seen under the microscope.[10] These two men were both Commissioners for Rebuilding the City of London, Wren was a representative for the King, and Hooke, for the City. Both men submitted plans. Wren mapped his concept into his most recent experience: the ideal of the European Baroque city which he had seen emerging in Parisian works and which had undoubtedly excited his abstract geometrical interest. Hooke mapped his idea of the City into the cells of the sponge, the cells which he had discovered under the microscope with their characteristic rectangular structure. Wren produced a radio-axial city: Hooke, the first consciously designed cellular city, a remarkable and singular invention. The city that was rebuilt was neither of these, but the one that everyone could grasp: excepting detail, of road width and constructional standards, the new London mapped, in almost one-to-one correspondence, into the city as it had been before the Fire.

So much for the fundamental importance of mapping in our thought processes and creative activities in general. In architectural and planning studies we have become habituated to at least three distinct ways of mapping buildings and urban space. We might say that we map the complex environment into separate sheets: one sheet maps the volumes,

Figure 1.18
Plan of London before the Great Fire

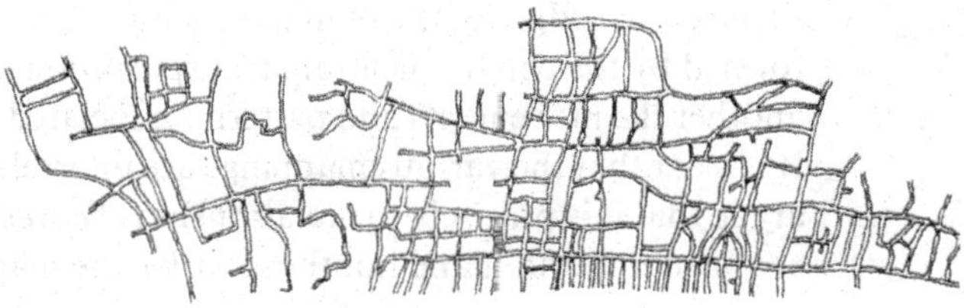

Sir Christopher Wren's proposal for the reconstruction

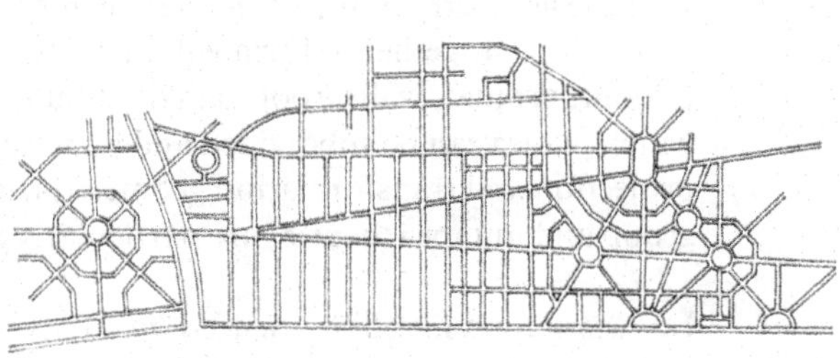

Dr Robert Hooke's proposal

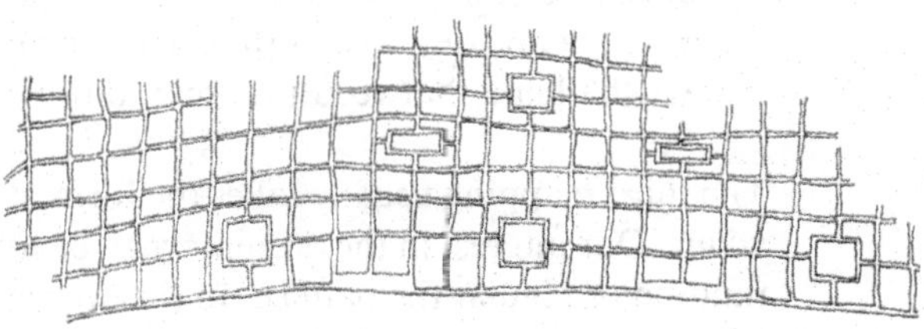

The cell structure of cork as seen through Hooke's microscope and illustrated in *Micrographia*

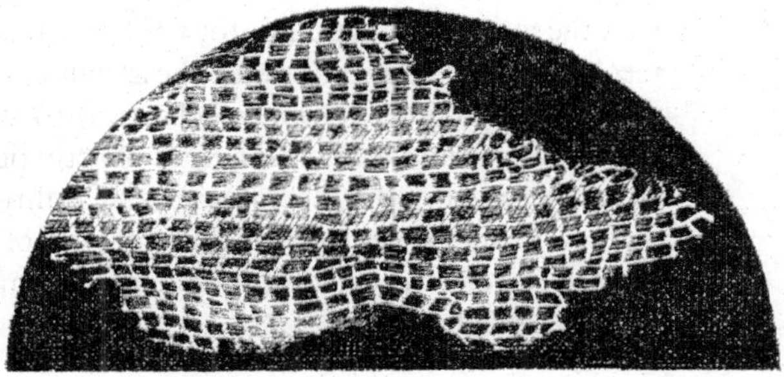

[11] For a discussion on taxonomic methods, see O. P. Tabor, *Traffic in Buildings 3; Analysis of Communication Patterns*, Working Paper 19, University of Cambridge, Land Use and Built Form Studies, 1970. And on stochastic and statistical theories see, for example, contributions in William L. Garrison and Duane F. Marble, eds., *Quantitative Geography*, Northwestern University Studies in Geography, no. 13, 1967, pp. 1–32; and Brian J. L. Berry and Duane F. Marble, *Spatial Analysis, a Reader in Statistical Geography*, Englewood Cliffs, Prentice-Hall, 1968. See also A. G. Wilson, 'A Statistical Theory of Spatial Distribution Models' in *Transportation Research*, vol. 1, no. 3, 1967, pp. 253–70.

surfaces and edges of the built environment, another the networks formed by the communication, transportation and service channels, and another the movements and patterns associated with human activities. It is clear that the various mappings are interrelated, and many would argue that it is precisely the relationships *between* these mappings that are most important and not those *within* the mappings themselves.

However that may be, the divisions do correspond to three broad mathematical interests. In the first mapping we are mainly concerned with the geometry of the *plane*, we are interested in the way in which planes come together to form volumes – floor, walls, ceilings – to create a room; ground, walls, roof to make a building – and in the ways in which planes can be subdivided and punctured. We shall find that projection and transformation geometries, isometry and symmetry, set, group and number theories are particularly relevant here.

For the second mapping we need *lines* in two or three dimensions. We map routes into lines, pipes into lines. The lines intersect at points which correspond to junctions or nodes. Often we are more interested in the relationship of lines and nodes than we are in any definite dimensional characteristics: we are then interested in graph theory. At other times we may be thinking of the direction and magnitude of the lines and we shall find that vector representation is convenient.

The third mapping usually takes the form of reducing activities to sets of *points*. Our interest in these points may be in their pattern, in how random or regular the pattern. In this we would need the statistical procedures of the ecologists and geographers. Our interest may be in grouping activities into clusters representing certain common attributes, in this we use various taxonomic techniques. Or we may be concerned with the movement of points, for which an analogy with statistical mechanics may be appropriate. The activities may be seen to be subject to random processes and stochastic theory may be relevant. For small groups and the micro-analysis of activities, set and graph theories have been employed by sociologists and anthropologists with some success. From the geometrical point of view of this book, however, these subjects have been considered too remote from our main theme to be included specifically.[11]

2 Translations, rotations and reflections

The nineteenth-century architect, Viollet-le-Duc, once wrote:

> 'Symmetry – an unhappy idea for which, in our homes, we sacrifice our comfort, occasionally our commonsense and always a lot of money.'[1]

The symmetry he had in mind was that imposed on architectural forms by his contemporaries, the academicians of the Beaux Arts. The plans of their building forms exhibited, almost without exception, *bilateral* symmetry of the whole and the parts: most plans could be resolved into a set of axes about which reflection was either total (major axes – usually, in fact, just one) or partial (minor axes). Out of this symmetry in plan there arose elevational reflected symmetries of the kind we have referred to in the previous chapter.

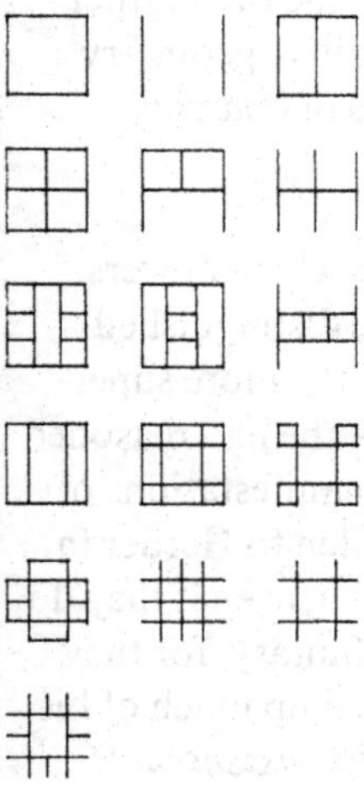

Figure 2.1
Combinations of axes within a square plan. After J. N. L. Durand

Figure 2.1 shows compositions of axes within a square plan from J. N. L. Durand's treatise *Leçons d'Architecture* published in 1819. Whenever a Beaux Arts architect had a planning problem to solve he would map the requirements into abstract generic structures of this kind until a satisfactory fit was achieved between the uses of various parts of the building and their geometric disposition, or *parti*. At the beginning the method was new and revolutionary. Here was a doctrine of architecture at once systematic and synthetic, owing little to precedents. But by mid-century, when Viollet-le-Duc wrote, architects were beginning to rely on precedents and examples from past styles of architecture for their invention, rather than on the 'style-free' abstract precepts of Durand. As Professor of Architecture at the same *École Polytechnique* as the mathematician Gaspard Monge, and in the scientific spirit encouraged by Napoleon in establishing the new school, Durand set out his *Leçons* like the axioms and theorems of a geometer. The treatise was concerned mainly with public buildings which the new Republic required and for which bilateral symmetry was often appropriate, and not so much with private dwellings which, Viollet-le-Duc felt, suffered undue corseting by symmetry.

Whilst other forms of symmetry were known to architects at the time, their full appreciation was dependent on developments in mineralogy,

[1] From *Discourses on Architecture*, vol. 2, translated from the French of Eugène-Emanuel Viollet-le-Duc by Benjamin Bucknall, New York, Grove Press, 1959, pp. 267–8.

[2] See F. C. Phillips, *An Introduction to Crystallography*, London, Longmans, Green, 1946. We illustrate examples of Haüy's cubelet approximations in Figure 5.27.

[3] For a commentary on nineteenth-century morphology see the contribution by the pioneer town-planner and biologist Patrick Geddes, 'Morphology' in *The Encyclopaedia Britannica*, vol. 16; 9th edition, 1898, pp. 837–46.

crystallography, morphology and eventually mathematics itself.[2] It was yet another Professor of *l'École Polytechnique*, René Just Haüy, who first formulated the 'law of symmetry' for crystal growth and who, as we shall see in Chapter 5, proposed a method of describing known crystal forms and creating synthetic, or hypothetical ones. His method was to study the symmetrical accretion of certain *molécules intégrantes*, which in themselves were invariant and incapable of further division, around definite axes of growth. Haüy's German translator, Christian Samuel Weiss, stressed the importance of the axis which is truly the line governing every figure round which the whole is uniformly disposed. 'All the parts look to it, and by it they are bound together as by a common chain and mutual contact.' This statement might well have been written by Durand about the axis in buildings. Haüy wrote *Traité de Minéralogie* in 1802 and in so doing earned himself the title 'father of crystallography'. And incidentally, Monge's descriptive geometry proved invaluable in the early graphic representations of complex crystal forms.

In 1813 A. P. de Condolle classified the symmetries to which flowers conformed in *Théorie Élémentaire de la Botanique*. He distinguished between the fundamental unity of structural type and the more superficial likeness of physiological adaptation. Earlier Goethe had reasoned that particular forms of flowers and plants were but manifestations of general abstract patterns. Knowing these, he had written to Herder in 1787, 'one could go on endlessly inventing plants which, even if they do not exist, might well do so without being just artistic fantasy, for they would have inner truth and necessity.' Goethe summed up much of his thought on the matter in his celebrated essay *Zur Naturwissenschaft überhaupt, besonders zur Morphologie* of 1817 in which he proposed the word 'morphology' for the study of unity of type of organic form. In his *Tagebuch* Goethe had written of 'architecture... being like mineralogy, botany and zoology', in as much as these subjects shared a common interest in spatial structure.[3]

Goethe was not a trained scientist and his cause was taken up by Auguste de Geoffroy Sainte-Hilaire who published *Morphologie Végétale* in 1841. Geoffroy was a notable biologist who had been responsible for interceding during the Revolution on Haüy's behalf when the latter was in prison and near to execution. Together with Darwin's work, the contribution made by Geoffroy to morphology inspired Ernst Heinrich Haeckel in his systematic study first of radiolarians, and then of organisms in general, from the point of view of symmetry; that is, the study of *promorphology*. Haeckel published *Generelle Morphologie der Organismen* in 1866, but his fame outside science rests upon the magnificently illustrated *Kunstformen der Natur* of 1899 – a work of great visual joy.

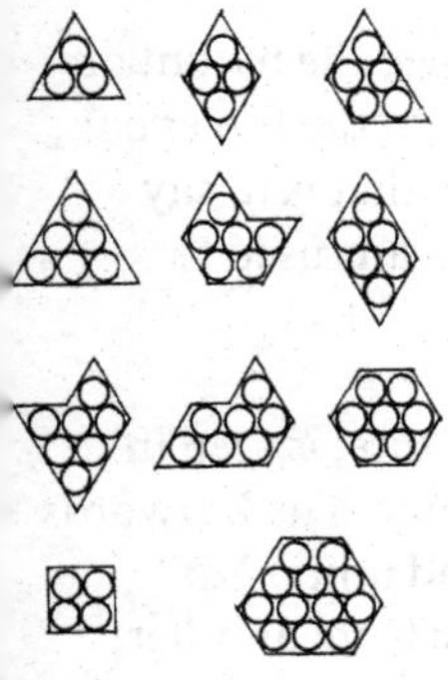

Figure 2.2
Solid pellets approximating simple forms. After Robert Hooke

By mid-century Haüy's crystallographic theories had been modified and extended by others, Weiss in particular, away from the solid cubelet *molécules*. First, an older notion of Robert Hooke's in *Micrographia*, which suggested that substances – and indeed organisms – might be considered to be made up of spherical pellets packed together in various ways, was revived (Figure 2.2); and then this physical model was dematerialized to become a geometrical system of points exhibiting certain symmetry properties: the abstract structure of modern crystallography. We have already met something similar to this in our analysis of three houses by Frank Lloyd Wright. There we represented each space by a point and its adjacency with other spaces by a line. In this example we were interested in demonstrating that the plans were topologically equivalent, but crystallographers are concerned with isometric equivalence, or *automorphism*, and their abstraction replaces solid crystalline units with a structural unit, or unit of pattern, preserving length but otherwise devoid of physical substance (Figure 2.3).

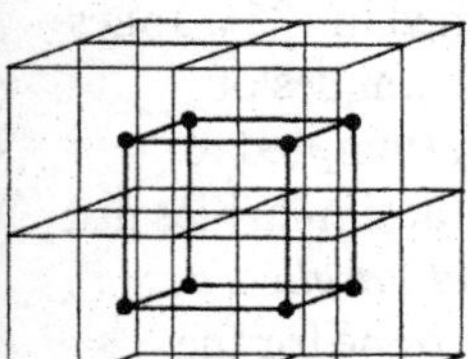

Figure 2.3
Development of a structural unit to replace solid cubelets. After F. C. Phillips

It was this unit of pattern and the laws governing its combination that Viollet-le-Duc wrote about in *Dictionnaire Raisonné de l'Architecture Française* published in 1866 when he argued that such symmetry as occurred in nature was not clapped on form, but was the very *principe* governing its growth, development and formation. In a remarkable passage he described mineralogical structure – he was not only an architectural innovator but also a sufficiently skilled geologist to be commissioned by the French authorities to carry out a survey of the Alps – and then he continued

> 'I have simply been trying to make it clear that the first creative event in the world we live in happened according to a rigorously applied *principe*, the only possible principle. If we follow all the phases of inorganic and organic creation, we shall soon recognize the logical order, in its most varied and even apparently different aspects, which results from a *principe*, from an *a priori* law, from which it never departs. It is from this method that all these acts

Figure 2.4
Crystal formations. After Viollet-le-Duc

[4] Translated by Lindsey March.

[5] Owen Jones. *The Grammar of Ornament*, London, Day and Son, Folio edition, 1856.

[6] See Introduction to the first edition of Le Corbusier's *Oeuvre Complète 1910–1929*.

[7] 'I got a packet of onion skin...and traced the multifold designs. I traced evenings and Sunday mornings until the packet of one hundred sheets was gone and I needed exercise to straighten up from this application.' Frank Lloyd Wright, *An Autobiography*, New York, Duell, Sloan and Pearce, 1943, p. 75.

> gain the *style* with which they are saturated. From the mountain down to the smallest crystal, from the lichen up to the forest oak, from the polyp to man himself, all possess *style*, that is to say a perfect harmony between the result and the means used to obtain it.'[4]

This preoccupation with *principe*, or in our terms structure, is seen in the works of many authors during the nineteenth century. The keywords of the century for poet and scientist were continuity and unity. No longer were events or objects looked at in isolation one from another, but now they were seen as part of a grand continuum governed by a finite and discoverable number of laws from which nature, or art for that matter, derived its essential unity. Typical of this approach, as applied to the decorative arts, is Owen Jones's monumental and exhilarating work *The Grammar of Ornament*[5] published in 1856. Jones was an English architect who had collected together examples of ornament from many different cultures and ages. He attempted to classify ornament according to its underlying structural similarities just as Durand had attempted for architecture in *Recueil et Parallèle des Édifices* of 1800; Goethe, Condolle and Geoffroy had done for the vegetable kingdom; Haüy and Weiss for crystal forms; and Haeckel was about to do for radiolarians. Years later, Jones's book excited both the young Le Corbusier who discovered it in L'Eplattenier's 'modest library, which contained all he judged necessary for our mental pabulum' at Chaux-de-Fondes;[6] and the youthful Frank Lloyd Wright[7] who worked through the nights tracing from it to exercise his skill at drafting before applying for a job with Louis Sullivan, 'the father of modern architecture', in Chicago. Both have written to say how deeply impressed they were by Jones's demonstration that beneath the superficial varieties of appearances lies the invariable logic of geometry (Figure 2.5.)

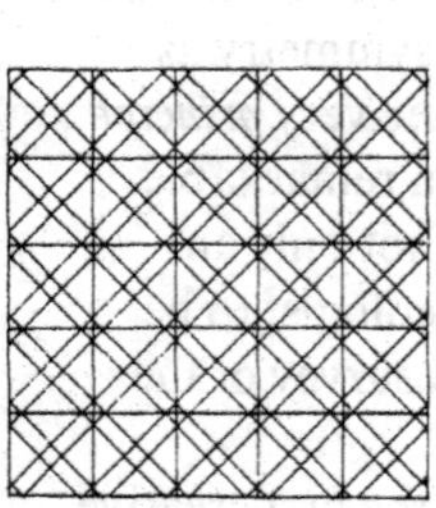

Figure 2.5
One of the structural systems underlying Moorish ornaments. 'The number of patterns that can be produced [based on such a system] would appear to be infinite.' After Owen Jones

It is well known that John Ruskin did much to promote the moral standing of modern architectural philosophy, but not too much has been said of his interest in crystal forms, which typically he bent to ethical purposes in a lugubrious set of lectures to schoolgirls on the elements of crystallization: *The Ethics of the Dust*, 1866. However, contrary to the other authors, Ruskin was interested in characterizing the idiosyncratic appearance of crystal forms rather than their underlying symmetry. He attacked Jones's concept of 'conventionalization' on the grounds that if it were true it would be possible for designers to produce ornamentation though they had 'no more brains than a looking glass or a kaleidoscope'.

In an essay, *Modern Manufacture and Design*, 1859, Ruskin tells the story of a friend who maintained that the essence of ornament consisted in three things: contrast, series and symmetry. Ruskin replies that none, nor all, would produce ornament. 'Here,' he says, making a smudge with his pen (Figure 2.6a), 'you have contrast: but it isn't ornament'; 'here – 1, 2, 3, 4, 5, 6'... 'you have series: but it isn't ornament; and here,' sketching a little stick-man (Figure 2.6b), 'you have symmetry; but it isn't ornament.' His friend replies, 'Your materials were not ornament because you did not apply them. I send them to you back, made up into a choice sporting neckerchief (Figure 2.6c): each figure is revolved on two axes, the whole opposed in contrasting series.'

Figure 2.6a

Figure 2.6b

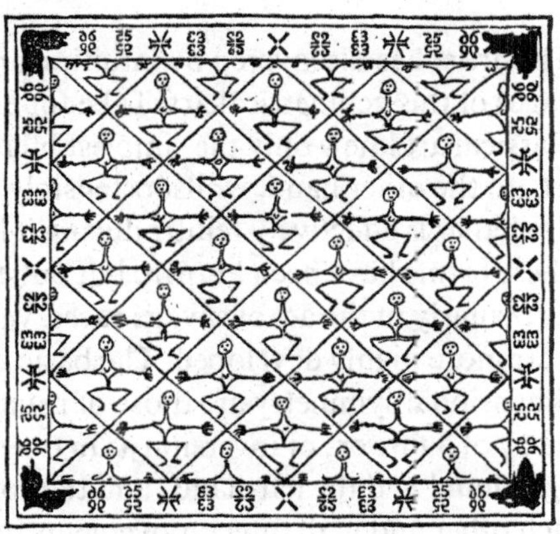

Figure 2.6c
'A choice sporting neckerchief.'

Ruskin comes back: 'Like all the people I have ever known who have your power of design, you are entirely unconscious of the essential laws by which you work, and confuse other people by telling them that the design depends on symmetry and series, when in fact, it depends entirely on your own sense and judgement.' Ruskin then argues that the designer would be best quit of the 'notion of formal symmetry'.

In a somewhat symmetrical way we have turned full circle. We started with Viollet-le-Duc's criticism of formal symmetry as applied by academic architects of his day, we then saw the growth of the idea of symmetry in various sciences and Viollet-le-Duc's understanding of this as a *principe* underlying appearances, and now here we have Ruskin im-

[8] J. L. Martin, Ben Nicholson, and N. Gabo, eds., *Circle; International Review of Constructive Arts*, New York, E. Weyhe, 1938, pp. 119–23.

patient with natural science, playing with the outward forms of crystals like an innocent schoolgirl, and airily dismissing symmetry as a 'sign of utterly bad, hopeless and base work'. Ruskin's stigma on formal symmetry became a battle cry of protagonists and polemicists of the Modern Movement in architecture, notably members of the Dutch De Stijl group, some theorists at the Bauhaus, and the Russian Constructivists. But it could be said that those who were the most successful innovators of architectural form, in particular Le Corbusier and Frank Lloyd Wright, were those who most understood symmetry as an abstract idea, Viollet-le-Duc's *principe*. Le Corbusier and Wright even bring new life to formal symmetry (Figure 2.7) in many of their projects.

Symmetry in the modern sense of the word was first made explicit to architects by the English crystallographer J. D. Bernal in an article 'Art and the Scientist' in *Circle: International Review of Constructive Art*, 1938.

> 'The artist has discovered by intuition and practice many of the stages of this geometrical art. Take for example symmetry. Classical art knew only the simplest bilateral symmetry. Modern art, on the other hand, whilst ostensibly rejecting symmetry altogether is effectively reintroducing it in more complex forms. These forms have been known, but outside the classical tradition, particularly in the art of savage races, where the sense of rhythm is far more highly developed. The basic concepts of the three-dimensional symmetry are those of rotation, such as the symmetry of a flower; of inversion as in the difference between the right-hand and the left-hand; and the combination of these with each other and with direct movements in space. This can be done only in a limited number of ways: 230 for three dimensions, 17 for two, but this is only for regular figures. By altering the scale a far larger number of internal harmonies, depending essentially on symmetry, can be introduced. Some of the more abstract artists have produced intuitively many of these complex rhythms. Architecture in particular gives great opportunities for symmetrical rhythms.'[8]

Ruskin's neckerchief displays each of the symmetry concepts mentioned by Bernal. The blobs and number series are handed or *reflected*, these rows are then *rotated* around the four sides of the scarf, and finally the stick-man motif is repeated or *translated*. The proof that the seventeen

Figure 2.7
The sparkling *poché* of the Darwin D. Martin House, Buffalo, New York, 1904, demonstrates Frank Lloyd Wright's mastery over the Beaux Arts tradition

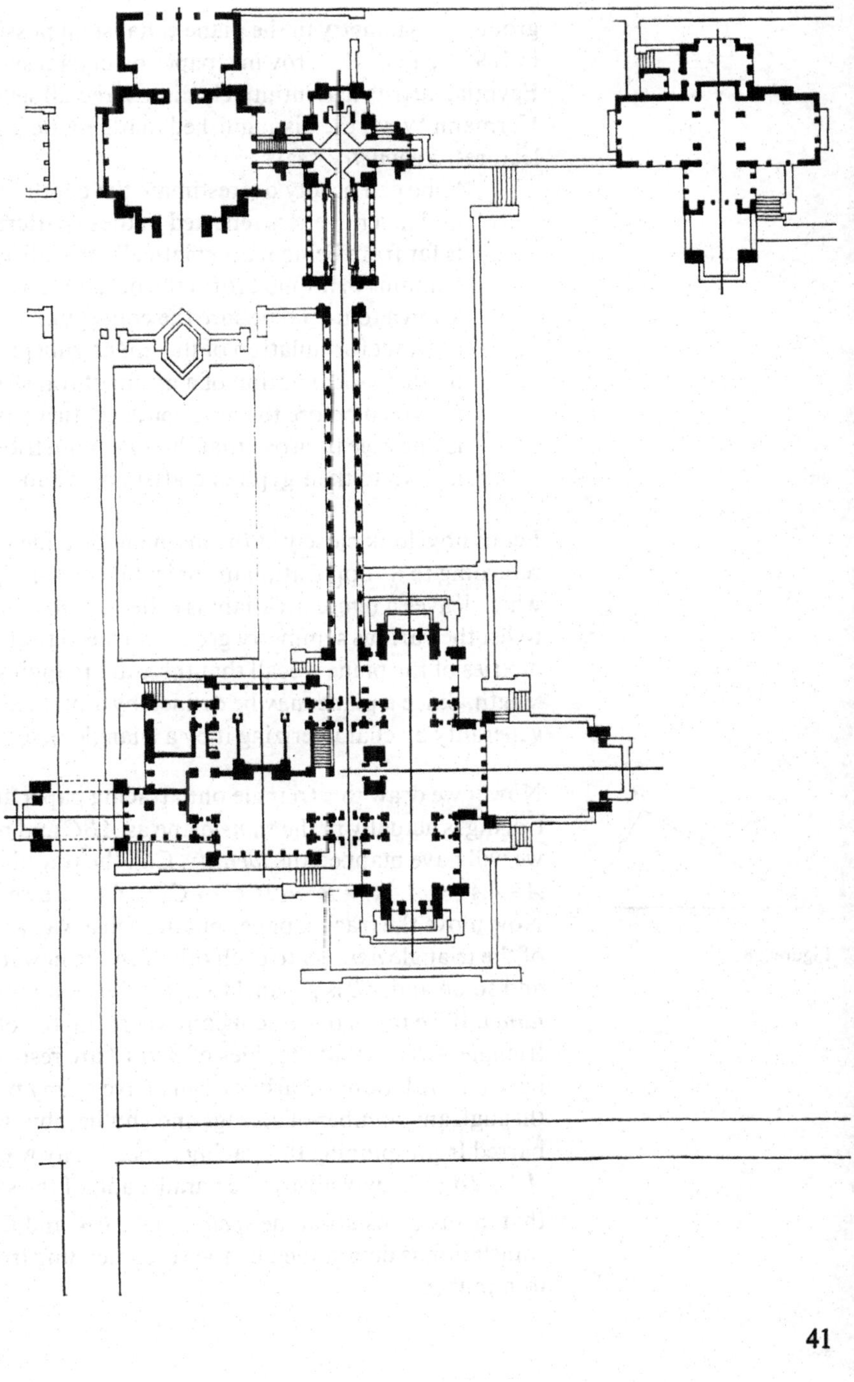

9 Hermann Weyl. *Symmetry*, Princeton University Press, 1952, pp. 103–4.

10 For illustrations see Chapter 3, Figures 3.8 to 3.12.

11 For a clear introduction see Max Jeger, *Transformation Geometry*, London, Allen and Unwin, 1966. Our treatment, however, follows H. S. M. Coxeter's excellent *Introduction to Geometry*, New York, Wiley, 1961.

groups of symmetry in the plane exhaust all possibilities was carried out in 1891 by E. S. Fedorov in a paper on crystal symmetry. However, Egyptian artists had intuitively discovered all seventeen groups. Hermann Weyl, the distinguished mathematician, in his excellent lectures, *Symmetry*,[9] says

> 'One can hardly overestimate the depth of geometric imagination and inventiveness reflected in these patterns. Their construction is far from being mathematically trivial. The art of ornament contains in implicit form the oldest piece of higher mathematics known to us. To be sure the conceptual means for a complete abstract formulation of the underlying problem, namely the mathematical notion of a group of transformations, was not provided before the nineteenth century; and only on this basis is one able to prove that the 17 symmetries already implicitly known to the Egyptian craftsman exhaust all possibilities.'[10]

Let us now look closely at the mathematical idea of symmetry before returning to its application in environmental design.[11] An *isometry* which leaves a figure invariant is called a *symmetry operation*. In order to list the various symmetry groups we need to study the different isometries of the plane: recall that these are mappings which preserve length. Since a plane may be defined by any three points in it we lose no generality by characterizing it by a triangle ABC.

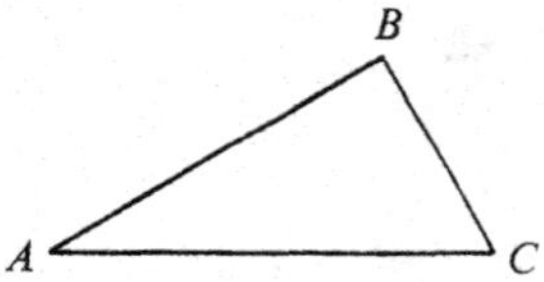

Figure 2.8

Now if we draw this triangle onto tracing paper and imagine that our tracing is actually in the *same* plane as ABC and not on a separate sheet we will have mapped the *identity*. Call the traced triangle $A'B'C'$ so that $A \to A' = A$, $B \to B' = B$, $C \to C' = C$ expresses the identity relation. Now move the tracing paper but in such a way as to keep the like sides of the triangle *parallel* to each other. In the new mapping $A \to A' \neq A$ and so on and AB is parallel to $A'B'$. This kind of mapping is a *translation*. If we move our tracing anywhere 'in' the plane of the original triangle ABC so that the sides of $A'B'C'$ are respectively parallel to it we have a translation. Clearly we can arrive at any particular position through any number of zigzags and shuffles, but for the moment we are barred from spinning the tracing paper in any way. The straight lines AA', BB', CC' will always be parallel and of the same length. We say that in any translation the *vectors* $\overline{AA'}$, $\overline{BB'}$ and $\overline{CC'}$ are all equal. A translation is determined by the vector leading from an origin A to a terminus A'.

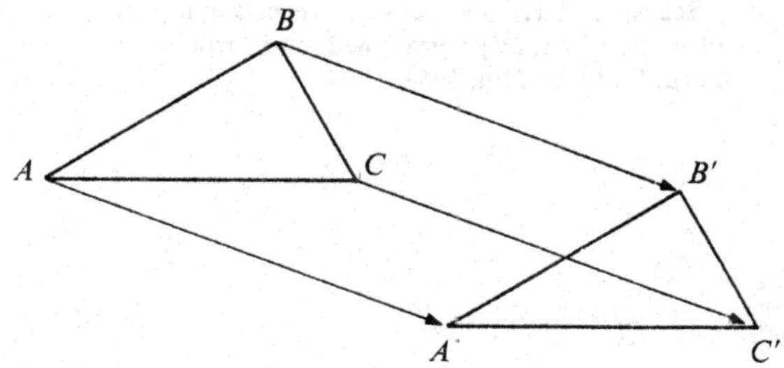

Figure 2.9

Let us translate ABC to a new position $A_1'B_1'C_1'$ and call this translation $\mathbf{T}_1$. Now from this position translate the triangle to another position $A_2'B_2'C_2'$ and call this move $\mathbf{T}_2$. It is apparent that we could have missed out the intermediate moves $\mathbf{T}_1$ and $\mathbf{T}_2$ and could have gone directly from the original position to $A_2'B_2'C_2'$. Let us call the translation that does this $\mathbf{T}_3$ (the subscripts are just little tags to distinguish one translation from another: we could use any mark we like to do this but usually it is sensible to use some sort of system).

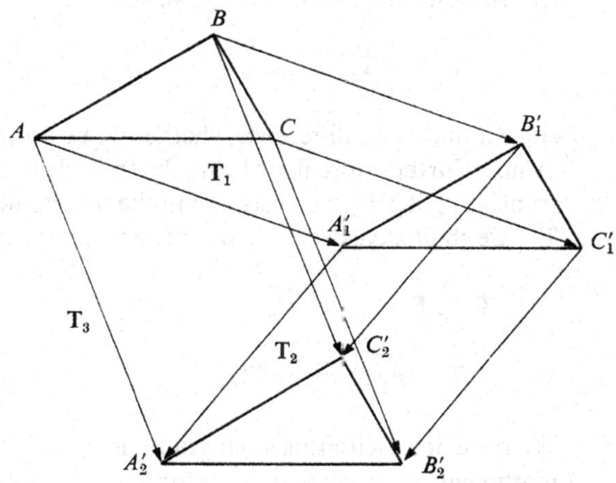

Figure 2.10

If we write $\mathbf{T}_2 . \mathbf{T}_1$ to mean first do operation $\mathbf{T}_1$ and then $\mathbf{T}_2$, we see that the *product* of two translations is the same as a third:

$$\mathbf{T}_2 . \mathbf{T}_1 = \mathbf{T}_3.$$

It happens to be true that we could have done $\mathbf{T}_2$ first and then $\mathbf{T}_1$ and still end up in the position given by the single translation $\mathbf{T}_3$. Thus

$$\mathbf{T}_2 . \mathbf{T}_1 = \mathbf{T}_3 = \mathbf{T}_1 . \mathbf{T}_2.$$

This isn't always so with other kinds of operations. For example if '**A**' is 'putting our socks on', and '**B**' is 'slipping on our shoes' we may write **B** . **A** for the combined operation, but **A** . **B** will gain us some odd looks when people see us wearing our socks over our shoes. We say translations are commutative: the puttings on of shoes and socks are not.

[12] See particularly Georges Papy's excellent introduction *Groups*, London, Macmillan, 1964. Also for fully programmed self-instruction, see Boyd Earl, *Groups and Fields*, New York, McGraw-Hill, 1963.

Further, if we make three translations $\mathbf{T}_1$, $\mathbf{T}_2$, $\mathbf{T}_3$, in sequence it will be noted that the resultant translation is the same whether we do $\mathbf{T}_1$ and $\mathbf{T}_2$ first and then $\mathbf{T}_3$, or $\mathbf{T}_2$ and $\mathbf{T}_3$ first and then $\mathbf{T}_1$. This may be written symbolically as

$$\mathbf{T}_3 . (\mathbf{T}_2 . \mathbf{T}_1) = (\mathbf{T}_3 . \mathbf{T}_2) . \mathbf{T}_1,$$

and we say that translations are *associative*. Finally it is clear that if our first 'translation' leaves $A'B'C'$ coincident with ABC we have the identity translation which we call **I** and we can write

$$\mathbf{T} . \mathbf{I} = \mathbf{T},$$

since it makes no difference whether we perform the identity operation or not. Furthermore if we bring the triangle back to ABC after first translating it **T** by a translation in the reverse direction which we label $\mathbf{T}^{-1}$ we shall have precisely what **I** maps, namely the identity, thus:

$$\mathbf{T}^{-1} . \mathbf{T} = \mathbf{I}.$$

We call $\mathbf{T}^{-1}$ the *inverse* of **T**.

We have dealt with this at length to demonstrate the properties of what mathematicians call a *group*,[12] for it is within the context of the group concept that they define symmetry. What are these properties? A system consisting of a set of elements $G = (a, b, c...)$ and an operation between any two elements called their *composition* and marked by some symbol, say †, is a group, providing four rules or *axioms* are satisfied.

Closure: For any two elements $a, b \in G$ there exists one element $c \in G$ such that

$$a \dagger b = c.$$

Associative: For any $a, b, c \in G$

$$(a \dagger b) \dagger c = a \dagger (b \dagger c).$$

Identity: There is an element $i \in G$ such that for all $a \in G$

$$a \dagger i = a.$$

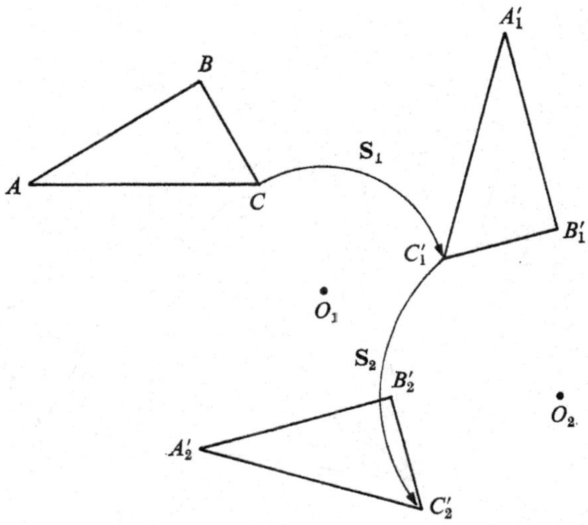

Figure 2.11

Inverse: For each element $a \in G$ and for each identity element i there is an element $a^{-1} \in G$ such that

$$a \dagger a^{-1} = i.$$

Now if we replace G by a set of translations $T = \{\mathbf{T}_1, \mathbf{T}_2, \mathbf{T}_3, \ldots\}$ in the plane and use the symbol . in place of †, calling the composition 'product', we see that translations obey these four axioms, thus forming a group with **I** as the identity. The same *algebraic structure* may appear in many different guises, with varying elements and compositions: the group structure is one of the most fundamental in mathematics.

Translations, however, also satisfy one more condition: the property of *commutativity*. This implies that going from $\mathbf{T}_1$ to $\mathbf{T}_2$ also implies going from $\mathbf{T}_2$ to $\mathbf{T}_1$. Such a group is distinguished from those that do not commute by the adjective 'Abelian' after the Norwegian mathematician N. H. Abel. Translations, then, form an *Abelian group*. If G is an Abelian group, a fifth axiom holds.

Commutativity: For any two elements $a, b \in G$

$$a \dagger b = b \dagger a.$$

To return to our triangle ABC and its tracing $A'B'C'$. Place $A'B'C'$ in the identity position coincident with ABC. Now take a pin and stick it through the tracing at random. We are no longer able to translate $A'B'C'$, but we may spin it about the pin to a position $A_1'B_1'C_1'$. Such a mapping is called a *rotation* (Figure 2.11). Take another pin and fix another point. Remove the first pin and spin $A'B'C'$ about the new *centre of rotation*. The new position $A_2'B_2'C_2'$ has been determined by first spinning the triangle about a centre of rotation O_1, and then about

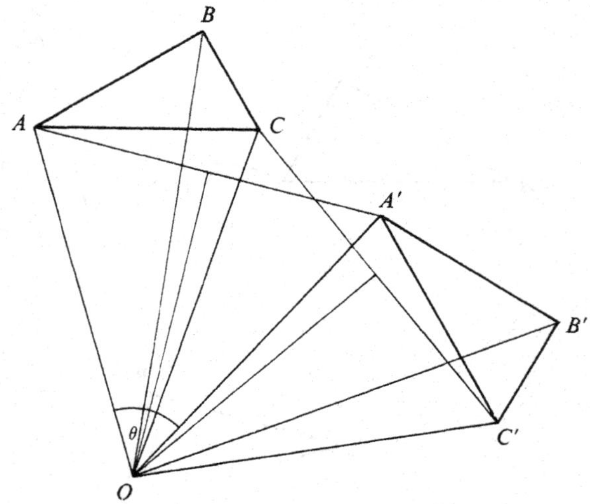

Figure 2.12

a second centre O_2. Call the first spin, through the angle ($\angle$) $CO_1\,C_1'$, $\mathbf{S}_1$ and the second through $\angle\, C_1'\,O_2\,C_2'$, $\mathbf{S}_2$. Then it can be shown that rotations form a group, but unlike translations it is not Abelian. In general

$$\mathbf{S}_1 \,.\, \mathbf{S}_2 \neq \mathbf{S}_2 \,.\, \mathbf{S}_1.$$

We can prove this as follows. It is obvious that the rotation $\mathbf{S}_1$ leaves O_1 unchanged:

$$\mathbf{S}_1(O_1) = O_1.$$

Thus when we operate $\mathbf{S}_2$ we see that

$$\mathbf{S}_2 \,.\, \mathbf{S}_1(O_1) = \mathbf{S}_2(O_1).$$

But

$$\mathbf{S}_1 \,.\, \mathbf{S}_2(O_1) = \mathbf{S}_1 \,.\, (\mathbf{S}_2(O_1))$$

and since $\mathbf{S}_1$ moves all points other than O_1 and in particular the point $\mathbf{S}_2(O_1)$ the two mappings of O_1 differ. Thus the rotation group is non-commutative.

All motion in a plane of this kind is either a translation or rotation. Indeed if we accept that a translation is a rotation through a zero angle about a suitable point at infinity we may say that every proper isometry of a plane is a rotation. The proof of this assertion is simple. Drop the tracing paper at random onto the sheet with ABC on it. We will have something like the situation above, providing the tracing paper does not turn over as it is dropped (that would be *improper*).

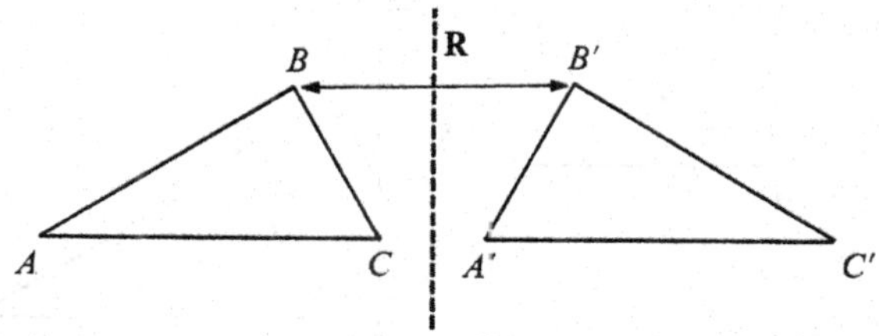

Figure 2.13

Join A to A' and B to B' and construct their perpendicular bisectors meeting in O. Draw in the lines OA, OA', OB, OB', and OC, OC'. By definition $OA = OA'$, $OB = OB'$ and, since by isometry we have $AB = A'B'$, triangles AOB and $A'OB'$ are congruent, thus $\angle AOB = \angle A'OB'$. By adding $\angle BOA'$ to each we see that $\angle AOA' = \angle BOB'$. Let us call this angle θ. A rotation θ about O carries AB into $A'B'$. Now, $\angle OAB + \angle BAC = \angle OA'B' + \angle B'A'C'$ by virtue of these congruences, and as $OA = OA'$ and $AC = A'C'$ we prove that triangle OAC and $OA'C'$ are congruent. It follows that $OC = OC'$ and $\angle COC' = \theta$. Thus the same rotation maps C onto C' which proves the assertion that every proper isometry of the plane is a rotation. The construction used in this proof is sufficient to determine any centre of rotation for such an isometry unless it degenerates into a translation.

If our tracing paper had turned over as we dropped it the motion would be described as *improper*. It is improper because we cannot change the handedness of our flat triangle without removing it from its two-dimensional world into the three-dimensional where we can turn it over and then put it back again as if nothing had happened. Since *we* are in a three-dimensional world it may seem perfectly reasonable to do this, but an analogy in three dimensions will make it obvious how improper the motion is. Occasionally we read of huge surpluses of army boots, not pairs, but boots just for the left foot, or right. To convert, say, a pile of left-footed boots into pairs would test the most imaginative quarter-master to his limit. He would need to dispatch half the pile to a destination out of this world and into four dimensions. There he would instruct whom? or what? to turn the boots over and to send them back to earth. Assuming that the boots survived the trip, the quartermaster would be in possession of a pile of matching pairs.

Nevertheless, taking our piece of tracing paper and turning it over, what do we get? If we hold our tracing so that the original triangle ABC and our copy $A'B'C'$ are coincident (identity), and then fold it over along any line we choose, we will have *reflected* the triangle along the line (Figure 2.13). Observe that, whereas we named the triangle ABC in a clockwise direction, its image is now described in an anti-clockwise manner, $C'B'A'$. Now join A to A', B to B' and C to C' with ruled lines. Note that they all cross the fold at right angles to it. Keeping the rest of the tracing paper firmly held, unfold the tracing so that $A'B'C'$ is again coincident with ABC. It will be immediately evident that the lines AA', BB' and

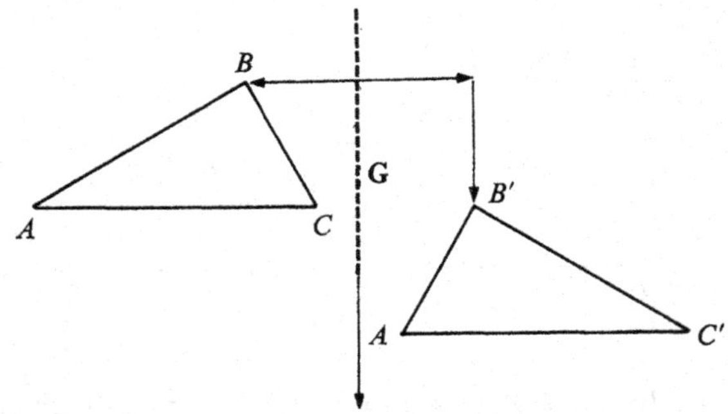

Figure 2.14

CC' are bisected by the line of the fold. This is precisely analogous to reflection in a mirror where the image X' of an object X is 'found' at a point on the perpendicular to the plane of the mirror from X and at a distance equally 'behind' the mirror as X is in front. For this reason the fold is often referred to as the mirror.

If we do not hold firmly to the tracing paper, and we merely turn it over haphazardly we obtain a *glide reflection*. The reason for this name becomes clear when we draw the lines AA', BB' and CC', join their midpoints together in a straight line, and fold the tracing back along this line. Run a pencil line against the edge of the fold and then glide the tracing paper along it until the two triangles are coincident. Unfolding the tracing without moving its position gives us a *reflection* as before, but to get the image of the triangle back to where it was when we originally threw it down, it is necessary to *glide* the fold of the tracing paper along the mirror plane: hence the name 'glide reflection' (Figure 2.14). An *improper*, or *opposite symmetry*, is either a reflection or a glide reflection. Such an isometry reverses sense, that is it turns left-handed figures into right-handed ones, and vice versa. Now, a translation is an isometry which leaves no point the same – we say it has no *invariant point* (although it is sometimes convenient to say that it leaves a point at infinity invariant); a rotation, as we have seen, is an isometry leaving one point invariant – the centre of rotation; while a reflection is an isometry which leaves one line (or plane), called its mirror, invariant. If an isometry has more than one invariant point, it must be either the identity (all points in the plane are invariant) or a reflection. A glide reflection, like a translation, has no invariant points.

As we have seen earlier with translations and rotations, when two operations follow one another we speak of their product. What then is the product of two reflections? One reflection 'hands' a set of points from left to right, and a second will 'hand' it back again so that the product is certainly a proper isometry. Furthermore if the two reflecting lines intersect in a point, O, it is easily shown that this isometry is a rotation about O through twice the angle between them (Figure 2.15). If OR_1 and OR_2 are the two reflecting lines and P_1 is the image of a point P in OR_1, and P_2 is the image of P_1 in OR_2 we have $2\angle R_1OP_1 = \angle POP_1$ and $2\angle P_1OR_2 = \angle P_1OP_2$ so that, adding, $2\angle R_1OR_2 = \angle POP_2$. Since $\mathbf{R}_2\mathbf{R}_1$ is a clockwise rotation, $\mathbf{R}_1\mathbf{R}_2$ is the corresponding

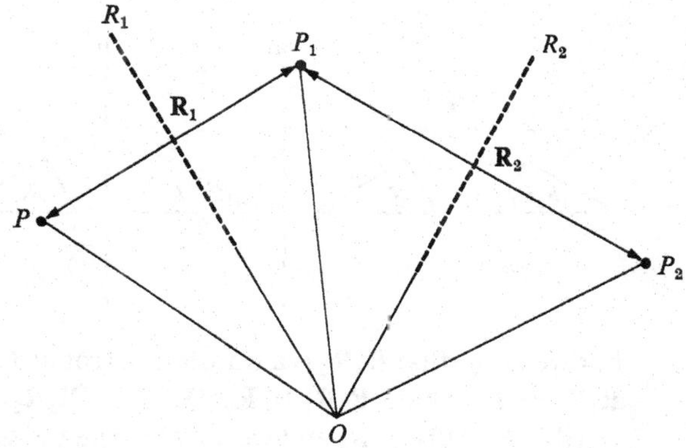

Figure 2.15

anti-clockwise rotation. Remembering that $\mathbf{R}_1^{-1}$ stands for the inverse operation to $\mathbf{R}_1$, that is, whereas the latter takes P to P_1, the former carries P_1 back again to P, we see that

$$\mathbf{R}_2\mathbf{R}_1 = \mathbf{R}_2^{-1}\mathbf{R}_1^{-1} = (\mathbf{R}_1\mathbf{R}_2)^{-1}$$

which is the same as $\mathbf{R}_1\mathbf{R}_2$ if the two mirrors are at right angles to one another in which case $(\mathbf{R}_1\mathbf{R}_2)^2 = \mathbf{I}$. Note that $(\mathbf{R}_1\mathbf{R}_2)^{-1} = \mathbf{R}_2^{-1}\mathbf{R}_1^{-1}$ and not $\mathbf{R}_1^{-1}\mathbf{R}_2^{-1}$. We put our socks on *first* and then our shoes, but we cannot take our socks off *first* before removing our shoes. If $\mathbf{R}_1\mathbf{R}_2 = \mathbf{I}$ then the two mirrors are the same, but if the two mirrors are parallel, then $\mathbf{R}_1\mathbf{R}_2$ is a degenerate rotation, or translation. This is familiar in rooms where two parallel walls are mirrored and we see ourselves disappearing into the distance, back to back, front to front, *ad infinitum*. The first image in a mirror m_1 is R_1, this is then reflected in a parallel mirror m_2 to become R_2R_1 behind us, this in turn is reflected in m_1 to become $R_1R_2R_1$ before us, then $R_2R_1R_2R_1$, $R_1R_2R_1R_2R_1$ and so on. At the same time our image is reflected behind us in m_2. This image R_2 is then reflected in m_1 to become R_1R_2 so that we see our back. Then the sequence continues as before, $R_2R_1R_2$, $R_1R_2R_1R_2$ and so on. The images may be divided into two sets: those *in* mirror m_1 in front of us, and those *in* m_2 behind. The sequence of images may be expressed as

$$m_1\text{:}\quad R_1, R_1R_2, (R_1R_2)R_1, \ldots, (R_1R_2)^n, (R_1R_2)^nR_1, \ldots$$

$$m_2\text{:}\quad R_2, R_2R_1, (R_2R_1)R_2, \ldots, (R_2R_1)^n, (R_2R_1)^nR_2, \ldots$$

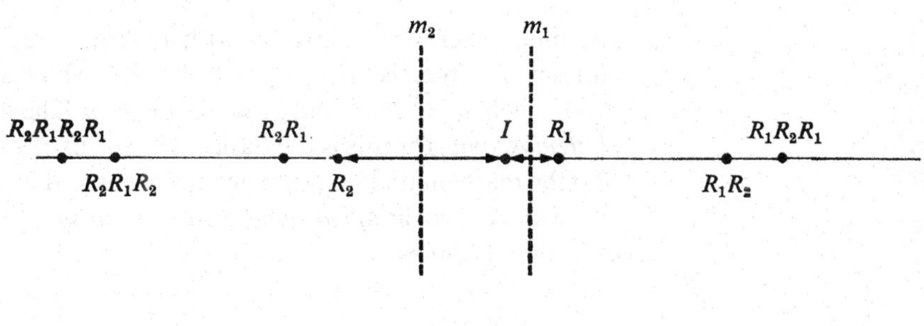

Figure 2.16

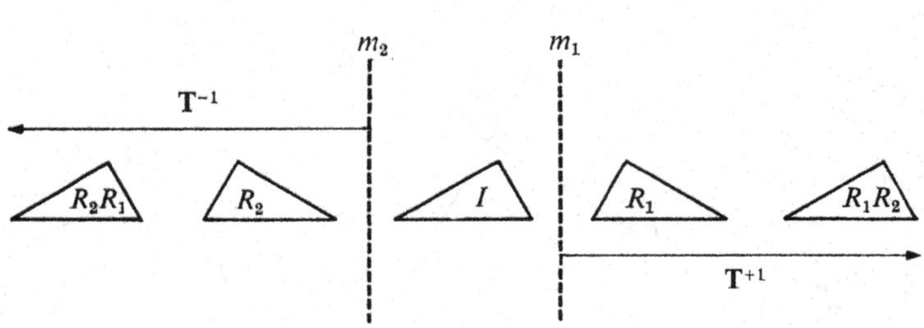

Figure 2.17

but we know that $\mathbf{R_1R_2}$ is a translation (rotation through 0°) so that if $\mathbf{R_1R_2 = T}$ we have $\mathbf{R_2R_1 = R_2^{-1}R_1^{-1} = (R_1R_2)^{-1} = T^{-1}}$ and $\mathbf{T^{-1}R_1 = (R_1^{-1}R_2^{-1})R_1 = R_2}$ and we may rewrite the sequences more simply as

m_1: $R_1, T, TR_1, \ldots, T^n, T^nR_1$

m_2: $T^{-1}R_1, T^{-1}, T^{-1}R_1, \ldots, T^{-n}, T^{-n}R_1$

showing that the double mirrors produce proper isometries – the translations – and improper – the reflections with translations. The product of reflections in two parallel mirrors is a translation through twice the distance between the mirrors.

These statements are neatly visualized by taking a piece of tracing paper and folding it along two parallel lines, call these m_1 and m_2. Draw a triangle between m_1 and m_2 when the paper is unfolded, label it I. Fold the paper about m_1 and trace the outline of I giving it the name R_1, now fold the paper about m_2, trace I through and mark it R_2. Next fold about m_1 keeping the paper between the 'mirrors' flat on the table all the time. Trace over the image of R_2 marking it R_1R_2. Unfold and crease along m_2 when the image R_1 will be seen. Trace R_1 and mark the new drawing R_2R_1. Continue folding and unfolding along m_1 and m_2 in order, marking each new tracing with an additional R_1 or R_2 in front of the name of the triangle being traced according to whether the fold is m_1 or m_2. Finally, unfold the tracing paper flat. The writing will be back to front, but the system should be perfectly clear (Figure 2.17).

The triangles marked with an even number of R_1s and R_2s are seen to be straight translations, $\mathbf{T}^{\pm n}$, of I through twice the distance between the mirrors, and those with an odd number are translations, $\mathbf{T}^{\pm n}$, of the first reflection R_1.

A similar exercise can be done with a piece of tracing paper folded along lines which intersect in one point at angles which are sub-multiples of 180°, such as 90°, 60°, and in general 180°/n. The images produced are *kaleidoscopic*, for this is the basic mechanism of this fascinating toy. To the mathematician, point groups generated by reflections of this kind are known as *dihedral* because of their bilateral symmetry about the mirror planes.

The group generated by three equiangular mirrors, $n = 3$, is D_3, by four, with $n = 4$, D_4 and so on. D_2 is the dihedral group formed by two perpendicular mirrors $n = 2$, and, for $n = 1$, D_1 is a straight reflection in a line. Take some tracing paper folded along two mutually perpendicular lines, m_1 and m_2, and draw a triangle in one quadrant. Call it I. Now fold about m_1 and trace a triangle R_1 over I. Unfold and crease again along m_2. Trace over I calling it R_2 and over R_1 naming it R_2R_1. Unfold and repeat the process taking m_2 first and then m_1 (Figure 2.18).

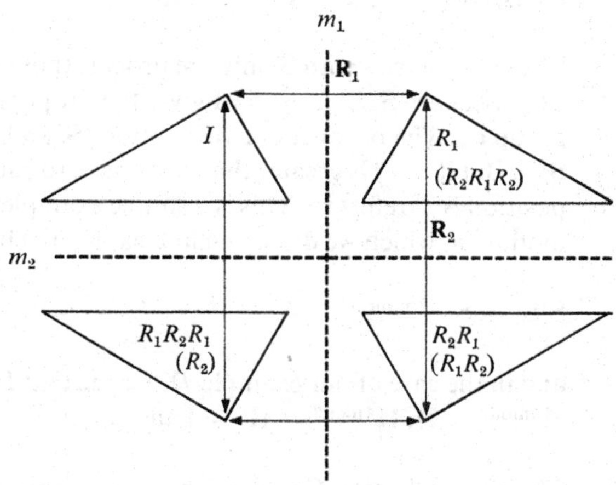

Figure 2.18

Note that we may now call the triangle R_1R_2 equally well R_2R_1. As we have seen, since $\mathbf{R}_1 = \mathbf{R}_1^{-1}$ and $\mathbf{R}_2 = \mathbf{R}_2^{-1}$ we may also write $\mathbf{R}_2\mathbf{R}_1 = \mathbf{R}_2^{-1}\mathbf{R}_1^{-1} = (\mathbf{R}_1\mathbf{R}_2)^{-1}$. These three equations, which may be expressed

$$\mathbf{R}_1^2 = \mathbf{R}_2^2 = (\mathbf{R}_1\mathbf{R}_2)^2 = \mathbf{I},$$

constitute what is known as the *abstract definition* of the symmetry group D_2. Similar definitions may be given for D_n:

$$\mathbf{R}_1^2 = \mathbf{R}_2^2 = (\mathbf{R}_1\mathbf{R}_2)^n = \mathbf{I}.$$

Looking again at our unfolded sheet we see that the triangles in diagonally opposite quadrants are proper isometries and can be effected by means of a rotation about the intersection of m_1 and m_2 through 180° (Figure 2.19). Let us call this rotation **H**, in view of the fact that it is a *halfturn*. Then we see that $\mathbf{H} = \mathbf{R}_1\mathbf{R}_2 = \mathbf{R}_2\mathbf{R}_1$ and $\mathbf{R}_2 = \mathbf{H}\mathbf{R}_1$ so that the four symmetry positions in D_2 are I, R_1, H and HR_1. In general, the

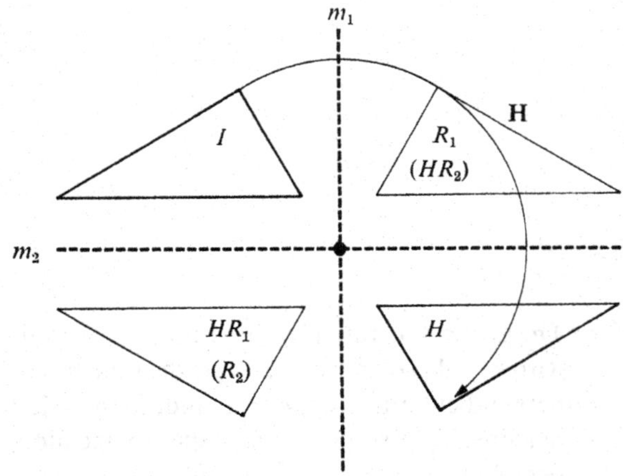

Figure 2.19

rotation, $\mathbf{R}_1\mathbf{R}_2 = \mathbf{S}$, is through $360°/n$ for D_n. D_n may be expressed entirely as products of $\mathbf{R}_1$ and $\mathbf{S}$:

$$\mathbf{R}_1, \mathbf{S}, \mathbf{S}\mathbf{R}_1, \mathbf{S}^2, \ldots, \mathbf{S}^{n-1}\mathbf{R}_1, \mathbf{S}^n = \mathbf{I}.$$

Those isometries with **S** only are proper, those with $\mathbf{R}_1$ are improper. The sequence $\mathbf{S}, \mathbf{S}^2, \ldots, \mathbf{S}^n = \mathbf{I}$ gives the proper isometries of the cyclic group C_n. The product of two rotations $\mathbf{S}^i$ and $\mathbf{S}^j$ from this sequence is $\mathbf{S}^{i+j}$. But if $i + j$ is greater than n we have to subtract n to find out what position $\mathbf{S}^{i+j}$ refers to. This is a simple example of the clock arithmetic, modulo n, which we discussed in Chapter 1. Thus,

$$\mathbf{S}^i\mathbf{S}^j = \mathbf{S}^{(i+j)\bmod n}$$

and in the case of our example D_2 we see that $\mathbf{H}^{2\bmod 2} = \mathbf{H}^0 = \mathbf{I}$, $\mathbf{H}^{3\bmod 2} = \mathbf{H}$, $\mathbf{H}^{4\bmod 2} = \mathbf{H}^0 = \mathbf{I}$ and so on.

The glide reflection **G** is, as we have seen, composed of a reflection and a translation. They evidently commute and we have

$$\mathbf{G} = \mathbf{RT} = \mathbf{TR}.$$

But a translation is also the product of two halfturns $\mathbf{H}_1\mathbf{H}_2$, or, as we have seen with the parallel mirrors, the product of two parallel reflections $\mathbf{R}_1\mathbf{R}_2$. A glide reflection may be expressed as the product of three reflections $\mathbf{R}\mathbf{R}_1\mathbf{R}_2$ (two perpendicular to the third), or of one halfturn and a reflection or a reflection and a halfturn. The latter statement is illustrated by Figure 2.20. We have a glide axis g, and two mirrors m_1 and m_2 perpendicular to g. The points of intersection of m_1 and m_2 with g we call O_1 and O_2. It is clear that the halfturn $\mathbf{H}_1$ is equivalent to $\mathbf{R}_1\mathbf{R}$ or $\mathbf{R}\mathbf{R}_1$ where **R** is reflection in the glide axis g, and that $\mathbf{H}_2$ is the same as $\mathbf{R}_2\mathbf{R}$ or $\mathbf{R}\mathbf{R}_2$. But

$$\mathbf{T} = \mathbf{H}_1\mathbf{H}_2 = \mathbf{R}_1\mathbf{R}\mathbf{R}\mathbf{R}_2 = \mathbf{R}_1\mathbf{R}_2$$

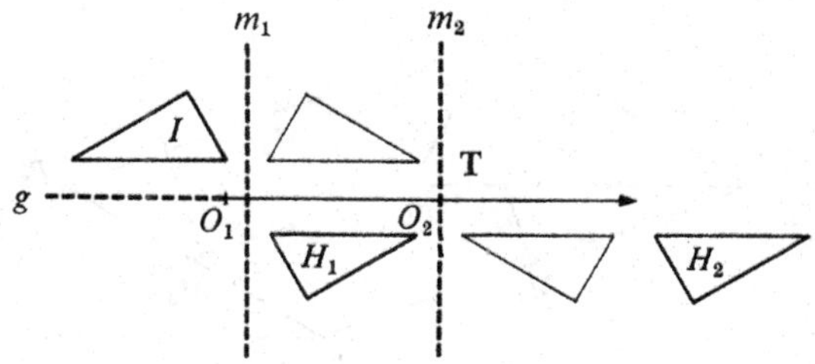

Figure 2.20

since $\mathbf{R}^2 = \mathbf{I}$. The glide reflection **G** is thus defined by

$$\begin{aligned}\mathbf{G} &= \mathbf{RT} = \mathbf{RR}_1\mathbf{R}_2 = \mathbf{H}_1\mathbf{R}_2\\ &= \mathbf{TR} = \mathbf{R}_1\mathbf{R}_2\mathbf{R} = \mathbf{R}_1\mathbf{H}_2.\end{aligned}$$

To summarize: we have been discussing transformations in one-to-one correspondence for the whole plane onto itself. Every point P maps onto a unique point P'. An isometry is a special kind of transformation: it is the transformation which preserves length. Isometries are of particular interest to architects and designers. Given, for example, an L-shaped room of fixed dimensions to be assembled with some other spaces it is likely that the architect will use most of the isometries we have discussed. He will certainly move the room around without changing its orientation in an attempt to find a desirable place for it. By doing so he *translates* it, **T**. He may *rotate* it, **R**, through an angle, particularly in a rectangular assembly through a right angle or a *halfturn*, **H**. If this does not satisfy his requirement he may 'flip' the room over to make it into a Γ-shaped room. He will have reflected it, **R**, and as he moves this new isometry into a suitable place he will be transforming his room by a *glide reflection*, **G**. These moves are, strictly speaking, *symmetry operations* with respect to the configuration: such an operation refers to a specific configuration – in our example the L-shaped room – and not the whole plane. A symmetry operation is an isometry that transforms a figure onto itself.

In the plane there are two distinct kinds of isometry. Those that preserve sense are called *proper*, those that are sense-reversing are said to be *improper*. Rotations, halfturns and translations are proper isometries: reflections and glide reflections are improper. The two general isometries are rotation and glide reflection, the others are special cases of these two. A halfturn is a rotation through 180°, and a translation is a rotation about a point at infinity. A reflection is the special case of a glide reflection when there is no translation (or glide). Figures 2.21–2.25 show each of the isometries and the constructions required to determine centres of rotation or lines of reflection. One interesting pattern emerges from this tabulation. If the isometry carries P_i onto P_i' and we note the arrangement of the midpoints M_i of P_iP_i' we see that with translations and rotations there is *no line* which can be struck through the points M_i, although in the case of the halfturn an *infinite number of lines* pass through M_i since all the midpoints are coincident and any number of

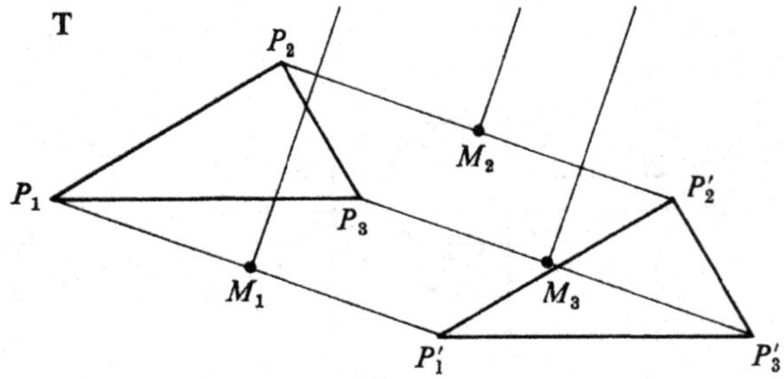

Figure 2.21

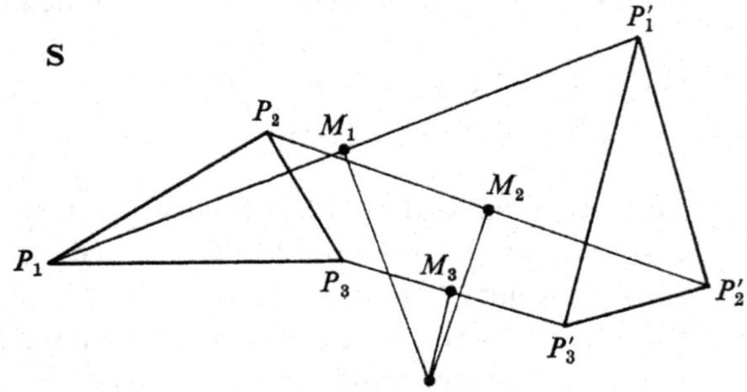

Figure 2.22

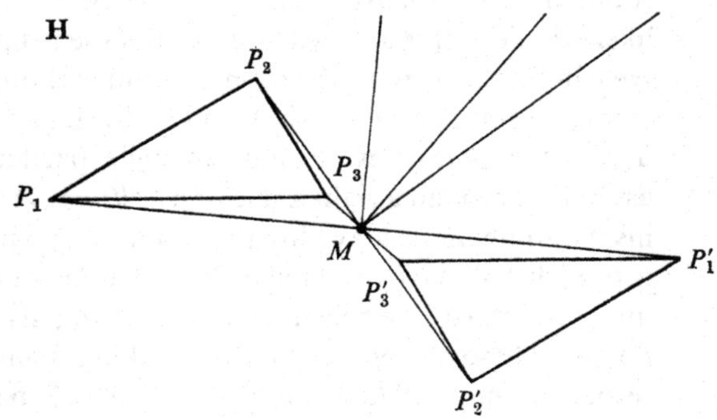

Figure 2.23

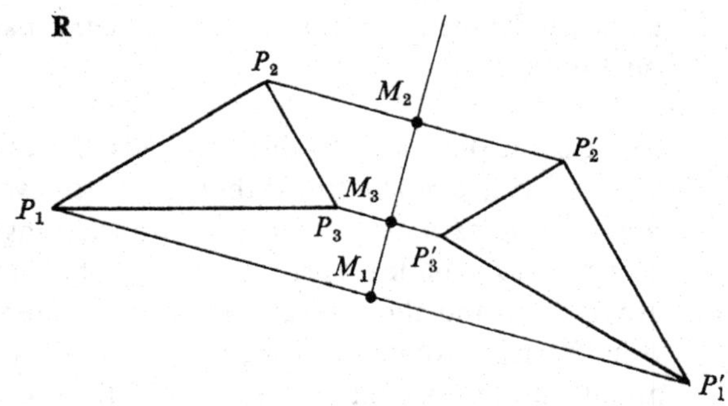

Figure 2.24

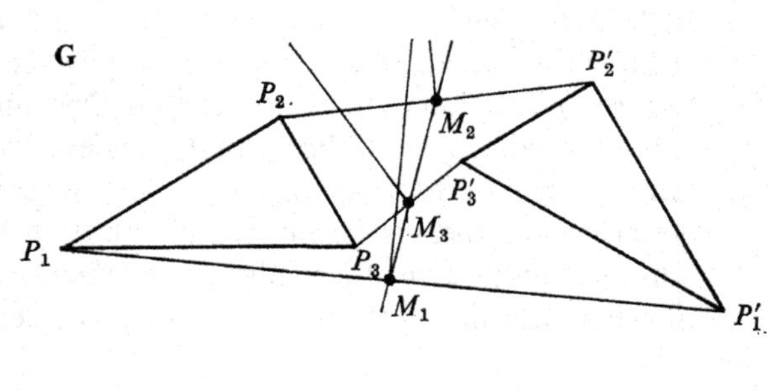

Figure 2.25

lines may be drawn through a point. On the other hand, with reflections and glide reflections the points M_i all lie on *one line* – the mirror or glide axis. When we look at the perpendiculars to P_iP_i' through M_i we find that with rotations and halfturns the perpendiculars all pass through *one point* and this is true for translations if we permit the convention that parallel lines intersect in a point at infinity. But with reflections we find that the perpendiculars all coincide along the mirror so that they are coincident with an *infinite number of points*. This is not so with glide reflections where the perpendiculars share *no point* in common. The pattern of this emerges very clearly when we tabulate the results:

	Proper		*Improper*	
Type of isometry	**T,S**	**H**	**G**	**R**
Midpoints collinear (number of lines)	0	∞	1	1
Perpendiculars concurrent (number of points)	1	1	0	∞

3 Symmetry groups in the plane

We are now in a position to enumerate the plane symmetry groups which may be classed according to their translational structure. There are three cases to be considered:

1 no translations present, the two finite point groups;
2 one translation present only, the seven frieze groups;
3 more than one direction of translation, the seventeen wallpaper groups.

Fejes Tóth has described these in some detail,[1] but perhaps a more approachable account is given by J. H. Cadwell in his book *Topics in Recreational Mathematics*.[2] Here we shall not repeat the mathematical discussion, but instead confine ourselves to illustrating the three classes of plane symmetry groups and to commenting on their application in architectural and environmental studies.

In Figure 3.1 we illustrate the two planar *point groups*. These are the only finite symmetry groups in the plane. The *cyclic group* has sense, or direction of spin, although this disappears if the figure itself, which is the subject of the symmetry operation, is bilaterally symmetrical about an axis through the centre of rotation. Properly speaking the group should then be called *dihedral*. The outstanding mathematician, Hermann Weyl, in his lectures on symmetry delivered at Princeton University on the eve of his retirement from the Institute of Advanced Studies in 1951, credited Leonardo da Vinci with the first tabulation of all the possible finite groups of rotation (proper and improper), namely

the cyclic group $C_1, C_2, C_3, \ldots, C_n, \ldots$
and the dihedral group $D_1, D_2, D_3, \ldots, D_n, \ldots$

where n is the period of the group, or the number of $360°/n$ rotations required to complete a full revolution. At one time, Leonardo had been engaged on a systematic study of the possible symmetries of a central building and how to attach chapels and niches without destroying the symmetry of the nucleus. In fact, like his architectural contemporaries, Leonardo essentially developed the dihedral group of symmetries for these proposals, avoiding the skewed 'asymmetry' of the cyclic group.

[1] Fejes Tóth. *Regular Figures*, Oxford, Pergamon Press, 1964.

[2] J. H. Cadwell. *Topics in Recreational Mathematics*, Cambridge University Press, 1966, pp. 112–29. For a more technical discussion see Paul B. Yale, *Geometry and Symmetry*, San Francisco, Holden-Day, 1968.

Figure 3.1
The two point groups: the cyclic group, C_n, consists of rotations about a single point, O, through $2\pi/n$, while the dihedral group, D_n, includes a reflection through O.

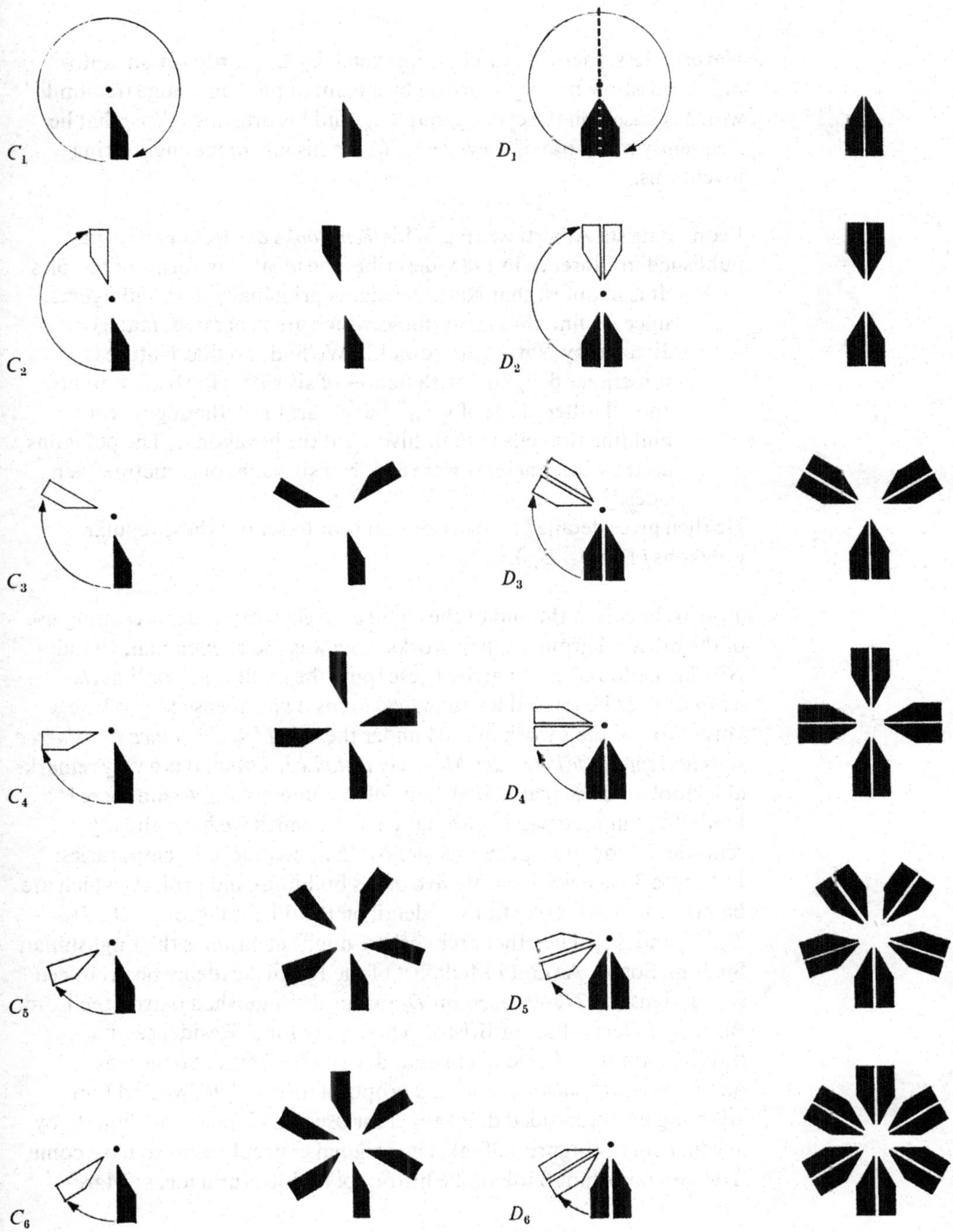

[3] This design is reproduced in A. E. Popham, *The Drawings of Leonardo da Vinci*, London, The Reprint Society, 1952, p. 303.

[4] Leone Battista Alberti. *Ten Books on Architecture*, facsimile of the Leoni edition by Joseph Rykwert, London, Tiranti, 1955, p. 138.

[5] For a discussion of the work of Ledoux see Emil Kaufmann, 'Three Revolutionary Architects, Boulleé, Ledoux, and Lequeu' in *Transactions of the American Philosophical Society*, vol. 42, part 3, 1952, pp. 431–564.

Nevertheless, there is a fascinating sketch by Leonardo for an 'automatic crossbow battery',[3] driven by a team of men on a huge treadmill which is based on the cyclic group C_{48}, and his drawings show that he frequently made use of skew gears, C_n, in his ingenious engineering inventions.

Leone Battista Alberti writing in his *Ten Books on Architecture*, first published in Florence in 1485, describes the ideal plan forms of temples:

> 'It is manifest that Nature delights principally in round figures, since we find that most things which are generated, made or directed by Nature are round.... We find too that Nature is sometimes delighted with figures of six sides; for bees, hornets, and all other kinds of wasps have learnt no other figure for building the cells in their hives, but the hexagon.... The polygons used by the Ancients were either of six, eight or sometimes ten sides.'[4]

He then gives detailed instructions on how to set out these regular polygons (D_6, D_8, D_{10}).

Two architects at the end of the eighteenth century made interesting use of the dihedral group in their works. One was the Frenchman, Claude-Nicolas Ledoux,[5] who narrowly escaping the guillotine – or '*l'hache nationale*', as he termed it – managed at his own expense to produce a volume of his life's work in 1804 under the title *L'Architecture Considérée sous le Rapport de l'Art, des Mœurs et de la Législation*. It is a very remarkable book and the plates are beautiful to contemplate. Again there is a kind of taxonomic passion about the work which we have already remarked upon in respect of his fellow 'Napoleonic' contemporaries. In Figure 3.2a–e we illustrate five of his buildings and projects which are based with minor exceptions of detail on the dihedral groups D_1, D_2, D_3, D_4 and D_{12}. The other architect we might mention is the Englishman Sir John Soane. As Gold Medallist of the Royal Academy he produced two designs in 1778–9 based on D_3 for his distinguished patron the Lord Bishop of Derry, Earl of Bristol. These were for a 'Residence of a Canine Family in Ancient Times and Modern Times'. Soane was obviously enthusiastic about this proposal for, in 1796, we find him adapting his three-sided dog kennel for use as a 'Sepulchral Church' by another client (Figure 3.2f–g). The original concept seems to have come from an important book in the history of architectural ideas: Marc-

Figure 3.2
Examples of point groups in architectural plans:
a, Montmorency Palace, D_1
b, De Witt House, D_2
c, Inn St. Marceau, D_3
d, Barrière de Picpus, D_4
e, House of Entertainment (circular colonnade and pavilions), D_{12}
all projects and buildings by Claude-Nicolas Ledoux:

f, Sepulchral Church, D_3
g, Kennels, D_3
both designs by Sir John Soane:

and projects and buildings by Frank Lloyd Wright
h, Greek Orthodox Temple, C_1
i, St Mark's Apartment Tower, C_2
j, Huntingdon Hartford Clubhouse, D_3
k, Suntop Homes, C_4
l, Daphne Mortuary, D_5

[6] From Arthur T. Bolton, *The Works of Sir John Soane, F.R.S., F.S.A., R.A. (1753–1837)*, London, The Sir John Soane Museum Publication, no. 8, (undated, *c.* 1924).

[7] James Fergusson. *The Illustrated Handbook of Architecture*, London, John Murray, 1859, p. 608.

[8] The English Renaissance architectural theorist, Sir Henry Wootton, condemned pentagons except in military (*sic*) architecture. He wrote that as pentagonal designs 'do more aim at rarity than commodity, so for my part I had rather admire than commend them'. From Sir Henry Wootton, *Elements of Architecture*, facsimile of 1624 edition published in London, 1903.

Antoine Laugier's *Essai sur l'Architecture* of 1752. In this Abbé Laugier describes a form for a church:

> 'In the triangle I inscribe a circle, which gives me the plan of the dome, which I carry up from the ground. At the three angles I construct three rotundas, which give me three sanctuaries, where I place three altars. On each of the three faces I open a door-way in the centre, so as to have three entrances, each facing an altar.'[6]

The need to repeat functional elements to justify the threefold symmetry appears absurd, and such buildings are rarely, if ever, built. While the plans of buildings based on the point groups are often pretty, they are not always of 'the sternest utility'. As Fergusson remarked of a triangular church at Planes in Provence: 'As a constructive puzzle it is curious, but it is doubtful how far any legitimate use could be made of such a *capriccio*.'[7]

In a rare instance of the use of D_5 in architecture, Frank Lloyd Wright projected a design for a mortuary in San Francisco on strictly functional grounds.[8] The client's brief was for five chapels of rest. Here there was a justifiable reason for having five entrances in order that mourners from different groups should not meet. The points symmetry made it possible for each chapel to share a single, central cremation core which served them (Figure 3.2l).

Frank Lloyd Wright also used other point groups in his work. Figure 3.2k illustrates the plan for Suntop Homes, built near Philadelphia, in which four identical houses are rotated through 90° to one another. The symmetry, C_4, is not unreasonable here, since each house is itself planned sensibly within its own area: of course, orientation to the sun differs for each house but with proper siting no house need be without east and west light. In a project for Huntingdon Hartford, Wright essentially used the dihedral group D_3 as a base for his design of the clubhouse for a luxury country club (Figure 3.2j). Three circular, bowl-like lounges and bars were to have been cantilevered out from a triangular base containing lifts and staircases. But overall, the exuberance of this project could not be bound, except in part, to the rigours of symmetry. Housing again provided a reason for cyclic symmetry in a project, in 1929, for apartment buildings in the Bouwerie, New York. Internally each dwelling is virtually identical (Figure 3.2i). However, a full C_4 symmetry as in the Suntop Homes would have meant four lifts

Figure 3.3
The seven frieze groups:

F_1, translation along one axis

F_1^1, reflection in the axis

F_1^2, reflection perpendicular to the axis

F_1^3, reflection and glide translation

F_2, halfturn

F_2^1, reflection in the axis

F_2^2, glide reflection, or reflection in pairs of axes

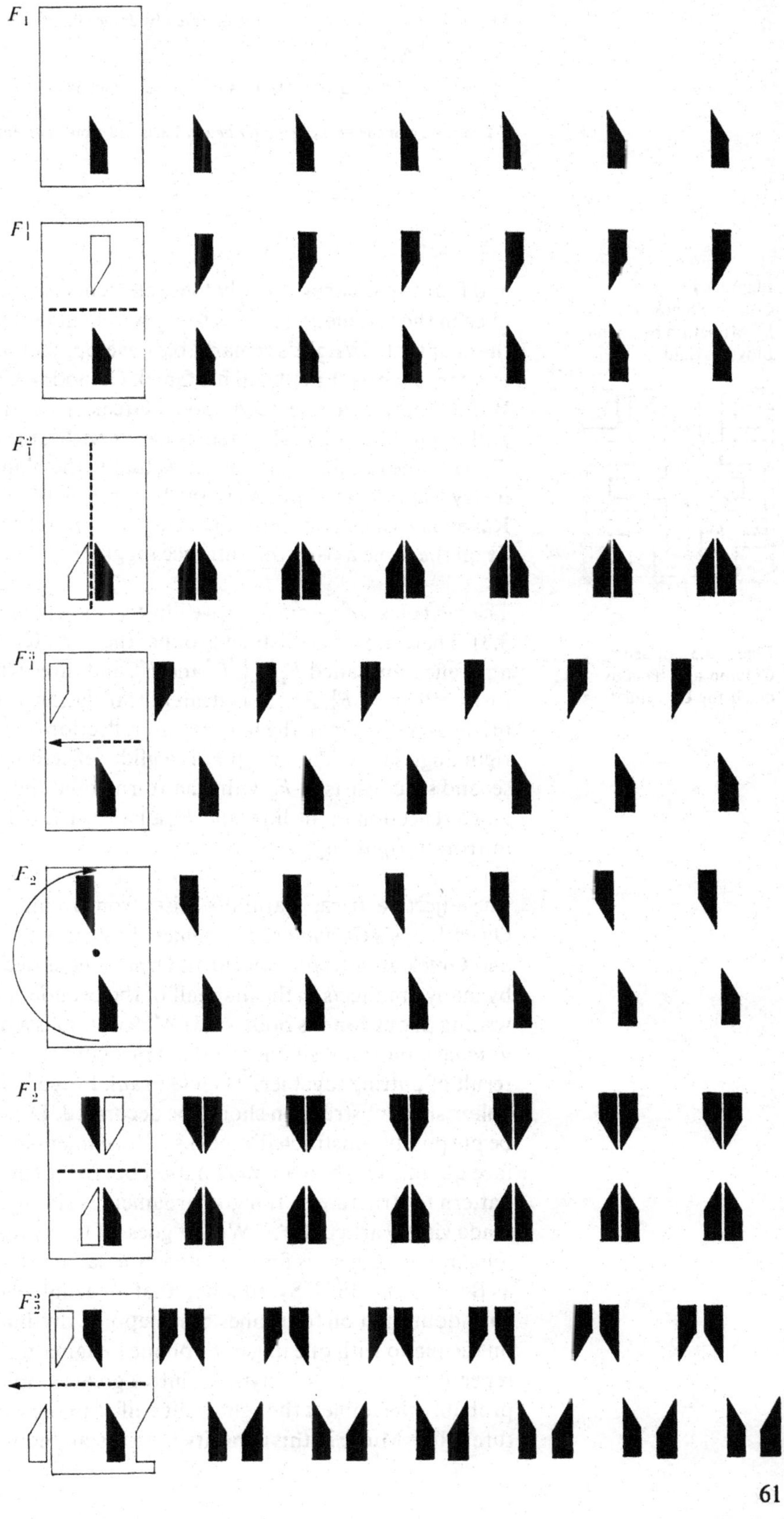

[9] From Le Corbusier, *My Work*, translated by James Palmes, London, The Architectural Press, 1960, p. 24.

[10] From Frank Lloyd Wright, *An Autobiography*, pp. 347–8.

[11] From Le Corbusier, *La Ville Radieuse*, Paris, Editions Vincent Fréal et Cie, 1964, p. 283.

Figure 3.4
Concrete block, D_4, for La Miniatura by Frank Lloyd Wright

Figure 3.5 opposite
Column and balcony detail for La Miniatura

and four fire-escapes. Clearly these elements can be shared and with two of each the symmetry reduces to C_2. We shall return towards the end of the chapter to Wright's remarkable development of this early project in later schemes. Finally, in his Greek Orthodox Church in Madison, Wright sets a concrete bowl upon a Greek cross structure (Figure 3.2h) Although the structural system is based on D_4, the plan itself reduces to D_1 with one dominant entrance. Actually, the plan has *no* dihedral symmetry – in a strict sense. As in the Byzantine church of San Vitale at Ravenna (which is structurally octagonal, D_4) the main altar does not lie on the same axis as the entrance loggia.

The next class of symmetries we illustrate are the *frieze groups* (Figure 3.3). There are seven distinct groups: there are four based on translation and reflection called F_1, F_1^1, F_1^2 and F_1^3, and three which permit half-turns called F_2, F_2^1, F_2^2. F_1 is straight translation along a line, F_1^1 involves reflection in the line, and F_1^2 reflection in a pair of mirrors at right angles to it. The group F_1^3 is a glide reflection along the line. The second set consists of F_2 with a halfturn about the line, F_2^1 a halfturn and a reflection in the line, and F_2^2 a halfturn and reflections in two mirrors at right angles to it.

The adjective 'frieze' naturally arises from architectural precedent. Owen Jones's *Grammar of Ornament* has many examples from Egyptian and Greek architecture onwards. Ornament as decoration was rejected by many architects in the first half of the twentieth century. Le Corbusier writing about Jones's book said: 'Decoration is a debatable topic, but *ornament* pure and simple is a thing of significance; it is a synthesis, the result of putting together.'[9] While Frank Lloyd Wright, echoing Jones's aphorism 'Construction should be decorated. Decoration should never be purposely constructed', speaks of 'ornament meaning not only surface qualified by human imagination but imagination giving natural pattern to structure.... Integral ornament is simply *structure-pattern* made visibly articulate'.[10] Wright goes on to say that what he calls 'integral ornament' is founded upon the same organic simplicities as Beethoven's Fifth Symphony, 'that amazing revolution in tumult and splendour built on four tones based upon a rhythm a child could play on the piano with one finger. Supreme imagination reared the four repeated tones, simple rhythms, into a great symphonic poem that is probably the noblest thought-built edifice in the world. And Architecture is like Music in this capacity for the symphony'. With this, Le

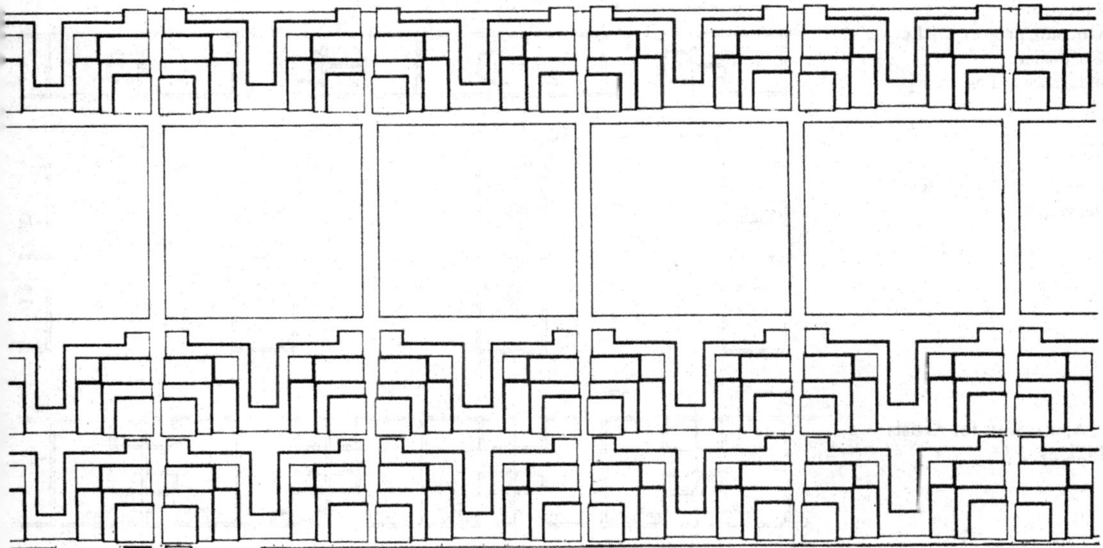

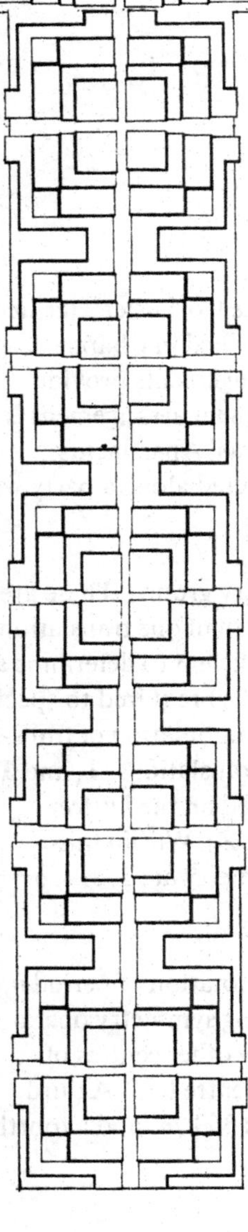

Corbusier agrees, as we see from his stirring statement on one of his grandest urban schemes, the development of the left bank of the Schelde at Antwerp: 'Il faut insister sur ce fait que *la diversité la plus totale règnera dans cette ville grâce aux infinies combinaisons des éléments simples qui la constituent...* celui des vastes ensembles architecturaux conjugués suivant une symphonie croissante.'[11] The problem about this kind of analogy at the city scale may be summed up by a question: 'Who wants to live *in* someone's symphony every day?'

In 1921 Frank Lloyd Wright built La Miniatura in Pasadena for Mrs George Millard. The house is built astride a small ravine between eucalyptus trees. It is constructed of concrete blocks reinforced with steel. The blocks were cast in wooden moulds and they are impressed by a design based on D_4 (Figure 3.4). The columns inside the house which support the balcony in the double height living-room are clad in half-blocks arranged as in the frieze group F_1^1 although, more properly, the symmetries of each block make this into F_2^1 (Figure 3.5). The frieze along the bottom edge of the balcony, oddly enough, does not represent a frieze group. This is so, since a half block on the top row is related to one on the bottom by a translation at 90° to the axis of the frieze, and no translations are permitted at an angle to the axis. Indeed this additional translation means that this particular frieze is a 'fragment' of the third class of symmetry groups – the wallpaper groups. However, by considering two half blocks – the top one over the bottom – as our unit element, the border can be read as frieze group F_1, or F_1^2 because of the reflected symmetry of each block. It seems reasonable to suppose that it is, in part, the ambiguity of multiple readings that makes pattern and symmetry so fascinating.

There are many examples of frieze group symmetries in mass housing where identical units are assembled into apartment buildings, terraces

Figure 3.6
Housing projects by Le Corbusier:
a, terrace at Pessac, F_1

b, semi-detached row at Pessac, F_1^1

c, Apartments for Artists 1928/29, F_1^2

d, terrace at Pessac, F_1^3

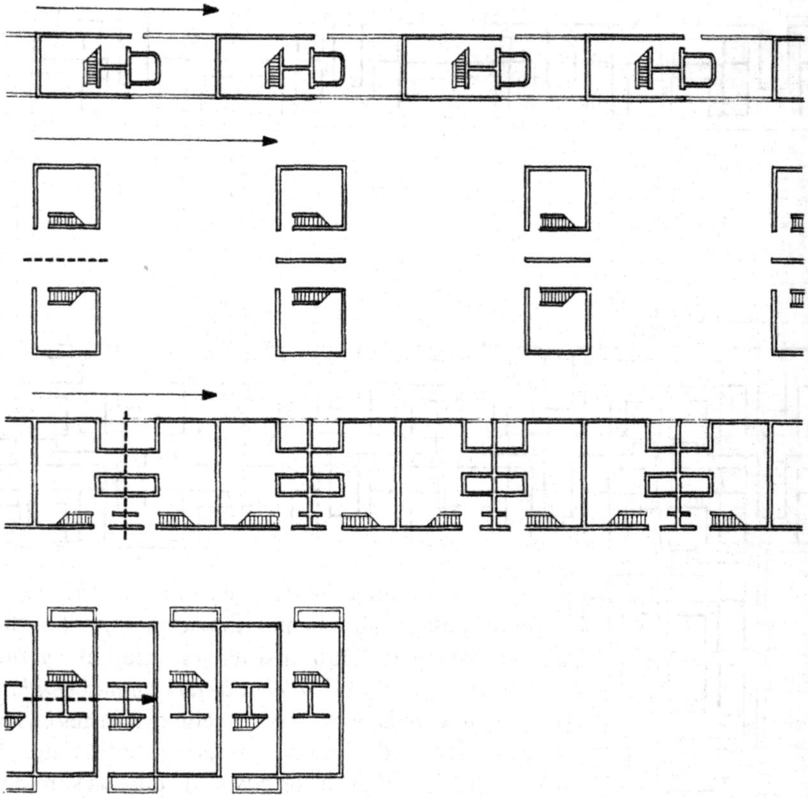

and other arrangements. The individual units of Le Corbusier's terraced and row housing at Pessac have already been described in Chapter 1. In Figures 3.6a, b and d we see that the block arrangements provide examples of the frieze groups F_1, F_1^1 and F_1^3. A frequent arrangement is illustrated by the F_1^2 pattern of the project for Artists' Apartments, Figure 3.6c, where individual apartments are reflected along a party wall containing the services to kitchens and bathrooms.

Let us now look at the third class of plane symmetry groups. These are the *wallpaper groups* which are generated by more than one translation. In two dimensions any two distinct vectors are sufficient to determine a third: this is analogous to the two coordinates (x, y) required to specify a point in a plane. In general, it follows that any combination of translations in a plane may be reduced to two distinct translations, $\mathbf{T}_1$ and $\mathbf{T}_2$. Any symmetry group of the plane will then be based on the *lattice generated* by $\mathbf{T}_1{}^i\,\mathbf{T}_2{}^j$ where i and j are integers $(i, j \in Z)$. Within this lattice the only possible periods of rotational symmetry are 2, 3, 4, 6. This is known as the crystallographic restriction.

The proof is elegantly simple. Let P be a centre of rotation of period n, that is the angle of rotation is $360°/n$. The remaining symmetry operations of the lattice transform P into infinitely many other centres of rotation of the same period. Let Q be the nearest centre to P. A third centre, P', is derived from Q by a rotation through $360°/n$; and a fourth,

12 Quoted from 'Plan by Frank Llyod Wright' in *City Residential Land Development Studies in Planning*, edited by Alfred B. Yeomans, Chicago, City Club of Chicago, 1915, pp. 96–102.

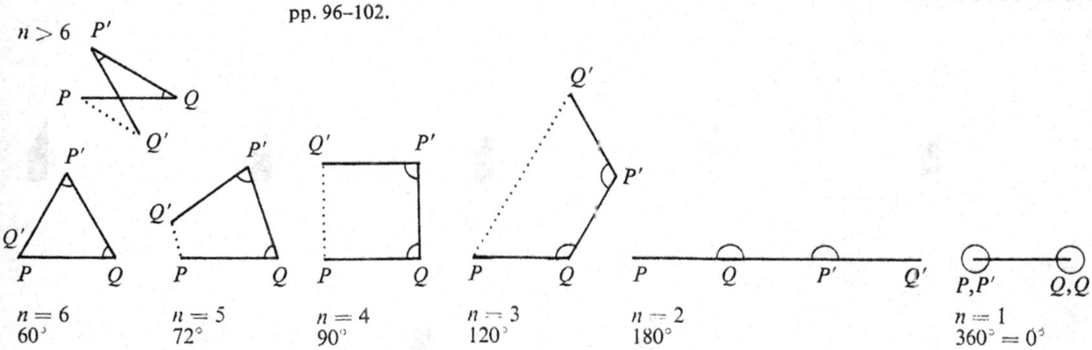

Figure 3.7

Q', by a rotation about P'. The segments PQ, QP' and $P'Q'$ are all equal. If n is greater than 6, the angle of rotation will be less than 60° and Q' will be nearer to P than Q which contradicts our assumption that Q was the nearest point. If $n = 6$ then P and Q' coincide. If $n = 5$, however, Q' will be closer to P than Q which again is a contradiction. Values of $n = 4, 3, 2$ are satisfactory. These angles – 60°, 90°, 120°, and 180° – are the bases of the wallpaper groups W_6, W_4, W_3, W_2. The wallpaper group W_1 is arrived at by direct translation with no rotations.

We illustrate the seventeen possible groups in Figures 3.8 to 3.12. In each case the unit cell, or element, of the pattern is delineated and the symmetry operations indicated. Fejes Tóth gives a full account of these symmetry groups in his book *Regular Figures*.

In 1913 Frank Lloyd Wright submitted *hors concours* a proposal in a competition for the development of a Chicago quarter-section (one-quarter mile square). Wright developed most of the property along the main perimeter roads as commercial buildings, making them

> '"background" buildings... continuously banked against the noisy city thoroughfare.... By thus drawing to one side all the buildings of this nature into the location they would naturally prefer, the greater mass of the subdivision is left quiet and clean... left intact as a residence park, developed according to the principle of the "quadruple block plan".... Each householder is automatically protected from every other householder. He is the only individual upon the entire side of his block. His utilities are grouped to the rear with his neighbors' utilities, and his yard, front or rear, is privately his own. His windows all look upon open vistas and upon no one's unsightly necessities. His building is in unconscious but necessary grouping with three of his neighbors', looking out upon harmonious groups of other neighbors, no two of which would present to him the same elevation even were they all cast in the same mould. A succession of buildings of any given length by this arrangement presents the aspect of well-grouped buildings in a park, *of greater picturesque variety than is possible where façade follows façade*.'[12]

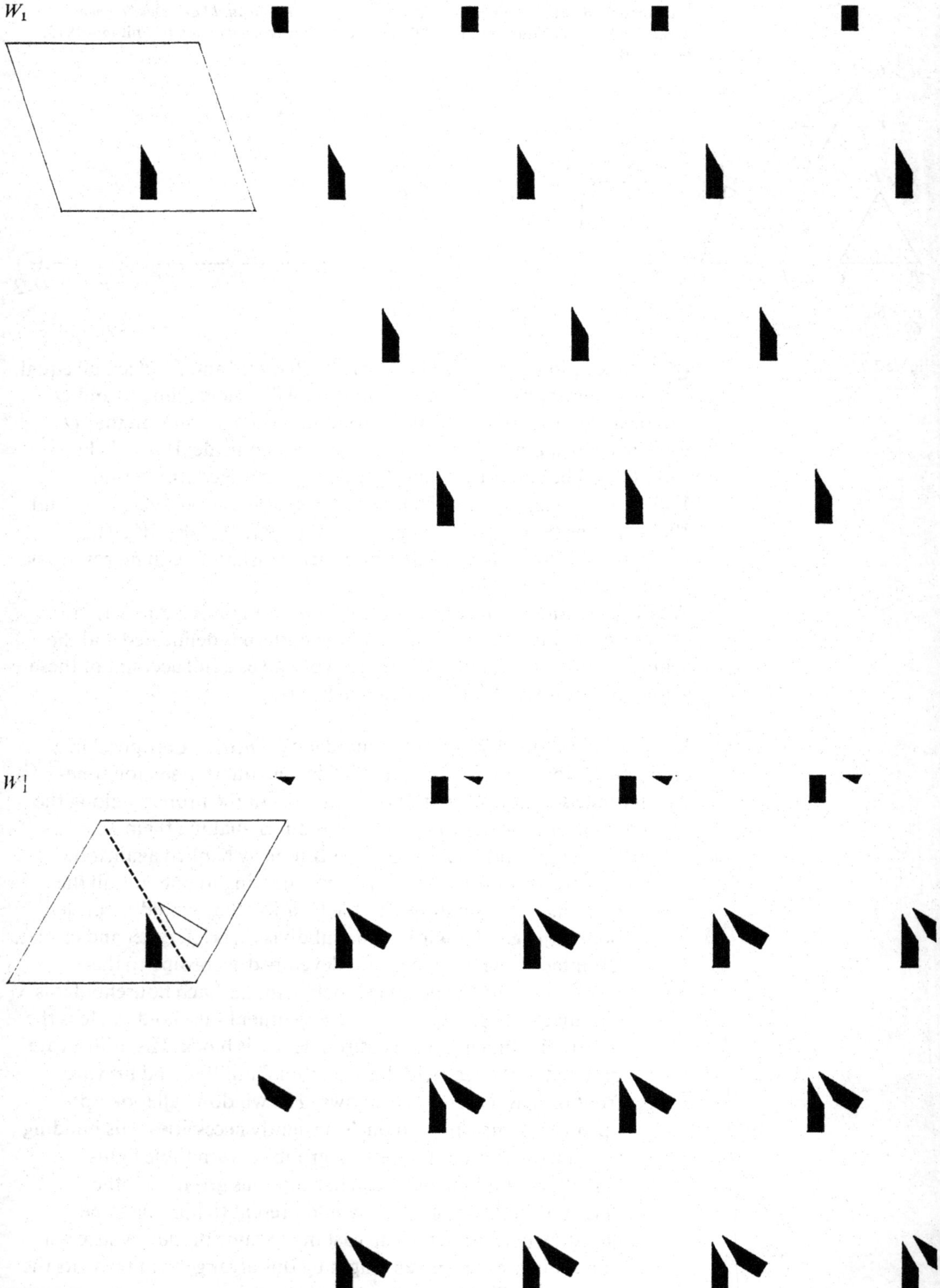

Figure 3.8

W_1^2

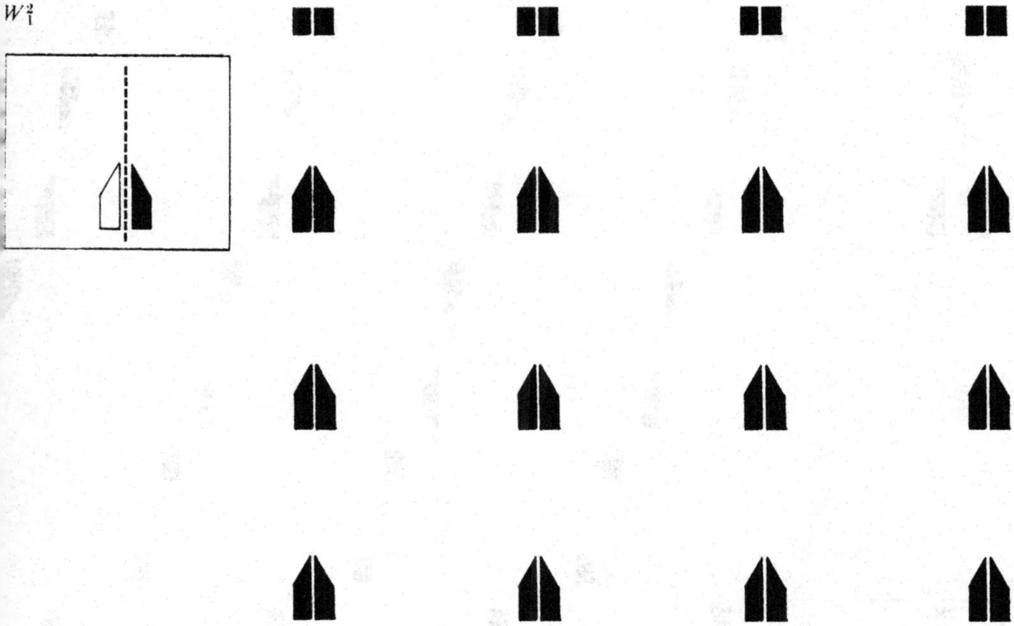

W_1^3

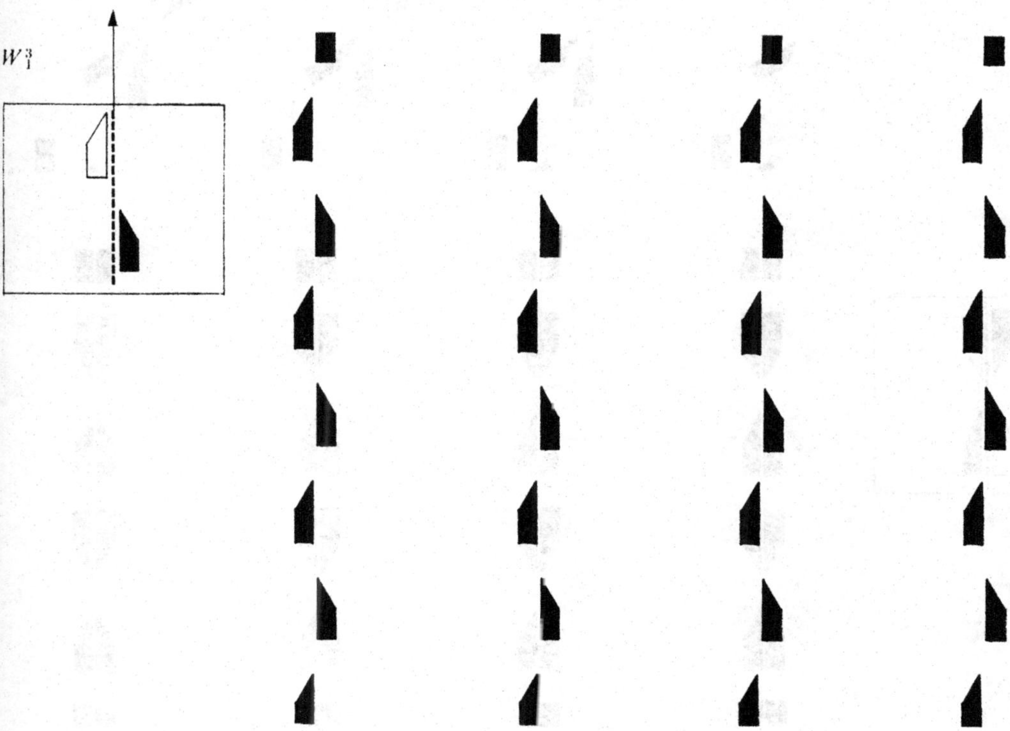

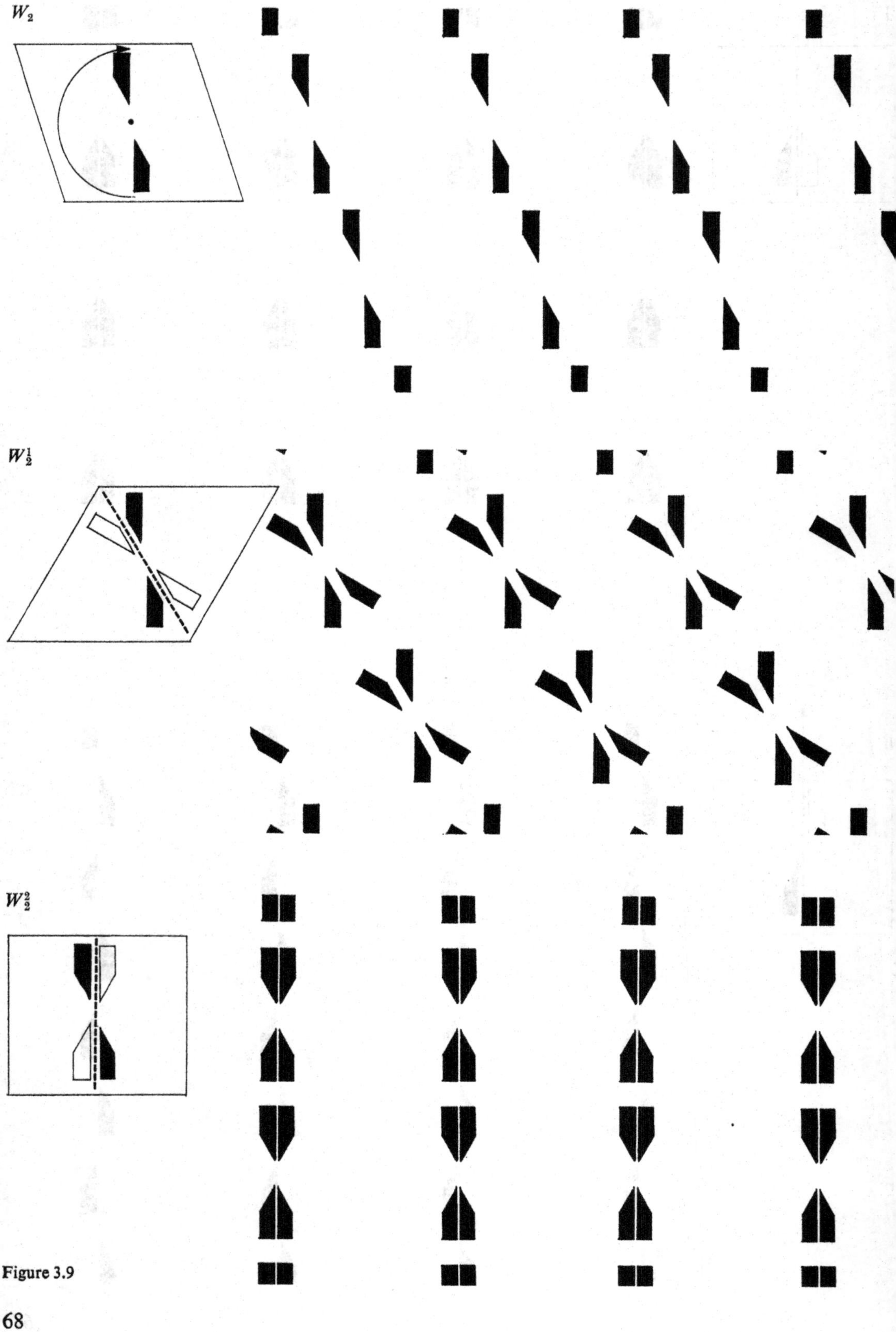

Figure 3.9

W_2^3

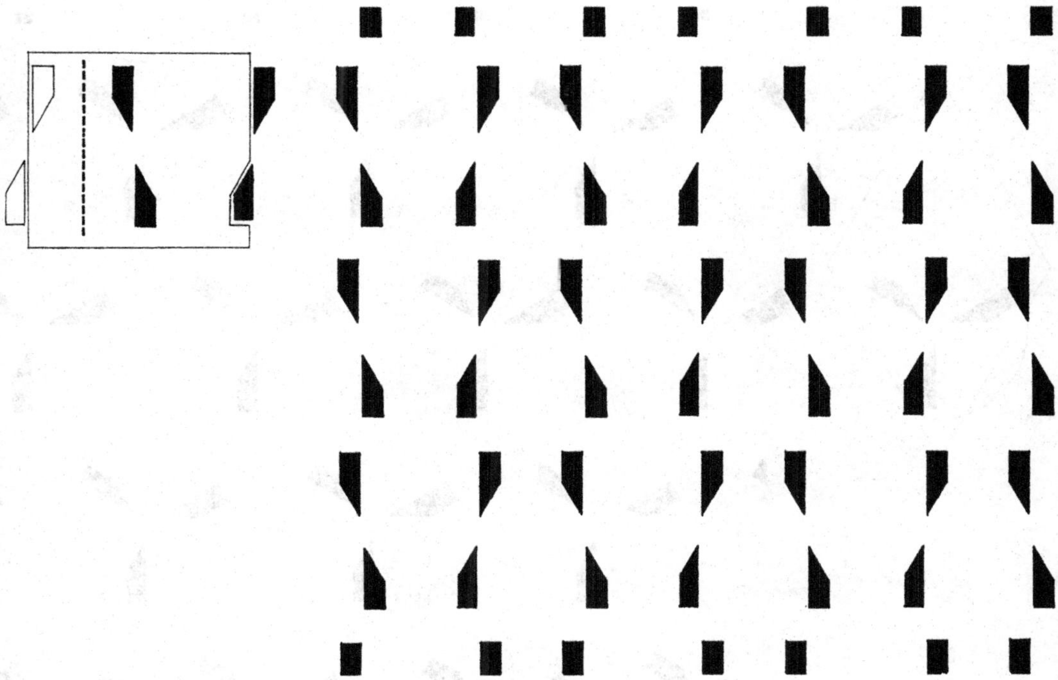

W_2^4

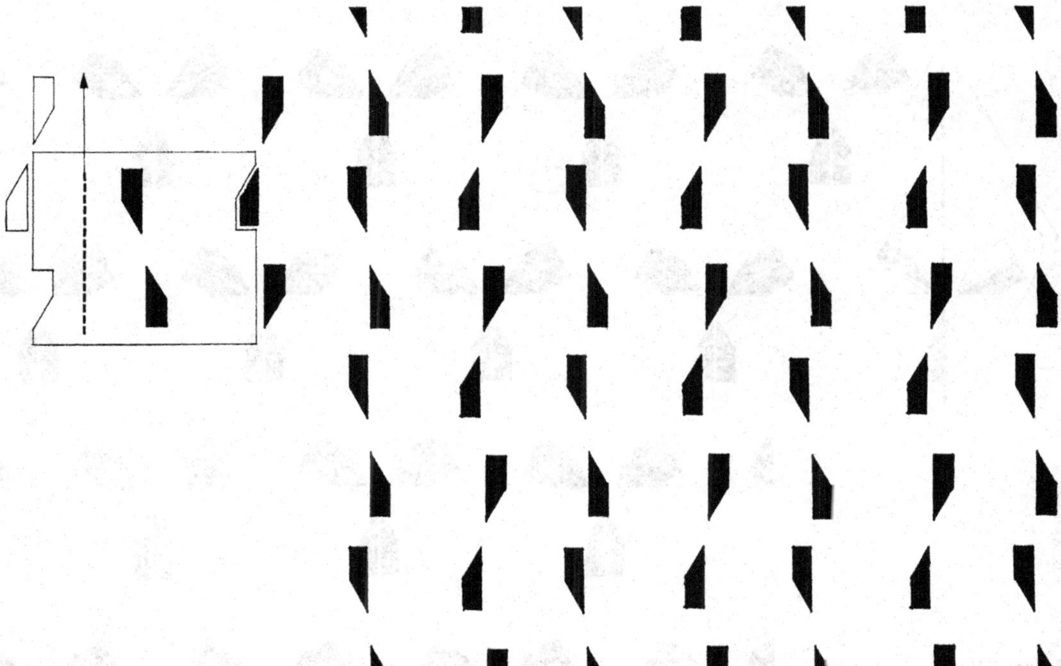

W_3

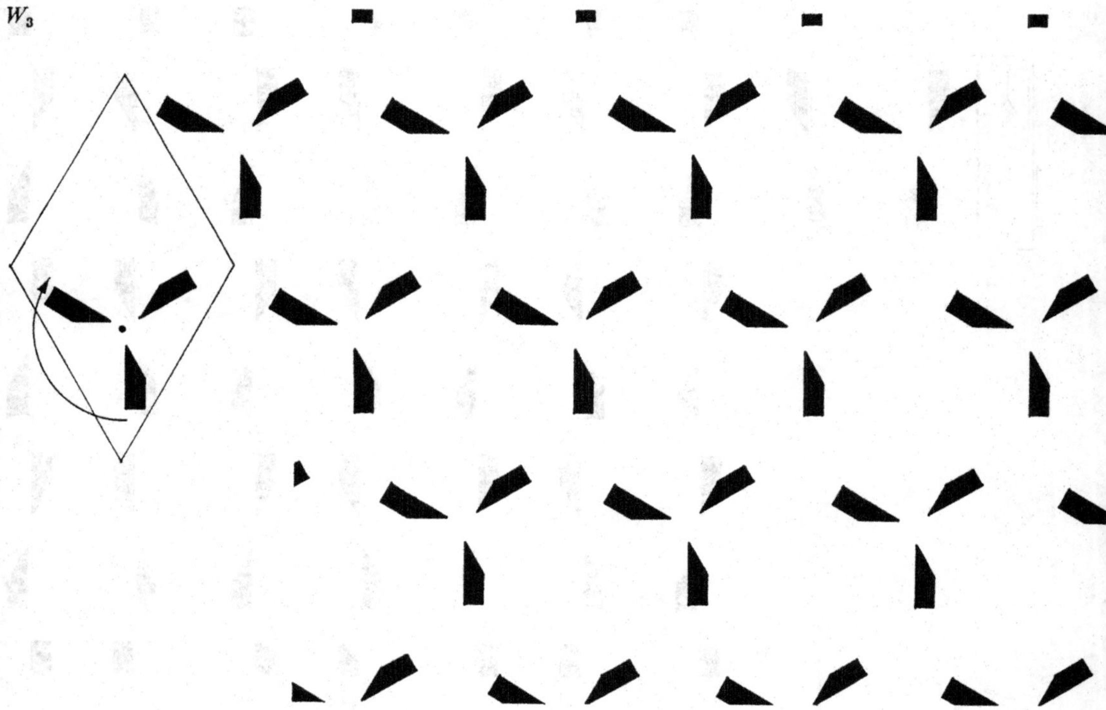

W_3^1

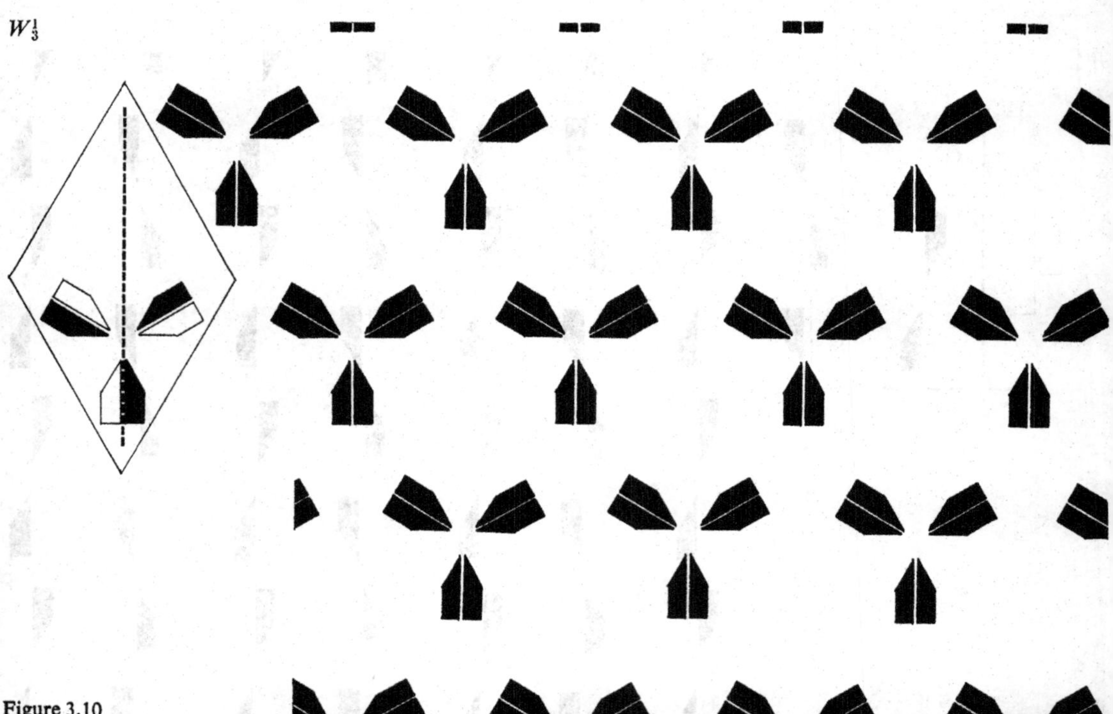

Figure 3.10

W_3^2

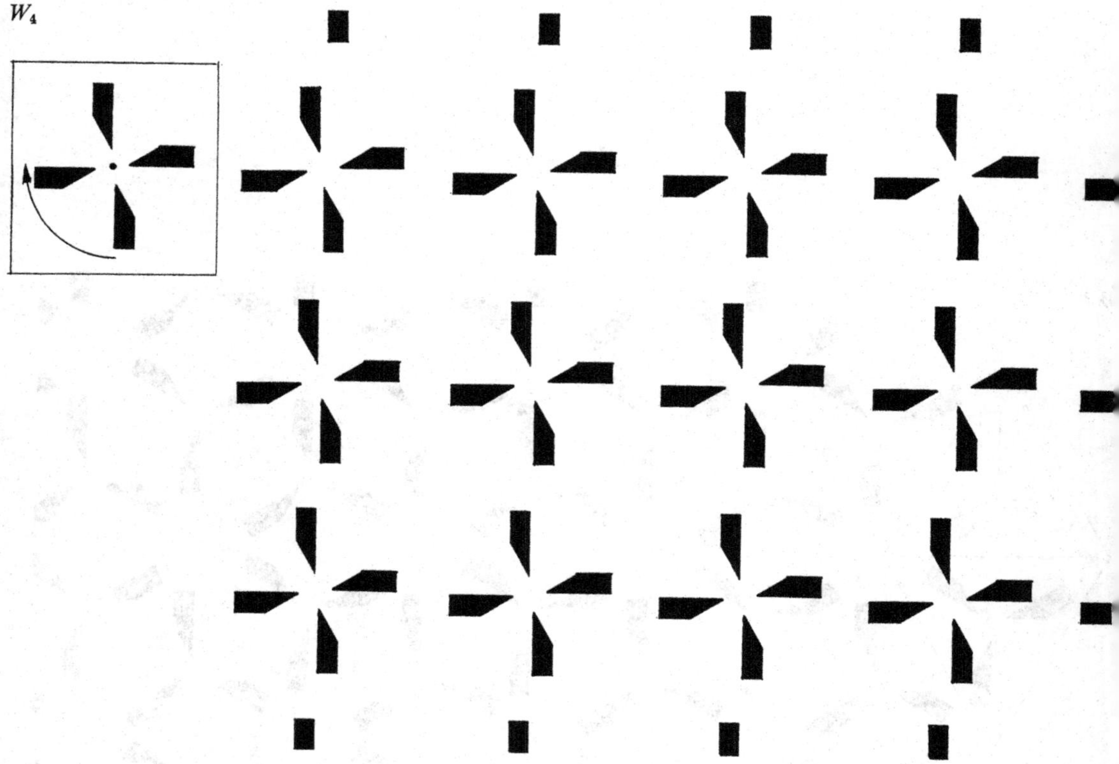

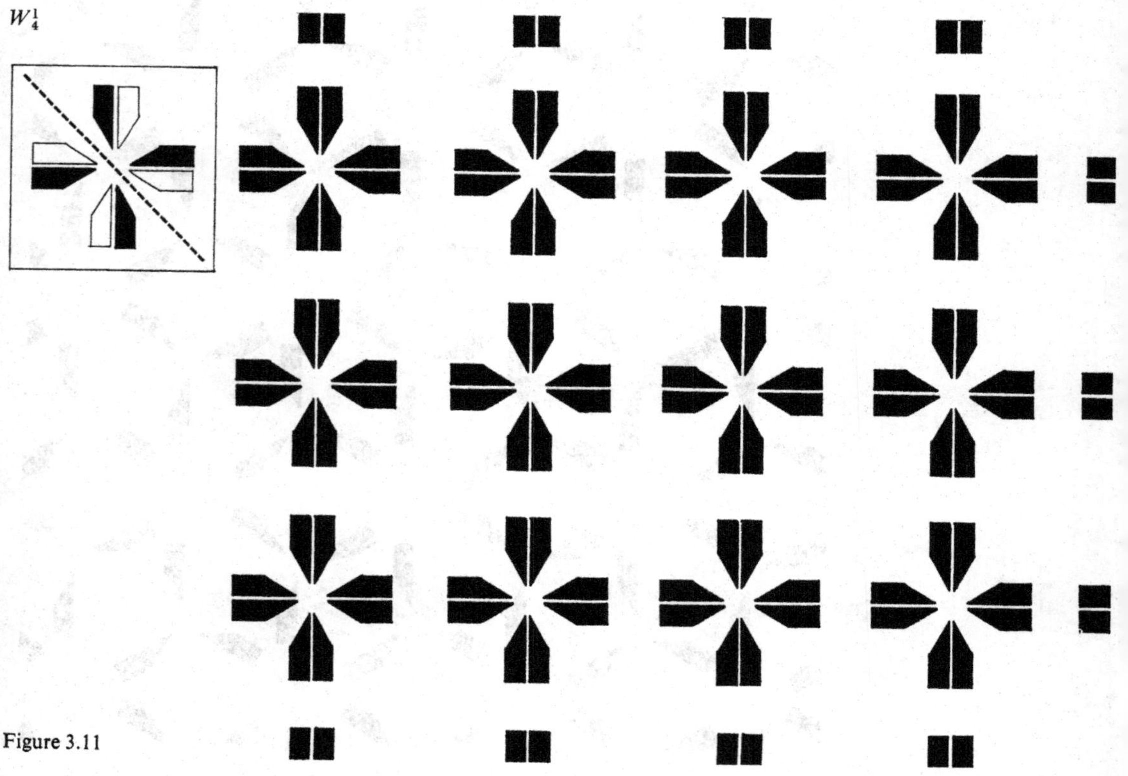

Figure 3.11

W_4^2

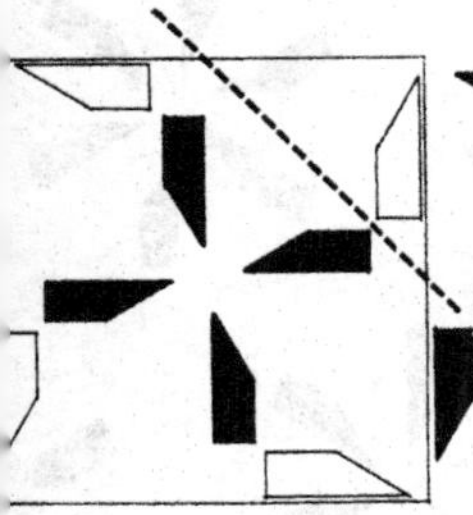

W_6

W_6^1

Figure 3.12

Our two examples of 'quadruple block' planning are, in essence, examples of W_4 and W_2^2 – each element being a cluster of four houses with point symmetry C_4 and D_2 respectively (Figure 3.13). As we have already mentioned, Frank Lloyd Wright's concrete block frieze in La Miniatura properly belongs to W_1^1, consisting as it does of a reflection (the bilateral symmetry of the half block) and a translation not in the

Figure 3.13
Housing layout by Frank Lloyd Wright exhibiting W_4 and W_2^2 patterns

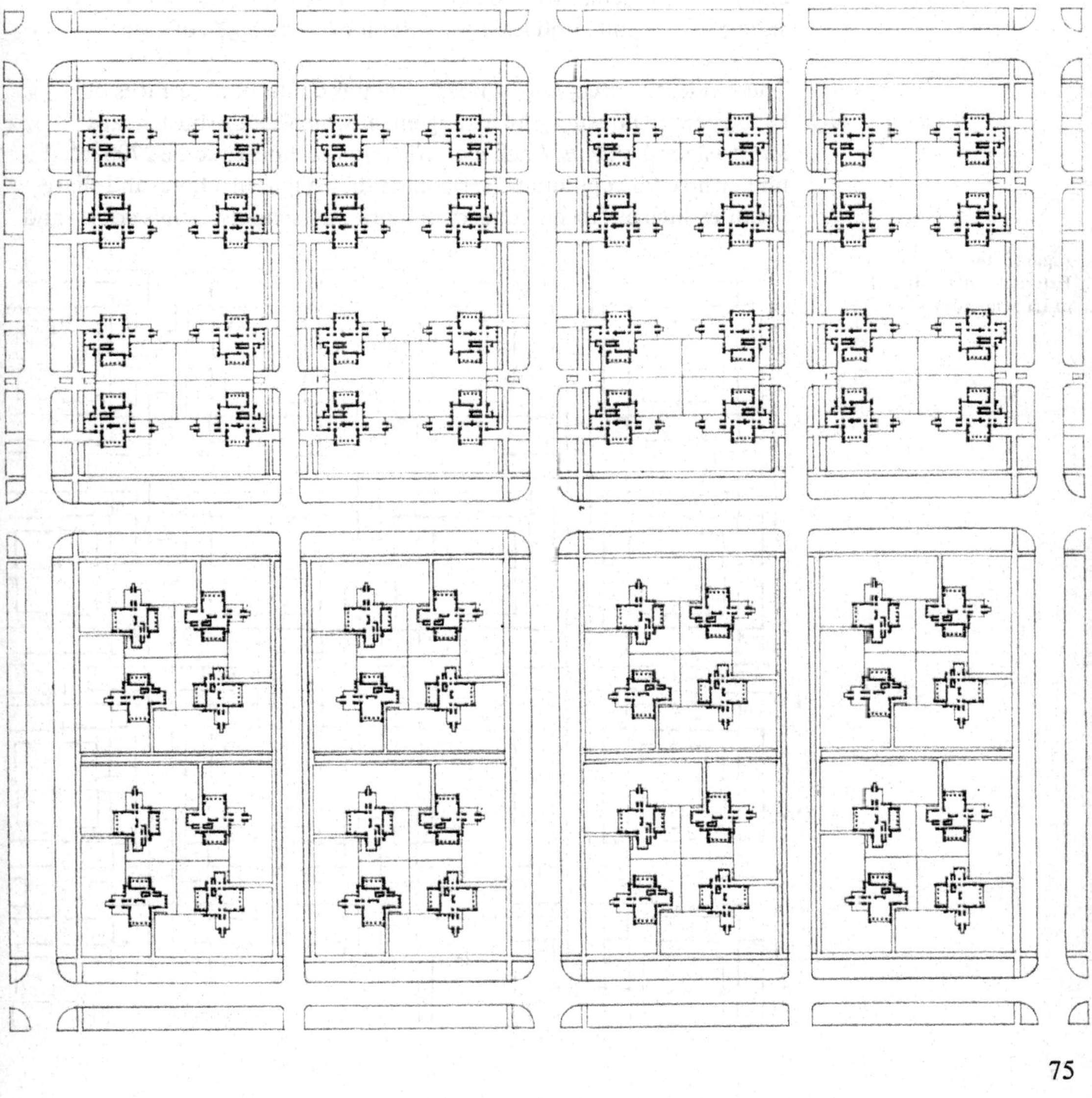

line of the mirror. On the other hand, the general wall treatment is W_4^2: each square block having symmetry, D_4 (Figure 3.14).

The project, in the late twenties, for La Ville Radieuse by Le Corbusier illustrates a number of symmetry groups (Figure 3.15). In the manufacturing section, which consists of *ateliers*, or flatted factories, we have an example of either W_1 or F_1 depending on our choice of unit (Figure 3.16). The particular form derives from the configuration of railway sidings which enter and leave an outer loop at an angle of 45°.

The residential area, *la ville verte*, contains continuous terraces of twelve-storey housing – an arrangement which Le Corbusier calls *lotissements à redents*. Each *unité d'habitation* houses some 2700 inhabitants who enjoy communal services as they might in a hotel and share facilities such as crèche, nursery and primary schools, tennis courts and

Figure 3.14
External wall treatment of La Miniatura

Figure 3.15
Le Corbusier's project for La Ville Radieuse

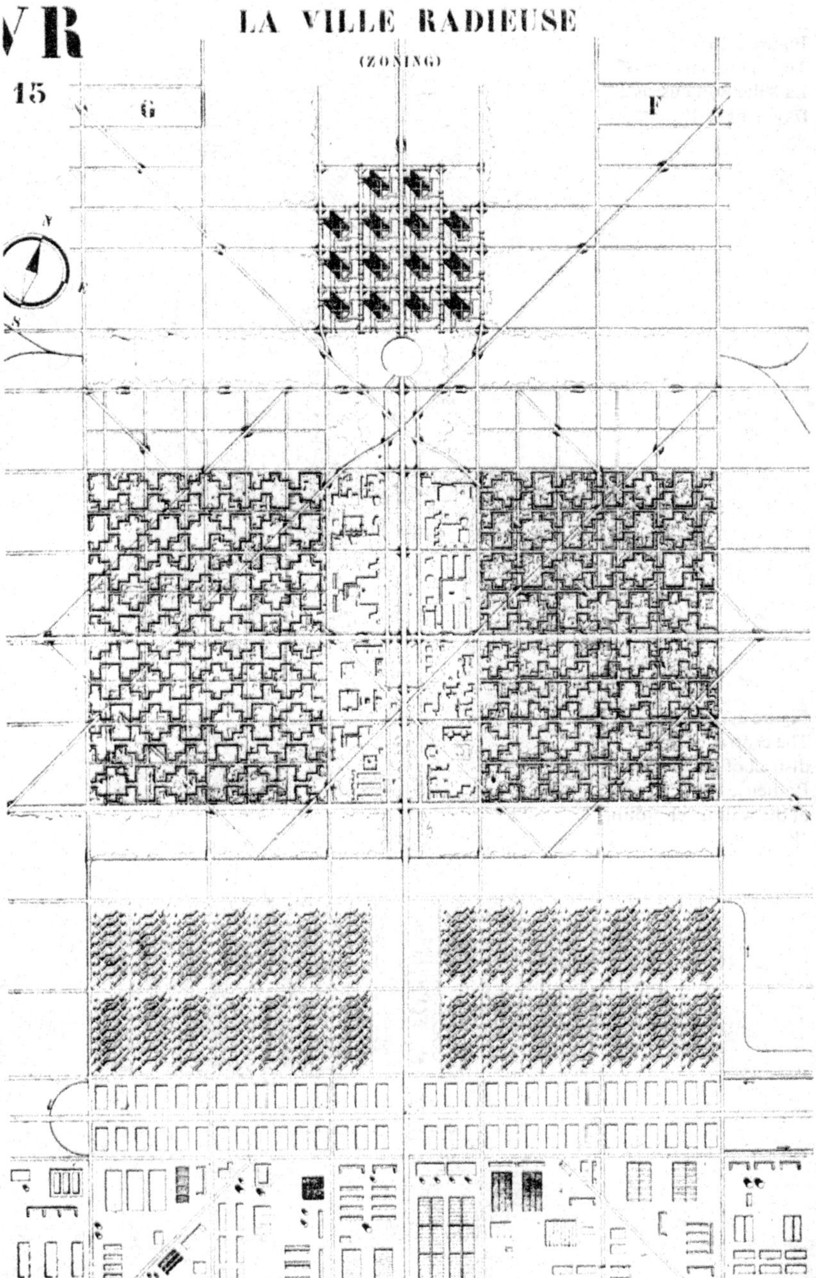

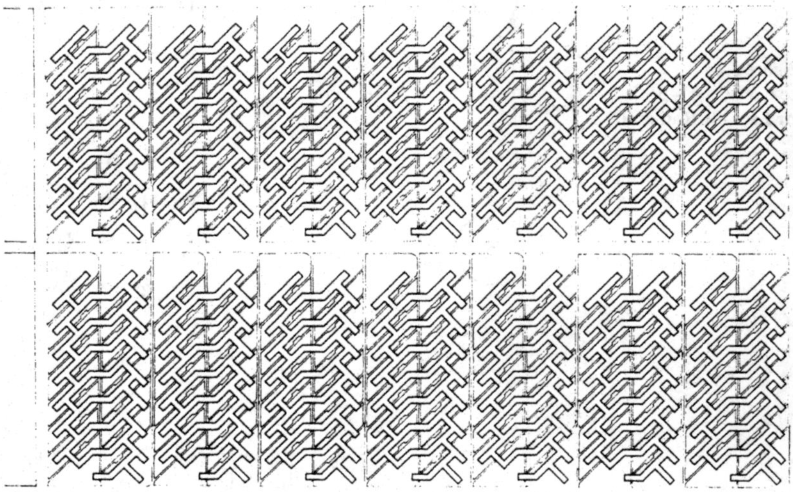

Figure 3.16
The industrial area of La Ville Radieuse is a fragment of W_1

Figure 3.17
The central business district of La Ville Radieuse is an example of the wallpaper group, W_4

13 For Le Corbusier's original description of the business district see his essay 'The Street' in *Oeuvre Complète 1910–1929*, pp. 118–19.

14 From Werner M. Moser, ed., *Frank Lloyd Wright; Sechzig Jahre Lebendige Architektur*, Winterthur, Switzerland, Buchdruckerei Winterthur AG, 1952, p. 27.

swimming pools within their own *lotissement*. The *autostrades* are based on a square grid 400 m × 400 m and are kept to the outside of the site, leaving a large traffic-free park within. Le Corbusier gives six examples of the many combinations which can be derived from a single architectural motif – 'une arabesque de redent susceptible d'assurer de grands spectacles architecturaux.' In these combinations there are examples of W_2, W^1_2 and W^2_2, but Le Corbusier mixes fragments of these together in his final composition. Note, however, that the north-east corner of *la ville verte* is a strict example of the wallpaper symmetry, W^1_2.

Finally, in the business section where cruciform towers are individually set within their own motorway box, we have the group W_4 (Figure 3.17). Each block is C_4, and each tower within the block is D_4. By using W_4, Le Corbusier maintains the neutrality of the square grid giving no preference to one route direction over another. The whole business section thus becomes a spectacular setting for one of those classic car chases of the silent cinema – with Harold Lloyd clutching at spindleberry[13] as he dangles over the side of the roof garden of a 60-storey-high skyscraper, while below the Keystone Cops appear and disappear in a spiralling maze of car parks and a helter-skelter of roads.

In 1939 Frank Lloyd Wright designed a project, the Crystal Heights Hotel, to be sited at the acute-angled intersection of Connecticut and Florida Avenues in Washington D.C. Werner Moser notes the essential features of the scheme:

> 'Clustered skyscraper group (application of the St Mark's Tower unit) containing a hotel, similar to 1930 project for Chicago with its towers linked in a line. South of this group Wright leaves a large open garden, enclosing it along the street contours with recessed 4 storey row-buildings, containing shops, garages, theatre and parking terraces on the flat roofs. The usual endless and unarticulated succession of apartments has been successfully avoided: Wright divided the whole block into many clearly accentuated groups, each containing 12 living units [per floor]. In this way the big block receives a rhythmical texture (light and shadow) on a comprehensible scale.'[14]

We have already illustrated the St Mark's-in-the-Bouwerie Tower as an example of the point group C_2. It is of some interest to study this generic form more closely. From 1929 until the end of his life, thirty years later, the 'quad' motif is one which recurs in numerous projects by Wright of which only one was built – the Price Tower in Bartlesville, Oklahoma. This sequence of variations provides us with the richest and most sustained application of symmetry in architecture that we know of.

The elementary unit of St Mark's Tower is an equilateral triangle. W_6^1 turns this single triangle into the regular plane tesselation {3, 6}. The Schläfli symbol {p, q} denotes a tesselation composed of *q* regular *p*-gons about each vertex. There are just three regular tesselations of the plane {6, 3}, {4, 4}, {3, 6}; that is to say, the hexagonal, the square and the triangular. In most of his buildings, Wright uses one or other of these tesselations as a base. For example, in the Sundt House he employs the hexagonal grid together with the triangular grid which contains it,

Figure 3.18
Behind the apparent romanticism of the Crystal Heights hotel, shops and theatre complex by Frank Lloyd Wright, lies a symmetrical structure of great simplicity

[15] Quoted from the Guide to the Price Tower, Bartlesville, Oklahoma.

in the Life Magazine project he uses a square grid as he does, surprisingly, as we see in Chapter 9, in the circular Jester House (Figure 1.13). For the tower, Wright describes his reasons for adopting the 60°-unit system as follows:

> 'The entire building is laid out on a 60° unit system. These units or modules consist of 60° parallelograms 2 ft 6 in across and 2 ft $10\frac{5}{8}$ in on a side. [The 2 ft 6 in dimension refers to the semi-diameter of the 60° rhombus, that is the height of the equilateral triangle which we take as the elementary unit in our discussion.] All walls, partitions, etc., fall on unit lines or subdivisions thereof. The unit basis of construction is made apparent in the lines inscribed in the floor slabs throughout the building. In laying out a building on such a unit system, a unity of design is achieved which would otherwise be unlikely; all shapes and forms grow out of or are related to this basic fabric as the pattern of a beautiful carpet grows from the warp and woof of the weaver's loom. In addition, the elimination of much of the customary dimensions on the drawing avoids confusion during construction since the unit system provides a constant reference point for the location of all component parts. The choice of the 60° unit system rather than the customary rectilinear results from the facility in which space relationships are coordinated in the former; not only does one area flow into and relate more naturally to human freedom and usage, but, if properly handled, the finished ensemble results in a much more open and facile plan.'[15]

F

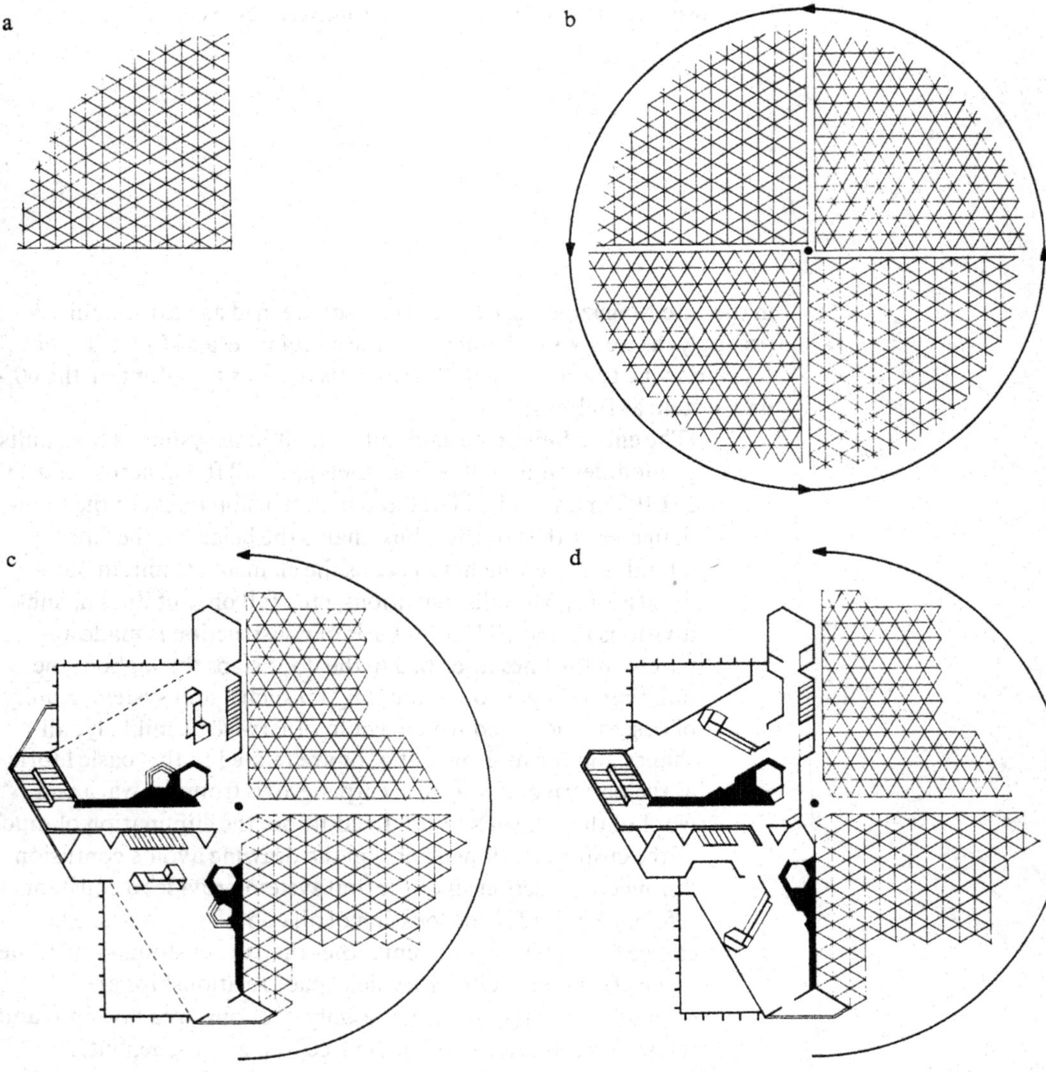

Figure 3.19
The geometrical development of the plans of an apartment building for St. Mark's-in-the-Bouwerie, New York, by Frank Lloyd Wright, C_2

In St Mark's Tower, Wright effectively takes a rectangular 'corner' of the triangular grid – one edge coinciding with an axis of the tesselation, the other at right angles bisecting a row of equilateral triangles (Figure 3.19a). This corner is then rotated through 90°, 180° and 270° about a point 10 inches from the corner. As an isometry transformation this may be represented by $\mathbf{S}_4{}^1$, $\mathbf{S}_4{}^2$, $\mathbf{S}_4{}^3$, $\mathbf{S}_4{}^4 = \mathbf{I}$ where $\mathbf{S}_4$ is a rotation of period 4. This rotation assembles the quadrants in pinwheel form (Figure 3.19b), leaving a 10-inch space between the four quarters whose dimensions in any given direction now clash irreconcilably: the perpendicular height of the triangular unit to its side being $\sqrt{3} : 2$. Wright, however, uses the gap so made for the four internal vertical shafts of reinforced concrete from which the slabs are cantilevered (Figure 3.20). This quintessential plan for the whole series of projects is C_4. But as we have already mentioned, in the St Mark's project the four apartments share two elevators

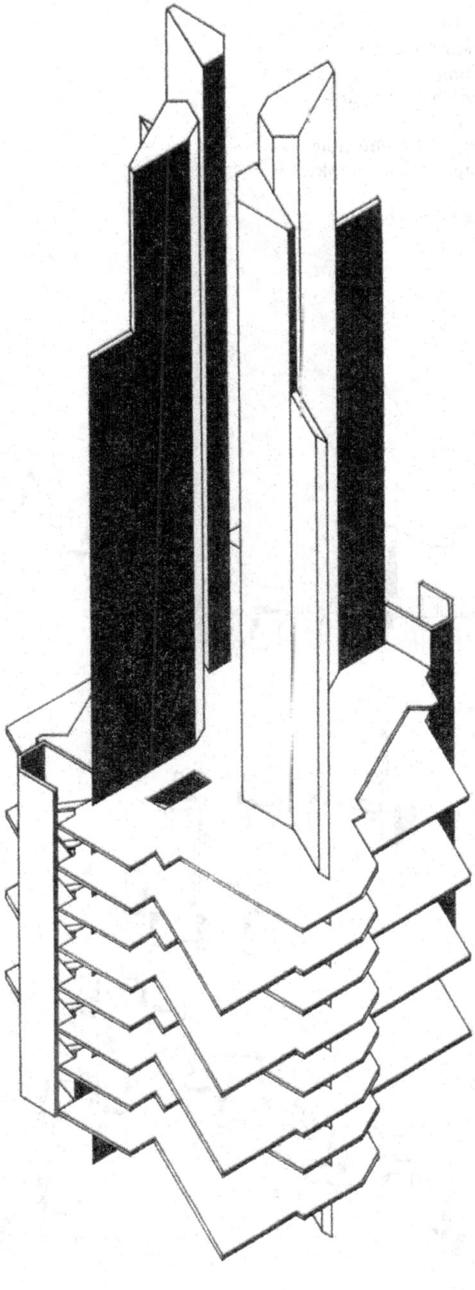

Figure 3.20
The structural system of the St. Mark's apartment building

and two escape stairs. This arrangement, together with a 10-inch translation to pull two of the quadrants yet further apart in order to increase the width of the escape corridor from 30 to 40 inches, makes the final symmetry of the floor plans of St Mark's Tower C_2 (Figure 3.19c–d).

A year later Wright designed a project for grouped apartment towers in Chicago (Figure 3.21b). Here Wright takes the St Mark's Tower plan (Figure 3.21a) as his element. He then rotates the plan through a half-turn, **H**. This produces a plan with six duplex dwellings and a good deal

Figure 3.21
The generic plans of building and projects by Frank Lloyd Wright:

a, the St. Mark's-in-the-Bouwerie apartment building
b, the Chicago apartments
c, Crystal Heights hotel and apartment complex; and
d, the Price Tower

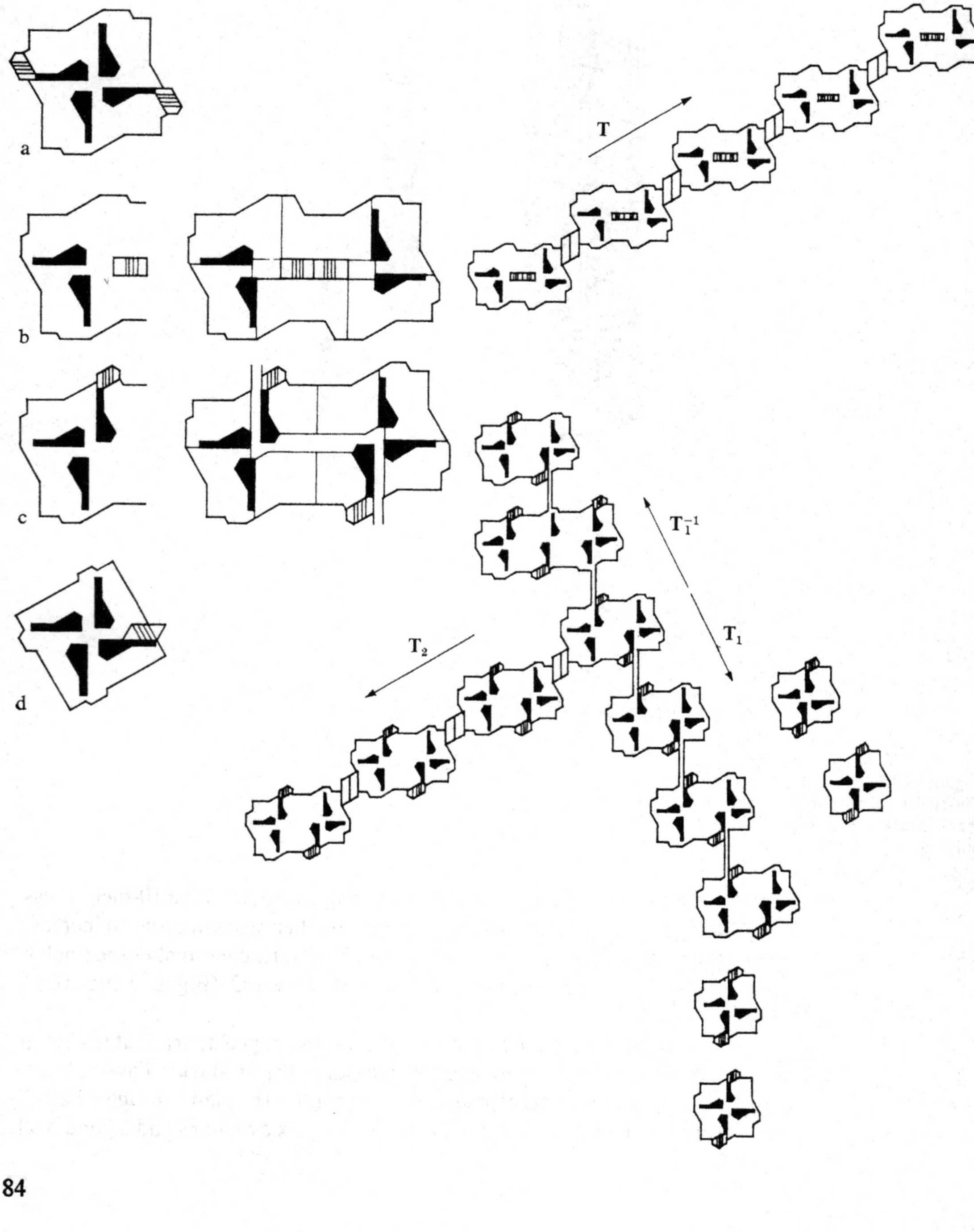

[16] All these schemes are illustrated in *The Drawings of Frank Lloyd Wright*, edited by Arthur Drexler, published for the Museum of Modern Art by Horizon Press, New York, 1962.

of redundancy in structure and fire-escapes. Wright removes two out of each cluster of four concrete shafts and arranges that the two required escape stairs back against one another in the centre of the building. The final plan exhibits C_2 symmetry, as does the structure. This new unit is then translated, **T**, along a line 30° to its main spine. The final plan consists of five units $I = T^0, T^1, T^2, T^3, T^4$ linked together by double-height garden balconies. The whole group may either be thought of as a fragment of F_1, or a lattice 'row' from W_2 which takes into account the halfturn symmetry of each unit.

The Crystal Heights Hotel project, by all outward appearances, is a great romantic composition, a cluster of towers arranged picturesquely to provide a telling skyline. But underlying this seemingly free form is a rigorous pattern-structure (Figure 3.21c): the culmination of the two previous studies. Again Wright spins the original quadruple unit through a halfturn, but this time the centre of rotation is even more eccentric and the resulting unit is more elongated than in the Chicago apartments. The plan now contains twelve hotel suites. The fire-escapes are external and one of the structural shafts is removed from each quadruple, thus making room for a central corridor. The symmetry of the unit is again C_2. This unit is essentially translated in two directions, $\mathbf{T}_1$ and $\mathbf{T}_2$. The development is $T_1^{-2}, T_1^{-1}, I = T_1^0, T_1^1, T_1^2$ and T_1^3 and $I = T_2^0, T_2^1, T_2^2, T_2^3$. There are then four free-standing towers, principally for the staff, and one additional halfturn on T_1^{-1} to form a triple element. The third tower of this new unit is some six storeys higher than the other towers. The units are linked to each other by glazed walkways. The quintessential symmetry pattern, ignoring the free-standing towers and the third tall tower, is again a fragment of W_2, but this time along both a 'row' and a 'column' of the lattice. This is undoubtedly one of Wright's finest works and one which, if it had been executed, would have been a valuable exemplar for urban development elsewhere.

The building that did get built on this principle was, like its progenitor, a single tower.[16] The Price Tower consists of a mixture of professional offices and duplex apartments (Figure 3.21d). While the structural symmetry C_4 is maintained, the requirement for six offices and one apartment to each two floors reduces the symmetry to C_1 – essentially no symmetry. This lack of rotational symmetry is clearly expressed in the exterior treatment of the single stair tower, the vertical copper sun blades for the apartments and the horizontal ones for the offices.

[17] From Owen Jones, *The Grammar of Ornament*, p. 156.

It seems clear to us that the two greatest form-makers of twentieth-century architecture – Frank Lloyd Wright and Le Corbusier – were able to innovate largely because of their appreciation and deep understanding of symmetry and pattern-structure. Both seemed to have taken to heart Owen Jones's proposition that 'the principles discoverable in the works of the past belong to us, not so the results. It is taking the end for the means. No improvement can take place in the Art of the present generation until all classes, artists, manufacturers, and the public, are better educated in Art, and the existence of general principles is more fully recognized'; and his advice that 'if a student in the arts, earnest in his search after knowledge, will only lay aside all temptation to indolence, will examine for himself the arts of the past, compare them with the works of nature, bend his mind to a thorough appreciation of the *principles* which reign in each, he cannot fail to be a creator, and to individualize new forms.'[17]

4 Matrices and vectors

When in 1925 Le Corbusier designed his housing scheme at Pessac, near Bordeaux, he was demonstrating the possibilities of mass produced housing.[1] He designed a number of distinct types of dwelling, each built of standardized elements. Le Corbusier used just three window sizes: a quadruple square window, a double square and a single square one. At some time he must have made what architects call a 'window schedule', a listing of the quantities of each window type in each building. This is most neatly done in the form of a table, or array of numbers, with the house types across the top and the window types down the side (Figure 4.1). The array of numbers has the characteristic that we may not change the positions of the numbers arbitrarily. Given a fixed order of house types and window types, each number in the array clearly has a fixed place. A *matrix* is a mapping of such an array onto an abstract mathematical structure, a system of mathematical pigeon-holes.[2]

Figure 4.1

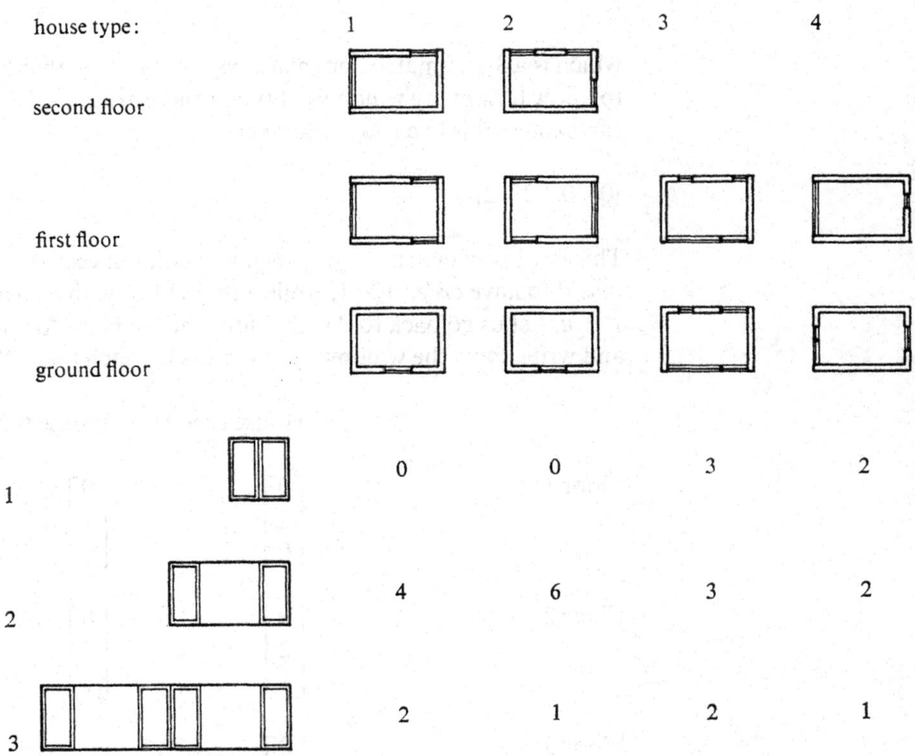

[1] For an interesting comment on the history of this development see Philippe Boudon, *Pessac de Le Corbusier, Etude Socio-architecturale, 1927–1967*, Paris, Dunod, 1969.

[2] A good introduction to matrices is A. C. Aitken, *Determinants and Matrices*, Edinburgh, Oliver and Boyd, 1939. A particularly lucid account is given by Jacob T. Schwartz, *Introduction to Matrices and Vectors*, New York, McGraw-Hill, 1961. Geometrical applications are discussed in D. C. Murdoch, *Analytic Geometry with an Introduction to Vectors and Matrices*, New York, Wiley, 1966.

The 3×4 matrix (three rows and four columns)

$$\begin{bmatrix} 0 & 0 & 3 & 2 \\ 4 & 6 & 3 & 2 \\ 2 & 1 & 2 & 1 \end{bmatrix}$$

is a representation of the number of windows of type 1 (first row), type 2 (second row), type 3 (third row) in each house: house type 1 (first column), house type 2 (second column) and so on. Thus the term in the ith row and jth column represents the number of windows type i in house type j. If we simply consider the windows in house type 1 we can do this in one column of numbers

$$\begin{bmatrix} 0 \\ 4 \\ 2 \end{bmatrix}$$

which is a 3×1 matrix, or *column vector*. On the other hand, if we want to know how many windows of type 1 there are in each house type, we can express this in a row of numbers

$$[0 \quad 0 \quad 3 \quad 2].$$

This is a 1×4 matrix, or *row vector*. A column vector with n terms in it is said to have *order* $n \times 1$, while a row vector with n terms has order $1 \times n$. Let us go back to the individual house plans for types 1 and 2 and write down the window type vectors for each floor. We have

	House type 1	House type 2
Floor 1:	$\begin{bmatrix} 0 \\ 1 \\ 0 \end{bmatrix}$	$\begin{bmatrix} 0 \\ 1 \\ 0 \end{bmatrix}$
Floor 2:	$\begin{bmatrix} 0 \\ 2 \\ 1 \end{bmatrix}$	$\begin{bmatrix} 0 \\ 2 \\ 1 \end{bmatrix}$
Floor 3:	$\begin{bmatrix} 0 \\ 1 \\ 1 \end{bmatrix}$	$\begin{bmatrix} 0 \\ 3 \\ 0 \end{bmatrix}$

On floors 1 and 2 the vectors for each house look the same, but they differ for floor 3. We say that two vectors are equal only if their like-positioned terms are equal. Clearly, this means that two vectors can only be compared for equality if they have the same number of terms, if, in fact, they have the same order.

House types 1 and 2 are actually two halves of a semi-detached house, the first two floors of which are reflected. How many windows of each type are there on floor 3 of house types 1 and 2? We add together all the windows of house type 1, then those of type 2 and finally type 3. This illustrates *vector addition:*

$$\begin{bmatrix}0\\1\\1\end{bmatrix}+\begin{bmatrix}0\\3\\0\end{bmatrix}=\begin{bmatrix}(0+0)\\(1+3)\\(1+0)\end{bmatrix}=\begin{bmatrix}0\\4\\1\end{bmatrix}.$$

Two vectors of the same order may be added together by adding each pair of like-positioned terms to form a third vector of the same order. If we now add up the windows of each type on floor 2 of house types 1 and 2 we obtain

$$\begin{bmatrix}0\\2\\1\end{bmatrix}+\begin{bmatrix}0\\2\\1\end{bmatrix}=\begin{bmatrix}(0+0)\\(2+2)\\(1+1)\end{bmatrix}=\begin{bmatrix}0\\4\\2\end{bmatrix}$$

but as, when we have two identical objects, a, it is natural to write $2a$ for $a + a$ so with vectors we write

$$2\begin{bmatrix}0\\2\\1\end{bmatrix}=\begin{bmatrix}0\\2\\1\end{bmatrix}+\begin{bmatrix}0\\2\\1\end{bmatrix}$$

which means, from our previous equation, that

$$2\begin{bmatrix}0\\2\\1\end{bmatrix}=\begin{bmatrix}0\\4\\2\end{bmatrix}.$$

This is an example of the multiplication of a vector by a *scalar*, or real number, which entails multiplying *every* term in the vector by the scalar. But can we multiply two vectors together?

Suppose the window types cost c_1, c_2, c_3 respectively. We can form a row vector with these as elements, their sequence corresponding to that which we have already used in the column vector of window types:

$$[c_1 \quad c_2 \quad c_3].$$

What is the total cost of windows in house type 1? The answer is, quite simply, $(c_1 . 0 + c_2 . 4 + c_3 . 2)$, that is to say, it is the sum of the products of each cost times the number of windows of each particular type. It would be useful, in more complicated situations, to have a definition of a vector product which gives us this result. We therefore define the *inner product* of two vectors of orders $1 \times n$ and $n \times 1$ as the sum of the product of terms in the row vector with their corresponding terms in the column vector, thus for house type 1

$$[c_1 \quad c_2 \quad c_3]\begin{bmatrix}0\\4\\1\end{bmatrix} = (c_1 . 0 + c_2 . 4 + c_3 . 1)$$

and for house type 2

$$[c_1 \quad c_2 \quad c_3]\begin{bmatrix}0\\6\\1\end{bmatrix} = (c_1 . 0 + c_2 . 6 + c_3 . 1).$$

Vectors are usually denoted by small bold letters, **a**, **b**, **c**,.... Their representative elements are a_i, b_i, c_i,..., where i takes the values 1, 2,..., n and where n is the *dimension* of, or number of terms in, the vector. Vectors may also be expressed $[a_i]$, $[b_i]$, $[c_i]$,..., when we wish to draw attention to the individual elements or *components*. By convention, unless otherwise stated, a vector is assumed to be a column vector

$$\mathbf{a} = \begin{bmatrix}a_1\\a_2\\\vdots\\a_n\end{bmatrix} = [a_i].$$

The row vector of these terms is called the *transpose* of $\mathbf{a}$ and is written

$$\mathbf{a}^T = [a_1 \quad a_2 \quad \dots \quad a_n] = [a_i]^T.$$

In future, in order to save space and since $(\mathbf{a}^T)^T$ is the same as $\mathbf{a}$, if we need to write the column vector out in full we will write it as $[a_1 \quad a_2 \quad \dots \quad a_n]^T$ on one line.

We are now in a position to state some rules of vector algebra but before we do so it is worth reminding ourselves of the operational rules of 'normal' algebra. This algebra deals with real numbers and two operations called *addition* (+) and *multiplication* (.) and it has been given the general name *number field* by mathematicians. A general *field* consists of a set of elements $F = a, b, c, \dots$ and *two compositions* +, . , and we write {F, +, . } to describe it. Recall that a *group* is a set of elements with just *one composition*. But to be a *field*, certain rules of operation as set out below must be satisfied.

Addition (+);
Multiplication (.) :

Closure: For all elements $a, b \in F$

$$a + b \in F$$
$$a \, . \, b \in F$$

Commutative:

$$a + b = b + a$$
$$a \, . \, b = b \, . \, a$$

Associative:

$$a + (b + c) = (a + b) + c$$
$$a \, . \, (b \, . \, c) = (a \, . \, b) \, . \, c$$

Identity: For every $a \in F$ there is an element $0 \in F$ such that

$$a + 0 = 0 + a = a.$$

For every $a \in F$ there is an element $1 \in F$, $1 \neq 0$ such that

$$a \, . \, 1 = 1 \, . \, a = a.$$

Inverse: For each $a \in F$ there is a unique element $-a \in F$ such that

$$a + (-a) = (-a) + a = 0.$$

For each $a \in F$, $a \neq 0$, there is a unique element $a^{-1} \in F$ such that

$$a \, . \, a^{-1} = a^{-1} \, . \, a = 1.$$

Distributive: $a \,.\, (b + c) = a \,.\, b + a \,.\, c$
$(a + b) \,.\, c = a \,.\, c + b \,.\, c$

The set of natural numbers $N = \{1, 2, 3, ...\}$ is *closed* under addition, since the addition of any two produces a third number in N, for example, $1 + 2 = 3$; but it does not possess an identity nor inverse for addition, that is $0, -1, -2, -3, ...$ are not in the set. The set of integers $Z = \{0, \pm 1, \pm 2, \pm 3, ...\}$ is not only closed under addition but it also satisfies all the rules 1 to 5 for addition: the integers form a *group* under addition. However, while Z is closed under multiplication it has no inverse element for multiplication; thus $\frac{1}{2}, \frac{2}{3}, \frac{3}{5}$, for example, are not in Z. We say that Z is not closed under division, but the set of rational numbers, Q, is, and the same is true for the set of real numbers R. Note that F is a commutative or *Abelian group* with respect to addition, and that non-zero elements of F also constitute such a group for multiplication.

Let us now summarize the laws of vectors in general. We shall give the laws for column vectors, but they also apply to row vectors. We define the following $n \times 1$ vectors

$$\mathbf{a} = [a_i] = [a_1 \quad a_2 \quad \ldots \quad a_n]^T$$
$$\mathbf{b} = [b_i] = [b_1 \quad b_2 \quad \ldots \quad b_n]^T$$
$$\mathbf{c} = [c_i] = [c_1 \quad c_2 \quad \ldots \quad c_n]^T$$

and the real numbers $\{a_i, b_i, c_i, r, s, ...\} \in R$. Then the following relations hold:

Equality:

$$\mathbf{a} = \mathbf{b} \Leftrightarrow a_i = b_i \text{ for all } i.$$

Thus, $\begin{bmatrix} a_1 \\ a_2 \end{bmatrix} = \begin{bmatrix} b_1 \\ b_2 \end{bmatrix} \Leftrightarrow \begin{cases} a_1 = b_1 \\ a_2 = b_2. \end{cases}$

Addition:

Closure:

$$\mathbf{a} + \mathbf{b} = [(a_i + b_i)].$$

The sum of two vectors is a third vector so that vectors obey the law of *closure* for addition. For example,

$$\begin{bmatrix} a_1 \\ a_2 \end{bmatrix} + \begin{bmatrix} b_1 \\ b_2 \end{bmatrix} = \begin{bmatrix} (a_1 + b_1) \\ (a_2 + b_2) \end{bmatrix}$$

Commutative:

but since the elements are real numbers and $(a_i + b_i)$ is the same as $(b_i + a_i)$ we have

$$\mathbf{a} + \mathbf{b} = [(a_i + b_i)] = [(b_i + a_i)] = \mathbf{b} + \mathbf{a},$$

thus demonstrating that vectors are *commutative* under addition. The *null vector* **0**, all of whose elements are zero, performs the role of the *identity* element since

Identity:

$$\mathbf{a} + \mathbf{0} = \mathbf{0} + \mathbf{a} = \mathbf{a}.$$

To illustrate:

$$\begin{bmatrix} a_1 \\ a_2 \end{bmatrix} + \begin{bmatrix} 0 \\ 0 \end{bmatrix} = \begin{bmatrix} (a_1 + 0) \\ (a_2 + 0) \end{bmatrix} = \begin{bmatrix} a_1 \\ a_2 \end{bmatrix}.$$

The vector $(-\mathbf{a}) = [-a_i]$ is the additive *inverse* of **a** since

Inverse:

$$\mathbf{a} + (-\mathbf{a}) = [a_i + (-a_i)] = [(a_i - a_i)] = \mathbf{0}$$

which is also evident from the example

$$\begin{bmatrix} a_1 \\ a_2 \end{bmatrix} + \begin{bmatrix} -a_1 \\ -a_2 \end{bmatrix} = \begin{bmatrix} (a_1 - a_1) \\ (a_2 - a_2) \end{bmatrix} = \begin{bmatrix} 0 \\ 0 \end{bmatrix}.$$

Finally, since

Associative:

$$(\mathbf{a} + \mathbf{b}) + \mathbf{c} = [(a_i + b_i) + c_i] = [a_i + (b_i + c_i)] = \mathbf{a} + (\mathbf{b} + \mathbf{c})$$

vector addition is *associative.*

Vectors under addition are seen to satisfy the five conditions for an Abelian group. Vectors of the same order are *closed* under addition; they possess an *identity*, the null vector **0**; each vector **a** has a unique *inverse* $-\mathbf{a}$; and the associative law holds. These four conditions suffice to make vectors under addition a *group*, but as we have seen with translations, the fifth property of being *commutative* makes the group Abelian.

Scalar product: Given any vector **a** and any scalar $r \in R$, then

Closure:

$$r\mathbf{a} = r[a_i] = [ra_i]$$

is also a vector so that vectors are *closed* under multiplication by scalars. For example,

$$r\begin{bmatrix} a_1 \\ a_2 \end{bmatrix} = \begin{bmatrix} ra_1 \\ ra_2 \end{bmatrix}.$$

Note especially $1\mathbf{a} = \mathbf{a}$ and $-1\mathbf{a} = -\mathbf{a}$. Also, since

Associative:

$$(rs)\mathbf{a} = [(rs)a_i] = [r(sa_i)] = r(s\mathbf{a})$$

the scalar product is *associative*. Further the *distributive* laws hold

Distributive:

$$r(\mathbf{a} + \mathbf{b}) = [(ra_i + rb_i)] = r\mathbf{a} + r\mathbf{b}$$
$$(r + s)\mathbf{a} = [(ra_i + sa_i)] = r\mathbf{a} + s\mathbf{a}.$$

The other operation which we illustrated earlier was the *inner product*. Vectors are not closed under this rule. So far, we have followed the convention that the inner product is performed on two vectors with the same number of terms but with the first a row vector and the second a column vector. This is not essential, but it leads to consistency of notation with matrix multiplication. Taking the column vectors **a** and **b**, it is usual to *transpose* one so that it becomes a row vector. An *inner product* satisfies the following properties:

Inner product: The inner product of two vectors is a real number. Vectors are not therefore closed under this product. The *inner product* is defined as

$$\mathbf{a}^T . \mathbf{b} = [a_i]^T [b_i] = \Sigma_i \, a_i b_i.$$

This is seen better in full

$$[a_1 \quad a_2 \quad \ldots \quad a_n]\begin{bmatrix} b_1 \\ b_2 \\ \vdots \\ b_n \end{bmatrix} = (a_1b_1 + a_2b_2 + \ldots + a_nb_n).$$

We then have the following relations

Distributive:

$$(\mathbf{a} + \mathbf{b})^T \,.\, \mathbf{c} = \mathbf{a}^T \,.\, \mathbf{c} + \mathbf{b}^T \,.\, \mathbf{c}$$
$$\mathbf{a}^T \,.\, (\mathbf{b} + \mathbf{c}) = \mathbf{a}^T \,.\, \mathbf{b} + \mathbf{a}^T \,.\, \mathbf{c}$$

Associative:

$$(r\mathbf{a}^T).\, \mathbf{b} = r(\mathbf{a}^T \,.\, \mathbf{b}) = \mathbf{a}^T \,.\, (r\mathbf{b})$$

Hermitian symmetry:

$$\mathbf{a}^T \,.\, \mathbf{b} = \mathbf{b}^T \,.\, \mathbf{a}$$

Positive definiteness:

$$\mathbf{a}^T \,.\, \mathbf{a} > 0 \text{ if } \mathbf{a} \neq \mathbf{0}$$

The last relation may appear a little surprising, but

$$\mathbf{a}^T \,.\, \mathbf{a} = [a_i]^T[a_i] = \Sigma_i\, a_i^2$$

and, since the squares of real numbers are always positive, it follows that a sum of squares is also positive. The inner product is only zero if $\mathbf{a} = \mathbf{0}$. For instance

$$[a_1 \quad a_2]\begin{bmatrix} a_1 \\ a_2 \end{bmatrix} = (a_1^2 + a_2^2)$$

which is greater than zero unless $a_1 = 0$ and $a_2 = 0$, in which case $[a_1 \quad a_2] = [0 \quad 0]$, the null vector.

Let us now return to our example of Le Corbusier's housing scheme at Pessac. The schedule of windows for the four dwelling types is given by the 3×4 matrix

$$\begin{bmatrix} 0 & 0 & 3 & 2 \\ 4 & 6 & 3 & 2 \\ 2 & 1 & 2 & 1 \end{bmatrix}$$

where each column contains the same ordered elements as the separate column vectors of window types for each house. Indeed, we may think of a matrix as a 'vector' of vectors

$$\left[\begin{array}{c:c:c:c} 0 & 0 & 3 & 2 \\ 4 & 6 & 3 & 2 \\ 2 & 1 & 2 & 1 \end{array}\right] \text{ or } \left[\begin{array}{cccc} 0 & 0 & 3 & 2 \\ \hdashline 4 & 6 & 3 & 2 \\ \hdashline 2 & 2 & 1 & 1 \end{array}\right].$$

Suppose now that we wish to form a matrix of costs of windows for each house type. This could be done separately as before, by taking the inner product of the cost vector $[c_1 \ c_2 \ c_3]$ and the window type vector for each house. But we can also multiply the matrix by the cost vector, treating each column of the matrix as if it were a separate vector and inserting the results of the inner products, in sequence, in a 1×4 vector, thus:

$$[c_1 \quad c_2 \quad c_3]\left[\begin{array}{c:c:c:c} 0 & 0 & 3 & 2 \\ 4 & 6 & 3 & 2 \\ 2 & 1 & 2 & 1 \end{array}\right]$$

$$= [(c_1 . 0 + c_2 . 4 + c_3 . 2) \quad (c_1 . 0 + c_2 . 6 + c_3 . 1) \quad (c_1 . 3 + c_2 . 3 + c_3 . 2) \quad (c_1 . 2 + c_2 . 2 + c_3 . 1)]$$

$$= [(4c_2 + 2c_3) \quad (6c_2 + c_3) \quad (3c_1 + 3c_2 + 2c_3) \quad (2c_1 + 2c_2 + c_3)].$$

In one go we have the costs for the four house types. Note that the 1×3 vector is abstracted from a table of (cost $\times$ window types) and the 3×4 matrix from a table of (window types $\times$ house types). The result is (cost $\times$ (window types $\times$ window types $\times$) house types), that is, the cost (of windows) per house type. This also gives us an insight into the pattern of matrix multiplication for $1 \times (3 \times 3 \times) 4 = 1 \times 4$ is the order of the resultant vector. In general, two matrices may only be multiplied if one is of order $m \times k$ and the other is $k \times n$ so that the resultant matrix is of order $m \times (k \times k \times) n = m \times n$. Such matrices are said to be *conformable for multiplication.*

At Pessac, eight type 1 houses were built, eight type 2, seven type 3 and seventeen type 4. How many windows of each type were required? Our main matrix is (window types $\times$ house types). If we form a matrix of (house types $\times$ number) and multiply these together we shall obtain (window types $\times$ (house types $\times$ house types $\times$) number) which is what we require. The 4×1 column vector

$$\begin{bmatrix} 8 \\ 8 \\ 7 \\ 17 \end{bmatrix}$$

is an abstract representation of the four house types (rows) and their number (column). The product

$$\begin{bmatrix} 0 & 0 & 3 & 2 \\ 4 & 6 & 3 & 2 \\ 2 & 1 & 2 & 1 \end{bmatrix} \begin{bmatrix} 8 \\ 8 \\ 7 \\ 17 \end{bmatrix}$$

is what we want, where, by forming the inner products of the three row vectors with the column vector we produce a $3 \times (4 \times 4 \times) 1 = 3 \times 1$ vector of quantities for each window type, namely

$$\begin{bmatrix} (0.8 + 0.8 + 3.7 + 2.17) \\ (4.8 + 6.8 + 3.7 + 2.17) \\ (2.8 + 1.8 + 2.7 + 1.17) \end{bmatrix} = \begin{bmatrix} 55 \\ 135 \\ 55 \end{bmatrix}$$

These products have resulted in aggregate costs and quantities. How can we obtain appropriate breakdowns of this information? Take costs first. We want a breakdown which will tell us the cost for each type of window in each type of house. This is an array of 3×4 items. If we *pre-multiply* our 3×4 matrix by a 3×3 matrix we will again obtain a $3 \times (3 \times 3 \times) 4 = 3 \times 4$ matrix. The matrix we are looking for is derived from a table of three separate elemental costs by window types:

	Window types		
	1	2	3
Cost of window type 1	c_1	·	·
Cost of window type 2	·	c_2	·
Cost of window type 3	·	·	c_3

Such a table is abstracted by a 3×3 matrix whose non-zero elements are on the diagonal. *Diagonal matrices*, such as this one, are always square – of order $n \times n$. Let us proceed to the multiplication

$$\begin{bmatrix} c_1 & \cdot & \cdot \\ \cdot & c_2 & \cdot \\ \cdot & \cdot & c_3 \end{bmatrix} \begin{bmatrix} 0 & 0 & 3 & 2 \\ 4 & 6 & 3 & 2 \\ 2 & 1 & 2 & 1 \end{bmatrix}.$$

How do we do it? The method is to partition the first matrix into its 1×3 row vectors and the second matrix into its 3×1 column vectors. We can then form the inner products of these vectors, making sure that we place the result of forming the inner product of the ith row vector and jth column vector in the ith row and jth column of the resultant matrix, thus

$$\begin{bmatrix} (c_1.0+0.4+0.2) & (c_1.0+0.6+0.1) & (c_1.3+0.3+0.2) & (c_1.2+0.2+0.1) \\ (0.0+c_2.4+0.2) & (0.0+c_2.6+0.1) & (0.3+c_2.3+0.2) & (0.2+c_2.2+0.1) \\ (0.0+0.4+c_3.2) & (0.0+0.6+c_3.1) & (0.3+0.3+c_3.2) & (0.2+0.2+c_3.1) \end{bmatrix}$$

$$= \begin{bmatrix} 0 & 0 & 3c_1 & 2c_1 \\ 4c_2 & 6c_2 & 3c_2 & 2c_2 \\ 2c_3 & c_3 & 2c_3 & c_3 \end{bmatrix}$$

which gives us the breakdown of costs for each window type in each house type. The method of multiplication we have shown applies to any pair of conformable matrices, that is, matrices of order $m \times k$ and $k \times n$ taken in *that* order, for $k \times n$ and $m \times k$ are not conformable taken in this sequence unless $n = m$. But diagonal matrix multiplication *before* a matrix simply entails, as we see from our example, multiplication of each row of the matrix by the scalar in the corresponding row of the diagonal matrix. This is equivalent to the scalar product of each row vector making up the matrix. For this reason the diagonal matrix is often called the *scalar matrix*.

To return to the other problem of providing a breakdown of quantities. By analogy with previous examples it seems reasonable to suppose that this will be given by the *post-multiplication* of our matrix by the scalar matrix whose diagonal elements are 8, 8, 7, 17. The product (window types $\times$ (house types $\times$ house types $\times$) quantities for each house type) gives us the $3 \times (4 \times 4 \times) 4$ matrix we are looking for.

$$\begin{bmatrix} 0 & 0 & 3 & 2 \\ 4 & 6 & 3 & 2 \\ 2 & 1 & 2 & 1 \end{bmatrix} \begin{bmatrix} 8 & \cdot & \cdot & \cdot \\ \cdot & 8 & \cdot & \cdot \\ \cdot & \cdot & 7 & \cdot \\ \cdot & \cdot & \cdot & 17 \end{bmatrix}.$$

As we might expect, if *pre-multiplication* by a diagonal matrix multiplies *rows* by scalars, *post-multiplication* by a diagonal matrix multiplies columns by corresponding scalars. The matrix of quantities is thus

$$\begin{bmatrix} 0 & 0 & 21 & 34 \\ 32 & 48 & 21 & 34 \\ 16 & 8 & 14 & 17 \end{bmatrix}.$$

The vector, **u**, all of whose elements are 1, is sometimes useful. For example, referring back to the cost matrix,

$$[1 \quad 1 \quad 1]\begin{bmatrix} 0 & 0 & 3c_1 & 2c_1 \\ 4c_2 & 6c_2 & 3c_2 & 2c_2 \\ 2c_3 & c_3 & 2c_3 & c_3 \end{bmatrix}$$

is a $1 \times (3 \times 3 \times) 4 = 1 \times 4$ vector

$$[(4c_2 + 2c_3) \quad (6c_2 + c_3) \quad (3c_1 + 3c_2 + 2c_3) \quad (2c_1 + 2c_2 + c_3)]$$

which is our original aggregate cost vector. Pre-multiplication by the 'all-one' vector, **u**, sums the columns, while post-multiplication by an 'all-one' vector of appropriate order sums the rows, as we see in the example of the matrix of quantities

$$\begin{bmatrix} 0 & 0 & 21 & 34 \\ 32 & 48 & 21 & 34 \\ 16 & 8 & 14 & 17 \end{bmatrix}\begin{bmatrix} 1 \\ 1 \\ 1 \\ 1 \end{bmatrix} = \begin{bmatrix} 55 \\ 135 \\ 55 \end{bmatrix}$$

the result of which accords with the previously calculated vector of aggregate quantities. What does the triple product

$$[c_1 \quad c_2 \quad c_3]\begin{bmatrix} 0 & 0 & 3 & 2 \\ 4 & 6 & 3 & 2 \\ 2 & 1 & 2 & 1 \end{bmatrix}\begin{bmatrix} 8 \\ 8 \\ 7 \\ 17 \end{bmatrix}$$

stand for? We have a 1×3 vector pre-multiplying a 3×4 matrix so that the result is a $1 \times (3 \times 3 \times) 4$ vector which then pre-multiplies a 4×1 vector to give a $1 \times (4 \times 4 \times) 1$ 'vector' or number. The 'meanings' of each matrix help us to understand what this number represents: cost × (window types × window types ×) (house types × house types ×) number. The triple product thus gives us the *total* cost of the windows in the scheme as a whole.

What would have happened to the breakdown of quantities if more of the Pessac scheme had been built: if, for example, five more of house type 4 were built? Our original 'quantities' matrix is

$$\begin{bmatrix} 0 & 0 & 21 & 34 \\ 32 & 48 & 21 & 34 \\ 16 & 8 & 14 & 17 \end{bmatrix}.$$

We need to add the windows due to the new houses. We may only add matrices if they are *conformable for addition*, a condition which requires that both matrices have the same order, $m \times n$. We must therefore construct a table for the new development as if it contained all the house types

	House types 1	2	3	4
Window type 1	0	0	0	2
Window type 2	0	0	0	2
Window type 3	0	0	0	1

The abstract matrix of this array multiplied by five for the five houses may now be added to our previous matrix. This is done by adding corresponding column (or row) vectors:

$$\left[\begin{array}{c:c:c:c} 0 & 0 & 21 & 34 \\ 32 & 48 & 21 & 34 \\ 16 & 8 & 14 & 17 \end{array}\right] + 5\left[\begin{array}{c:c:c:c} 0 & 0 & 0 & 2 \\ 0 & 0 & 0 & 2 \\ 0 & 0 & 0 & 1 \end{array}\right]$$

$$= \left[\begin{array}{c:c:c:c} (0+5\cdot 0) & (0+5\cdot 0) & (21+5\cdot 0) & (34+5\cdot 2) \\ (32+5\cdot 0) & (48+5\cdot 0) & (21+5\cdot 0) & (34+5\cdot 2) \\ (16+5\cdot 0) & (8+5\cdot 0) & (14+5\cdot 0) & (17+5\cdot 1) \end{array}\right]$$

$$= \begin{bmatrix} 0 & 0 & 21 & 44 \\ 32 & 48 & 21 & 44 \\ 16 & 8 & 14 & 22 \end{bmatrix}.$$

Clearly, this is an elaborate way to achieve this particular result, but it does illustrate the method of matrix addition and of the multiplication of a matrix by a scalar.

These exercises have introduced most of the essential rules of matrix algebra which we may now summarize. Matrices are usually denoted by a bold capital letter. Let **A**, **B**, **C**,... be matrices of general order $p \times q$ to be specified more precisely as required. We may write **A** in terms of its typical element, a_{ij}, in the ith row and jth column, and similarly for **B**, **C**,.... Thus, for matrices such as

$$\mathbf{A} = \begin{bmatrix} a_{11} & a_{12} & \dots & a_{1q} \\ a_{21} & a_{22} & \dots & a_{2q} \\ \vdots & \vdots & & \vdots \\ a_{p1} & a_{p2} & \dots & a_{pq} \end{bmatrix} = [a_{ij}]$$

and the real numbers $\{a_{ij}, b_{ij}, c_{ij}, r, s, \dots\} \in R$ the following definitions and relations hold:

Definitions:

$\mathbf{A}_{mn}$, a *matrix of order* $m \times n$. Where the order is otherwise understood the suffices are dropped.
$\mathbf{A}_{1n}$, a *row vector* of order $1 \times n$
$\mathbf{A}_{m1}$, a *column vector* of order $m \times 1$
$\mathbf{A}_{nn}$, a *square matrix* of order $n \times n$
$\mathbf{A}_{nn} = [a_{ij}]$ is, (1), *triangular* if and only if, for $i \geqslant j$, $a_{ij} = 0$
is, (2), *strictly triangular* if and only if, for $i > j$, $a_{ij} = 0$
is, (3), *diagonal* if and only if, for $i \neq j$, $a_{ij} = 0$
is, (4), *symmetric* if, and only if, $a_{ij} = a_{ji}$
is, (5), *skew-symmetric* if, and only if, $a_{ij} = -a_{ji}$.
Examples of these matrices are

$$\underset{(1)}{\begin{bmatrix} \cdot & f & g \\ \cdot & \cdot & h \\ \cdot & \cdot & \cdot \end{bmatrix}} \quad \underset{(2)}{\begin{bmatrix} a & f & g \\ \cdot & b & h \\ \cdot & \cdot & c \end{bmatrix}} \quad \underset{(3)}{\begin{bmatrix} a & \cdot & \cdot \\ \cdot & b & \cdot \\ \cdot & \cdot & c \end{bmatrix}} \quad \underset{(4)}{\begin{bmatrix} a & f & g \\ f & b & h \\ g & h & c \end{bmatrix}} \quad \underset{(5)}{\begin{bmatrix} \cdot & f & g \\ -f & \cdot & h \\ -g & -h & \cdot \end{bmatrix}}$$

A diagonal matrix is like a 'sloping' vector, and since it is completely specified by the statement 'the diagonal matrix **A** whose non-zero elements correspond to the elements of the vector **a**' it is frequently abbreviated $\hat{\mathbf{a}}$. Thus if $\mathbf{a} = [a_1 \quad a_2 \quad a_3]^T$, then

$$\hat{\mathbf{a}} = \begin{bmatrix} a_1 & \cdot & \cdot \\ \cdot & a_2 & \cdot \\ \cdot & \cdot & a_3 \end{bmatrix}.$$

Equality: Two $m \times n$ matrices **A** and **B** are *equal* if their corresponding elements are equal

$$\mathbf{A} = \mathbf{B} \Leftrightarrow a_{ij} = b_{ij} \text{ for all } i, j. \text{ Thus}$$

$$\begin{bmatrix} a_{11} & a_{12} \\ a_{21} & a_{22} \end{bmatrix} = \begin{bmatrix} b_{11} & b_{12} \\ b_{21} & b_{22} \end{bmatrix} \Leftrightarrow \begin{cases} a_{11} = b_{11} \\ a_{21} = b_{21} \\ a_{12} = b_{12} \\ a_{22} = b_{22}. \end{cases}$$

Addition: Two $m \times n$ matrices **A** and **B** are *conformable for addition* and their sum is also a $m \times n$ matrix.

Closure: $\mathbf{A} + \mathbf{B} = [(a_{ij} + b_{ij})]$ so that conformable matrices are closed under addition. The following rules also pertain:

Commutative: $\mathbf{A} + \mathbf{B} = [(a_{ij} + b_{ij})] = [(b_{ij} + a_{ij})] = \mathbf{B} + \mathbf{A}$

Associative: $\mathbf{A} + (\mathbf{B} + \mathbf{C}) = [a_{ij} + (b_{ij} + c_{ij})] = [(a_{ij} + b_{ij}) + c_{ij}] = (\mathbf{A} + \mathbf{B}) + \mathbf{C}$

Identity: $\mathbf{A} + \mathbf{0} = \mathbf{0} + \mathbf{A} = \mathbf{A}$ where $\mathbf{0}$ is an $m \times n$ matrix with all zero elements.

Inverse: $\mathbf{A} + (-\mathbf{A}) = [a_{ij} + (-a_{ij})] = [(a_{ij} - a_{ij})] = \mathbf{0}$

Matrices of order $m \times n$ thus constitute an Abelian group under addition. We also have similar rules for scalar multiplication as we have for vectors.

Scalar product: Given any matrix **A** and any scalar $r \in \mathbf{R}$, then

Closure: $r\mathbf{A} = r[a_{ij}] = [ra_{ij}]$

which is another $m \times n$ matrix.

Associative: $(rs)\mathbf{A} = [(rs)]a_{ij}] = [r(sa_{ij})] = r(s\mathbf{A})$

Distributive: $r(\mathbf{A} + \mathbf{B}) = [(ra_{ij} + rb_{ij})] = r\mathbf{A} + r\mathbf{B}$
$(r + s)\mathbf{A} = [(ra_{ij} + sa_{ij})] = r\mathbf{A} + s\mathbf{A}$

Multiplication: Two matrices are *conformable for pre-multiplication* if the *rows* of the first matrix contain as many elements as the *columns* of the second. Thus two matrices $\mathbf{A}_{mp}$ and $\mathbf{B}_{pn}$ are conformable and their product is an $m \times n$ matrix:

$$\mathbf{AB} = [a_{ik}][b_{kj}] = [\Sigma_k\, a_{ik}b_{kj}]$$

which may also be represented in terms of the inner product of the rows of $\mathbf{A}$, the $1 \times p$ vectors $\mathbf{a}_i^T$, and the columns of $\mathbf{B}$, the $p \times 1$ vectors $\mathbf{b}_j$, thus

$$\begin{bmatrix} \mathbf{a}_1^T \\ \mathbf{a}_2^T \\ \vdots \\ \mathbf{a}_m^T \end{bmatrix} [\mathbf{b}_1 \quad \mathbf{b}_2 \quad \dots \quad \mathbf{b}_n] = [\mathbf{a}_i^T.\mathbf{b}_j].$$

The products of two matrices are not usually commutative and indeed if $\mathbf{AB}$ exists there is no reason why, in general, $\mathbf{BA}$ should exist, as we have already discussed. Matrices of the form $\mathbf{A}_{nm}$ and $\mathbf{B}_{mn}$ are conformable for pre- and post-multiplication, as are square matrices. Usually $\mathbf{AB} \neq \mathbf{BA}$. However, the following rules apply for conformable matrices:

Associative: $\mathbf{A}(\mathbf{BC}) = (\mathbf{AB})\mathbf{C}$

Distributive: $\mathbf{A}(\mathbf{B} + \mathbf{C}) = \mathbf{AB} + \mathbf{AC}$
$(\mathbf{A} + \mathbf{B})\mathbf{C} = \mathbf{AC} + \mathbf{BC}$

Identity: The *identity matrix*, $\mathbf{I}$, of order n is the $n \times n$ diagonal matrix

$$\mathbf{I} = [\delta_{ij}] \quad \text{where } \delta_{ij} = \begin{cases} 0 \text{ if } i \neq j \\ 1 \text{ if } i = j. \end{cases}$$

Then for any $n \times n$ square matrix $\mathbf{A}$

$$\mathbf{AI} = \mathbf{IA} = \mathbf{A}$$

and for any $m \times n$ matrix **B**

$\mathbf{BI} = \mathbf{B}$

and any $n \times m$ matrix **C**

$\mathbf{IC} = \mathbf{C}$.

For example,

$$\begin{bmatrix} a_{11} & a_{12} \\ a_{21} & a_{22} \end{bmatrix} \begin{bmatrix} 1 & \cdot \\ \cdot & 1 \end{bmatrix} = \begin{bmatrix} 1 & \cdot \\ \cdot & 1 \end{bmatrix} \begin{bmatrix} a_{11} & a_{12} \\ a_{21} & a_{22} \end{bmatrix} = \begin{bmatrix} a_{11} & a_{12} \\ a_{21} & a_{22} \end{bmatrix}$$

$$\begin{bmatrix} a_{11} & a_{12} & a_{13} \\ a_{21} & a_{22} & a_{23} \end{bmatrix} \begin{bmatrix} 1 & \cdot \\ \cdot & 1 \end{bmatrix} = \begin{bmatrix} a_{11} & a_{12} & a_{13} \\ a_{21} & a_{22} & a_{23} \end{bmatrix}$$

$$\begin{bmatrix} 1 & \cdot \\ \cdot & 1 \end{bmatrix} \begin{bmatrix} a_{11} & a_{12} \\ a_{21} & a_{22} \\ a_{31} & a_{32} \end{bmatrix} = \begin{bmatrix} a_{11} & a_{12} \\ a_{21} & a_{22} \\ a_{31} & a_{32} \end{bmatrix}.$$

Inverse: Only square matrices have inverses and then they have to satisfy the condition that their *determinant* is not zero. They are then said to be *non-singular*.

[The determinant of $n \times n$ matrix **A** is a number given by the expression

$\det \mathbf{A} = \Sigma_N \, \epsilon_{j_1 j_2 \ldots j_n} a_{1j_1} a_{2j_2} \ldots a_{nj_n}$

where $\epsilon_{j_1 j_2 \ldots j_n} = +1$ if $j_1 j_2 \ldots j_n$ is an *even* permutation of $1, 2, \ldots, n$ and -1 if $j_1 j_2 \ldots j_n$ is an *odd* permutation of $1, 2, \ldots, n$; and where the summation extends over all $N = n!$ permutations $j_1 j_2 \ldots j_n$ of the integers $1, 2, \ldots, n$. The determinant of a square matrix **A** of order 2 is the number $(a_{11}a_{22} - a_{12}a_{21})$: $j_1 j_2$ are the two permutations of 1 and 2, namely 1, 2 and 2, 1 so that $a_{1j_1} a_{2j_2} = a_{11}a_{22}$ for the even permutation 1, 2; and $a_{1j_1} a_{2j_2} = a_{12}a_{21}$ for the odd permutation 2, 1 (it takes an odd number of pairwise interchanges to produce 2,1 from 1, 2).]

The multiplicative *inverse* of a square non-singular matrix is defined by

$\mathbf{AA}^{-1} = \mathbf{A}^{-1}\mathbf{A} = \mathbf{I}$.

For example, the 2×2 matrix $\begin{bmatrix} a & b \\ c & d \end{bmatrix}$ has an inverse if its *determinant* defined as $(ad - bc)$, is not zero. This inverse is given by the matrix

$$\frac{1}{ad - bc} \cdot \begin{bmatrix} d & -c \\ -b & a \end{bmatrix}.$$

Let us test this by direct multiplication

$$\frac{1}{ad - bc} \cdot \begin{bmatrix} d & -b \\ -c & a \end{bmatrix} \begin{bmatrix} a & b \\ c & d \end{bmatrix} = \frac{1}{ad - bc} \cdot \begin{bmatrix} (\ ad - bc) & (db - bd) \\ (-ca + ac) & (ad - bc) \end{bmatrix}$$

$$= \begin{bmatrix} 1 & \cdot \\ \cdot & 1 \end{bmatrix}.$$

This brief survey of the basic rules of matrix algebra merely touches on some of its complications, but it is clear that matrices do not usually form a multiplicative group. Only square non-singular matrices do so, and then the group is non-commutative.

Figure 4.2

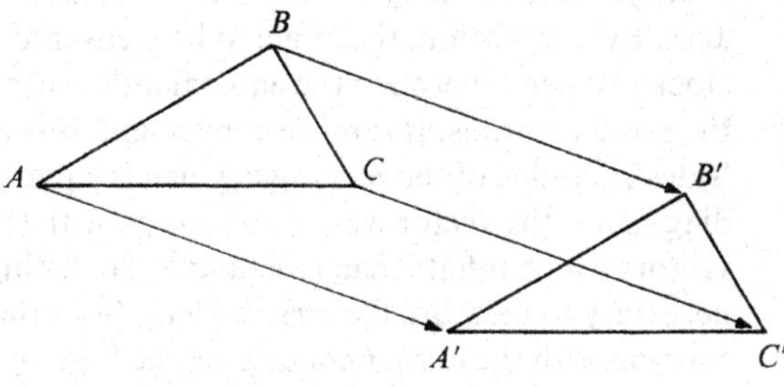

So far, in this chapter, we have discussed vectors as ordered 'lists' of numbers, but now we shall give them some geometric substance. In Chapter 2 we referred to the translation, **T**, which carries every point P to P' in such a way that if A, B, C are any three non-collinear points and A', B', C' are their corresponding images under translation then the vectors $\overline{AA'}$, $\overline{BB'}$, $\overline{CC'}$ are all equal (Figure 4.2). By this we mean that the length between any P and its image P' is equal to the length between any other point, Q, and its image, Q'. We also mean that the *direction* of any two vectors is the same, even though their positions differ. A vector in this geometric sense has just two properties, *length* and *direction*, in

Figure 4.3

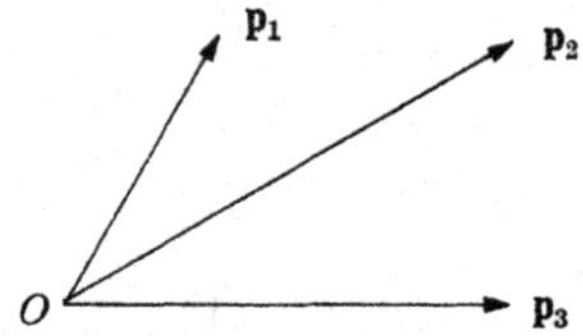

Figure 4.4

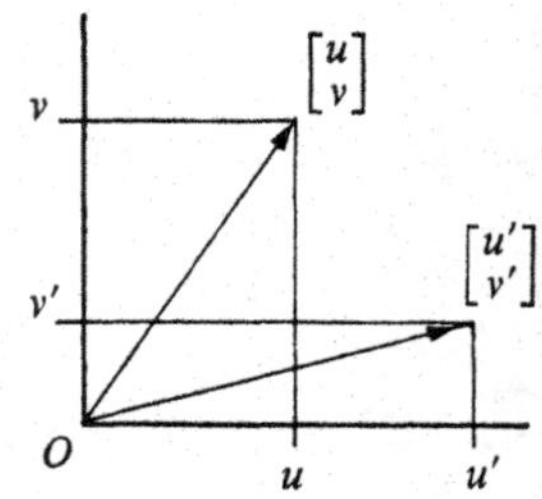

Figure 4.5

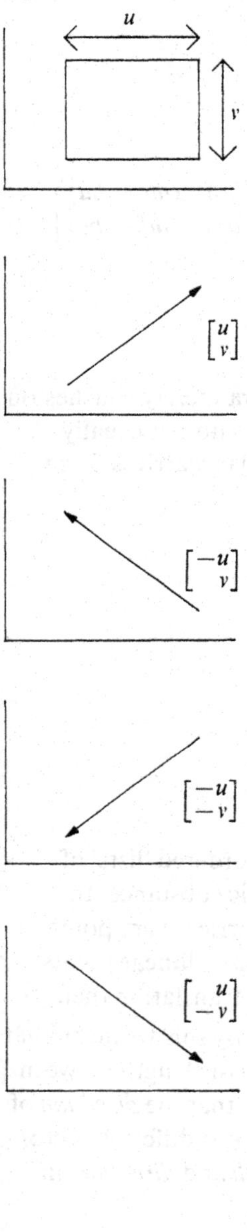

fact $\overline{AA'}$, $\overline{BB'}$, $\overline{CC'}$, *as vectors*, are indistinguishable and may be given the single name **p**, let us say. The *line segments* AA', BB', CC' are, of course, distinguishable because they possess the additional characteristic of *position* which vectors do not have. Sometimes, however, it is useful to give *vectors* a precise location; we then refer to them as *position vectors* (Figure 4.3). The most frequently used position vectors are those which have their 'tail' tied to a fixed reference point, the origin O.

Consider a plane marked with an origin and two axes which intersect at right-angles at the origin. (The stipulation about being a right angle is arbitrary, and any non-zero angle would do. It just happens that we habitually measure things, like the slope of a hill for instance, along perpendicular axes.) How can we specify a vector **p** with reference to the origin and axes? We could measure its length which we denote by $\| \mathbf{p} \|_2$ signifying its *magnitude* or *absolute value*. We would also need to specify its direction by the angle the vector makes with respect to the axes. By convention, the angle to be measured is the angle in an anti-clockwise sense between the equivalent position vector at the origin and the positive x-axis. It is more convenient, however, to measure the 'sides', u and v, of the rectangle which is square to the axes and whose diagonal is the vector we are concerned with (Figure 4.4). There are four vectors which inhabit each rectangle. To distinguish between them it is necessary to measure the *directed* lengths of the sides: that is to say, starting with the corner containing the 'tail' of the vector if we measure u in the same direction as the positive x-axis we assume it is positive, but if we measure u in the opposite direction we prefix a minus sign, thus, $-u$. Similarly with v in regard to the y-axis. Since it is important to know which of the two numbers is measured in the x-direction and which in the y we represent this geometric vector with the algebraic vector notation. The four vectors on the diagonals of the rectangle of sides u and v are shown in Figure 4.5. Notice that the pairs of 'equal but opposite' vectors are related one to another by the scalar product with -1, that is,

$$-1\begin{bmatrix} u \\ v \end{bmatrix} = \begin{bmatrix} -u \\ -v \end{bmatrix} \quad \text{and} \quad -1\begin{bmatrix} u \\ -v \end{bmatrix} = \begin{bmatrix} -u \\ v \end{bmatrix}.$$

Figure 4.6

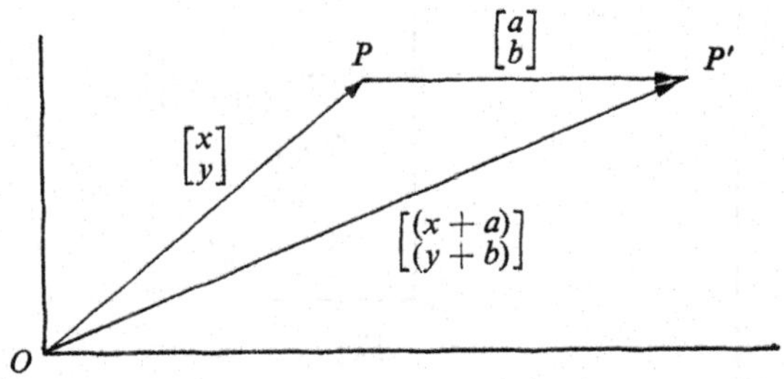

Geometrically, how shall we interpret vector addition? Suppose that a point P is defined by its position vector $[x \quad y]^T$ and that we then add the vector $[a \quad b]^T$. By the law of vector addition we obtain a new vector $[a+x \quad b+y]^T$ which as a *position* vector determines another point P' in the plane (Figure 4.6). We then have the well-known 'parallelogram of forces' diagram in which $\overline{OP} + \overline{PP'} = \overline{OP'}$. Now it is clear that whatever point we choose in the plane, $P, Q, \ldots$, the vector $[a \quad b]^T$ represents the shift from that point to the image. We thus have a one-to-one correspondence between a *vector* and a *translation*. Indeed, vectors and translations provide an example of what mathematicians call *isomorphism:*

	Translations $\{T, .\}$	Vectors $\{V, +\}$
Closure:	$\mathbf{T}_1\mathbf{T}_2 \in T$	$\mathbf{a} + \mathbf{b} \in V$
Commutative:	$\mathbf{T}_1\mathbf{T}_2 = \mathbf{T}_2\mathbf{T}_1$	$\mathbf{a} + \mathbf{b} = \mathbf{b} + \mathbf{a}$
Associative:	$\mathbf{T}_1(\mathbf{T}_2\mathbf{T}_3) = (\mathbf{T}_1\mathbf{T}_2)\mathbf{T}_3$	$\mathbf{a} + (\mathbf{b} + \mathbf{c}) = (\mathbf{a} + \mathbf{b}) + \mathbf{c}$
Identity:	$\mathbf{TI} = \mathbf{IT} = \mathbf{T}$	$\mathbf{a} + \mathbf{0} = \mathbf{0} + \mathbf{a} = \mathbf{a}$
Inverse:	$\mathbf{TT}^{-1} = \mathbf{T}^{-1}\mathbf{T} = \mathbf{I}$	$\mathbf{a} + (-\mathbf{a}) = (-\mathbf{a}) + \mathbf{a} = \mathbf{0}$

Isomorphic groups have the same *structure* although they may differ in respect of notation and the nature of their elements. In this case both vectors under addition, and translations under multiplication, have the structure of an Abelian, commutative, group. Notice, too, that whereas translations have the identity **I**, vectors have the identity **0**. An everyday example of isomorphism, which parallels this, is to be found in the use of additive logarithms for multiplying real numbers together. If $r_i \in R$ and $\ln r_i \in L$, we then have

$$r_i \leftrightarrow \ln r_i, \text{ for all } r_i,$$

and

$$r_i r_j \leftrightarrow \ln r_i + \ln r_j.$$

Especially, we have

$$r_i^n \leftrightarrow n \ln r_i.$$

which is similar to the relationship between successive translations $\mathbf{T}^n$ and the addition of equal vectors $n\mathbf{a}$: in the one case we have a *scalar exponent* rule, and in the other a *scalar product* rule.

Consider a pair of rectangular axes in the plane with an origin O. If the y-axis were moved 3 units to the right it is clear that every y-com-

Figure 4.7

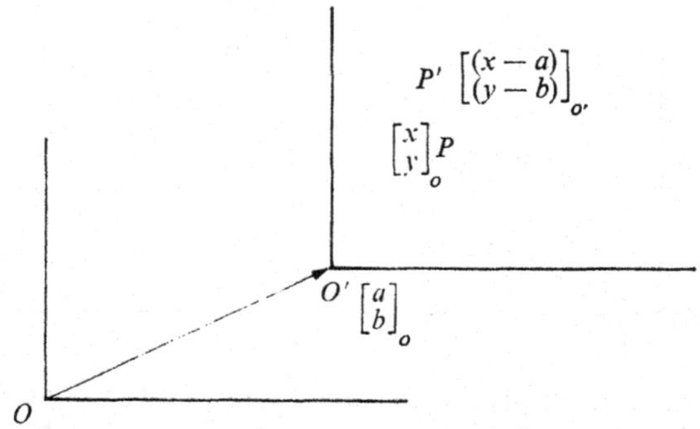

ponent of the position vector of a typical point would be decreased by 3, while a similar move to the left (O moves to O' which is $[-3 \quad 0]^T$ on the old system) also 'decreases' the y-component of a typical position vector by -3. The same effect occurs with shifts up and down with respect to the x-components. In general, if the coordinate axes (the base vectors) are moved to a new origin O', or the point $[a \quad b]^T$ measured from O, then the new name which the point P, $[x \quad y]^T$, acquires is P', $[x-a \quad y-b]^T$, with respect to the new origin O' (Figure 4.7). Such a transformation involving the *translation of axes*, where points remain fixed but the reference system changes, is called an *alias* transformation since points acquire new *names*. The previous example involved a mapping of one set of points, P, Q,..., onto a set of points P', Q',..., at a new *address* with regard to a fixed reference system. Such a translation is called an *alibi* (Figure 4.6).

We have been loosely referring to *coordinate axes*. Strictly speaking we should speak of *base vectors*. The vector $[x \quad y]^T$ may be expressed in terms of the sum of two scalar products, thus

$$\begin{bmatrix} x \\ y \end{bmatrix} = x\begin{bmatrix} 1 \\ 0 \end{bmatrix} + y\begin{bmatrix} 0 \\ 1 \end{bmatrix}$$

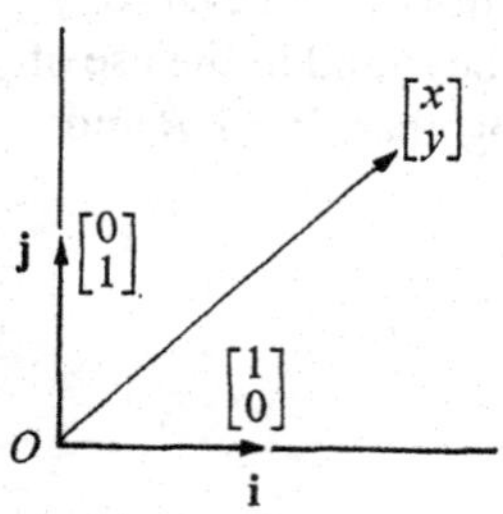

Figure 4.8

any vector may be expressed in this way. The zero-one vectors $[1 \quad 0]^T$ and $[0 \quad 1]^T$ are called the base vectors which, if they are mutually perpendicular and they need not be, are usually abbreviated **i** and **j**. The scalars x and y are the respective i- and j- components of the vector $[x \quad y]^T$ (Figure 4.8). There is a one-to-one correspondence between the traditional x- and y-axes with units of measure along each and these *unit vectors*, and since the former is certainly more familiar we shall continue to refer to the x- and y- components of a vector, rather than the i- and j- components.

We see now that the scalar product $x\mathbf{i}$ gives *length* x to the vector in the direction of **i**, and $y\mathbf{j}$ gives *length* in the direction of **j**. In general, if **p** is any vector, $m\mathbf{p}$ is a vector in the same direction as **p** but m times as long. But what geometrical interpretation can be given to the *inner product* of two vectors? For example, what meaning can we attribute to the results

Figure 4.9

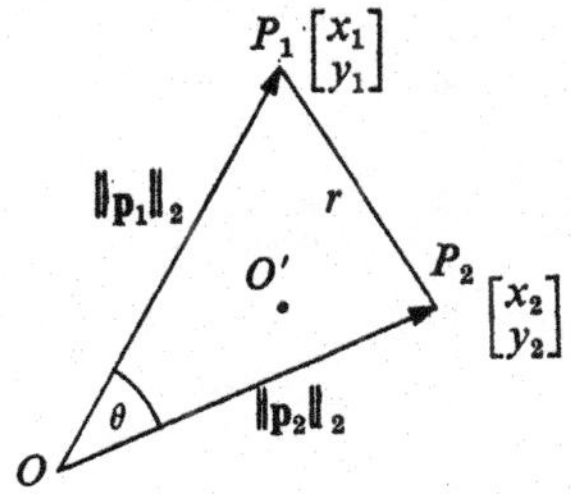

$$\mathbf{i}\,.\,\mathbf{j} = 0$$
$$\mathbf{i}\,.\,\mathbf{i} = (\mathbf{i}^2) = 1$$
$$\mathbf{j}\,.\,\mathbf{j} = (\mathbf{j}^2) = 1.$$

Consider first the general case of two position vectors $\overline{OP_1}$ ($\mathbf{p}_1$) and $\overline{OP_2}$ ($\mathbf{p}_2$) with components x_1, y_1 and x_2, y_2 respectively. The inner products

$$\mathbf{p}_1 \cdot \mathbf{p}_1 = (\mathbf{p}_1{}^2) = x_1{}^2 + y_1{}^2$$
$$\text{and} \quad \mathbf{p}_2 \cdot \mathbf{p}_2 = (\mathbf{p}_2{}^2) = x_2{}^2 + y_2{}^2$$

look remarkably familiar. They are vector representations of Pythagoras's theorem concerning the square of the hypotenuse being equal to the sum of the squares on the other two sides. Thus $\mathbf{p}\,.\,\mathbf{p}$ is the square of the length of the vector $\mathbf{p}$. This length is usually denoted by $\|\mathbf{p}\|_2 = (\mathbf{p}\,.\,\mathbf{p})^{\frac{1}{2}}$. We see that the unit vectors $\mathbf{i}$ and $\mathbf{j}$ have length $\|\mathbf{i}\|_2 = \|\mathbf{j}\|_2 = 1$. But what geometric meaning can we give to $\mathbf{p}_1\,.\,\mathbf{p}_2 = x_1x_2 + y_1y_2$?

Let the angle $\angle P_1OP_2$ be θ, and let the length of the vector $\overline{P_1P_2}$ be r. Then by the *law of cosines* for the triangle P_1OP_2 (Figure 4.9) we have

$$r^2 = \|\mathbf{p}_1\|_2{}^2 + \|\mathbf{p}_2\|_2{}^2 - 2\,\|\mathbf{p}_1\|_2 \cdot \|\mathbf{p}_2\|_2 \cos\theta$$

but in the triangle $P_1O'P_2$ we see that $r^2 = (x_1 - x_2)^2 + (y_1 - y_2)^2$ so that on expansion

$$\begin{aligned} r^2 &= (x_1{}^2 + x_2{}^2) + (y_1{}^2 + y_2{}^2) - 2(x_1x_2 + y_1y_2) \\ &= \|\mathbf{p}_1\|_2{}^2 + \|\mathbf{p}_2\|_2{}^2 - 2\mathbf{p}_1 \cdot \mathbf{p}_2 \end{aligned}$$

whence combining the two equations for r^2 we have the result that

$$\mathbf{p}_1 \cdot \mathbf{p}_2 = \|\mathbf{p}_1\|_2\|\mathbf{p}_2\|_2 \cos\vartheta.$$

When $\theta = \frac{\pi}{2}$, $\cos\theta = 0$, so that

$$\mathbf{i}\,.\,\mathbf{j} = 0$$

confirms that the two unit vectors are perpendicular as we defined them. The angle between any two vectors is then given by its cosine

$$\cos\theta = \frac{\mathbf{p}_1 \cdot \mathbf{p}_2}{\|\mathbf{p}_1\|_2\|\mathbf{p}_2\|_2} = \frac{x_1x_2 + y_1y_2}{{}_{+}\sqrt{(x_1^2 + y_1^2)} + {}_{+}\sqrt{(x_2^2 + y_2^2)}}.$$

The angle we choose between the two lines is the *interior*, or enclosed, angle which satisfies the conditions $0 \leqslant \theta \leqslant 180°$, so that there is no ambiguity since arc cos θ is then unique.

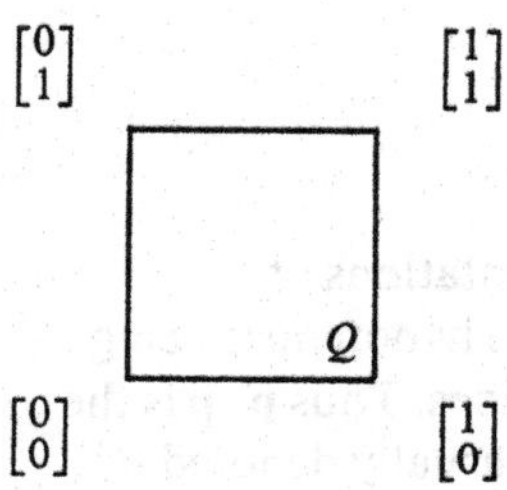

Figure 4.10

Consider the square Q in Figure 4.10. Its vertices are located by the position vectors $[0 \quad 0]^T$, $[0 \quad 1]^T$, $[1 \quad 1]^T$ and $[1 \quad 0]^T$. If we then apply the translation $[4 \quad 0]^T$ to these points we move the square to a new position – four units along the x-axis – with vertices

$$\begin{bmatrix}0\\0\end{bmatrix} + \begin{bmatrix}4\\0\end{bmatrix} = \begin{bmatrix}4\\0\end{bmatrix}$$

$$\begin{bmatrix}0\\1\end{bmatrix} + \begin{bmatrix}4\\0\end{bmatrix} = \begin{bmatrix}4\\1\end{bmatrix}$$

$$\begin{bmatrix}1\\1\end{bmatrix} + \begin{bmatrix}4\\0\end{bmatrix} = \begin{bmatrix}5\\1\end{bmatrix}$$

$$\begin{bmatrix}1\\0\end{bmatrix} + \begin{bmatrix}4\\0\end{bmatrix} = \begin{bmatrix}5\\0\end{bmatrix}.$$

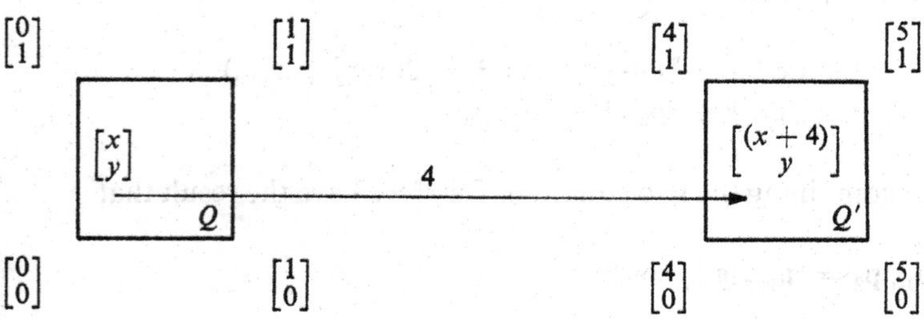

Figure 4.11

At the same time all the points within the square $[x \quad y]^T \in Q$ are translated to $[(x + 4) \quad y]^T$ as in Figure 4.11. If, however, we apply the translation $[\frac{3}{2} \quad \frac{3}{2}\sqrt{3}]^T$, the square is moved at 30° to the previous translation, but now with a three-unit shift in the direction of the translation (Figure 4.12).

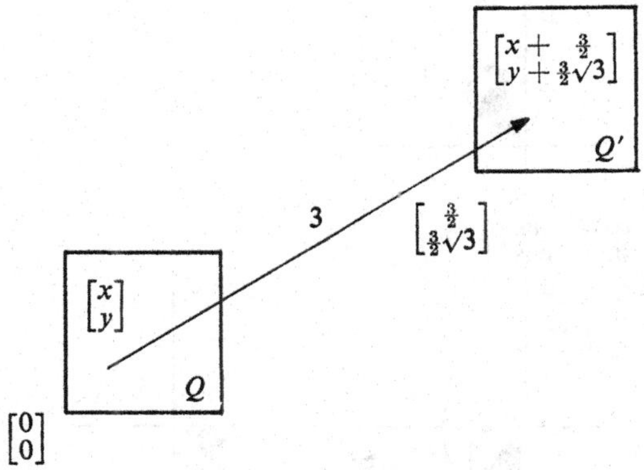

Figure 4.12

In Le Corbusier's scheme at Pessac two of the house types are arranged in rows. These are the semi-detached dwellings, house types 1 and 2 (Figure 1.6), and the terrace unit, type 3 (Figure 1.5). Consider the set of points (position vectors) $\mathbf{s} = [x \quad y]^T$ which make up the plan of the semi-detached house S_0 in Figure 4.13, so that we may write $\{\mathbf{s} \mid \mathbf{s} \in S_0\}$; and the set of points (position vectors) $\mathbf{t} = [x \quad y]^T$ within the terrace house T_0 in the same figure, $\{\mathbf{t} \mid \mathbf{t} \in T_0\}$. We can represent the next house in the row of semi-detached houses by a translation represented by a vector $\mathbf{a} = [4 \quad 0]^T$ if our unit of measure is the length of side of Le Corbusier's square module. Thus the set of points in

$$\begin{aligned} &S_1 \text{ is } \{\mathbf{s} + \mathbf{a}\} \\ &S_2 \text{ is } \{\mathbf{s} + 2\mathbf{a}\} \\ &\vdots \qquad \vdots \\ &S_i \text{ is } \{\mathbf{s} + i\mathbf{a}\}. \end{aligned}$$

Hence, if $[x \quad y]^T$ is the position vector of a point of the ground plan in S_0, then the corresponding point in the ith pair of semi-detached dwellings S_i is $[(x + 4i) \quad y]^T$.

The terrace is set at 30° to the row of semi-detached dwellings. The shift between one house and the next in the line of translation is now 3 units. As above, in the example of the square, the vector $\mathbf{b} = [\frac{3}{2} \ \frac{3}{2}\sqrt{3}]^T$ represents the required translation. The ith house in the terrace T_i is then described by the set of vectors $\{\mathbf{t} + i\mathbf{b}\}$.

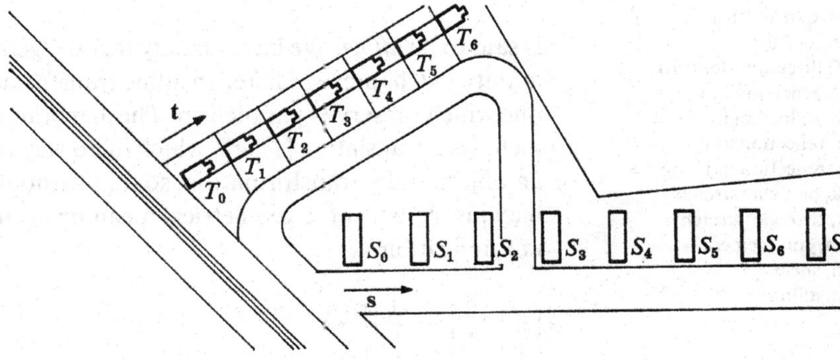

Figure 4.13
Le Corbusier's layout for houses at Pessac

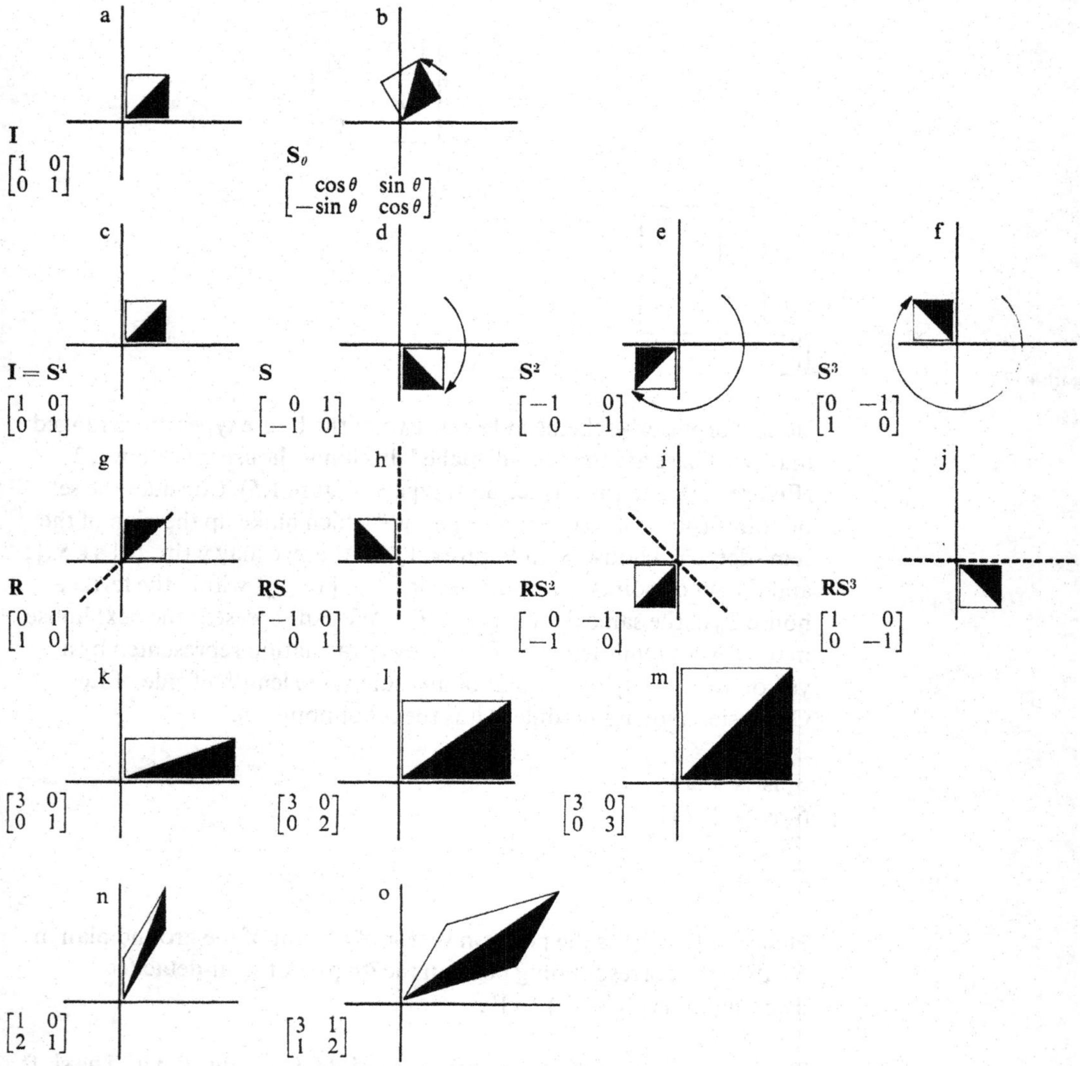

Figure 4.14
Examples of matrices of some common transformations:
a, identity
b, rotation through θ
c, wholeturn (identity)
d, quarterturn
e, halfturn
f, three-quarter turn
g, reflection in $y = x$
h, reflection in $x = 0$
i, reflection in $y = -x$
j, reflection in $y = 0$
k, one-way stretch
l, two-way stretch
m, enlargement
n, shear
o, affinity

Translations as we have seen in Chapter 2 are just one of a number of isometry transformations. The general transformation

$$\begin{bmatrix}x\\y\end{bmatrix} \leftrightarrow \begin{bmatrix}a & b\\c & d\end{bmatrix}\begin{bmatrix}x\\y\end{bmatrix} + \begin{bmatrix}f\\g\end{bmatrix}$$

is said to be *affine*. We have already met the geometrical concept of affinity in Chapter 1. There, an affine transformation was described as one which preserved parallelism. The algebraic expression above includes a translation $[f \quad g]^T$ which in no way changes the shape of an object under transformation, so that without loss of generality we may just consider the geometrical meaning of the 2×2 matrix pre-multiplication

$$\begin{bmatrix}x\\y\end{bmatrix} \leftrightarrow \begin{bmatrix}a & b\\c & d\end{bmatrix}\begin{bmatrix}x\\y\end{bmatrix}.$$

An affine transformation is reversible so that the *inverse* $\begin{bmatrix} a & b \\ c & d \end{bmatrix}$ must exist and, as we have seen, this requires the determinant $(ad - bc)$ to be non-zero. Figure 4.14 shows the transformations induced by various 2×2 matrices.

The general *rotation* matrix is

$$\mathbf{S}_\theta = \begin{bmatrix} \cos\theta & \sin\theta \\ -\sin\theta & \cos\theta \end{bmatrix}$$

whose determinant is $\cos^2\theta + \sin^2\theta = 1$. Matrices of this kind form a multiplicative Abelian group with the properties that $\mathbf{S}_{\theta_1}\mathbf{S}_{\theta_2} = \mathbf{S}_{\theta_1+\theta_2}$, $\mathbf{S}_\theta \mathbf{S}_\theta^{-1} = \mathbf{I}$ and, especially, $\mathbf{S}_\theta{}^n = \mathbf{S}_{n\theta}$ (de Moivre's theorem). Notice that $\mathbf{S}_\pi = \begin{bmatrix} -1 & 0 \\ 0 & -1 \end{bmatrix}$ is a halfturn, $\mathbf{H}$, and is equivalent to $-\mathbf{I}$ which 'explains' our previous discovery that $[u \quad v]^T$ and $-[u \quad v]^T$ are opposite in sense, but equal in length. The rotation through one right angle forms a cyclic group of period 4:

$$\mathbf{S}_{\frac{\pi}{2}} = \begin{bmatrix} 0 & 1 \\ -1 & 0 \end{bmatrix}$$

$$\mathbf{S}_{\frac{\pi}{2}}{}^2 = \begin{bmatrix} 0 & 1 \\ -1 & 0 \end{bmatrix}\begin{bmatrix} 0 & 1 \\ -1 & 0 \end{bmatrix} = \begin{bmatrix} -1 & 0 \\ 0 & -1 \end{bmatrix} = \mathbf{S}_\pi \, (= \mathbf{H})$$

$$\mathbf{S}_{\frac{\pi}{2}}{}^3 = \begin{bmatrix} 0 & 1 \\ -1 & 0 \end{bmatrix}\begin{bmatrix} -1 & 0 \\ 0 & -1 \end{bmatrix} = \begin{bmatrix} 0 & -1 \\ 1 & 0 \end{bmatrix} = \mathbf{S}_{\frac{3\pi}{2}}$$

$$\mathbf{S}_{\frac{\pi}{2}}{}^4 = \begin{bmatrix} 0 & 1 \\ -1 & 0 \end{bmatrix}\begin{bmatrix} 0 & -1 \\ 1 & 0 \end{bmatrix} = \begin{bmatrix} 1 & 0 \\ 0 & 1 \end{bmatrix} = \mathbf{S}_{2\pi} = \mathbf{I}.$$

The *reflection* matrices do not form a group. The product of two reflections is, as we know from Chapter 2, a rotation, so that reflection matrices are not closed under multiplication. The general reflection matrix is

$$\begin{bmatrix} -\sin\theta & \cos\theta \\ \cos\theta & \sin\theta \end{bmatrix} = \begin{bmatrix} 0 & 1 \\ 1 & 0 \end{bmatrix}\begin{bmatrix} \cos\theta & \sin\theta \\ -\sin\theta & \cos\theta \end{bmatrix}$$

so that a reflection may always be compounded of a reflection about $y = x$ and a proper rotation. A *glide reflection* reintroduces the translation which for the moment we are ignoring. The determinant of a reflection matrix is $-(\cos^2\theta + \sin^2\theta) = -1$. Together the four reflection matrices and the four rotation matrices form a *non-commutative* multiplicative group with the following multiplication table:

	I	**S**	**S²**	**S³**	**R**	**RS**	**RS²**	**RS³**
I	**I**	**S**	$\mathbf{S}^2$	$\mathbf{S}^3$	**R**	**RS**	$\mathbf{RS}^2$	$\mathbf{RS}^3$
S	**S**	$\mathbf{S}^2$	$\mathbf{S}^3$	**I**	$\mathbf{RS}^3$	**R**	**RS**	$\mathbf{RS}^2$
$\mathbf{S}^2$	$\mathbf{S}^2$	$\mathbf{S}^3$	**I**	**S**	$\mathbf{RS}^2$	$\mathbf{RS}^3$	**R**	**RS**
$\mathbf{S}^3$	$\mathbf{S}^3$	**I**	**S**	$\mathbf{S}^2$	**RS**	$\mathbf{RS}^2$	$\mathbf{RS}^3$	**R**
R	**R**	**RS**	$\mathbf{RS}^2$	$\mathbf{RS}^3$	**I**	**S**	$\mathbf{S}^2$	$\mathbf{S}^3$
RS	**RS**	$\mathbf{RS}^2$	$\mathbf{RS}^3$	**R**	$\mathbf{S}^3$	**I**	**S**	$\mathbf{S}^2$
$\mathbf{RS}^2$	$\mathbf{RS}^2$	$\mathbf{RS}^3$	**R**	**RS**	$\mathbf{S}^2$	$\mathbf{S}^3$	**I**	**S**
$\mathbf{RS}^3$	$\mathbf{RS}^3$	**R**	**RS**	$\mathbf{RS}^2$	**S**	$\mathbf{S}^2$	$\mathbf{S}^3$	**I**

If the determinant of the affine matrix $\mathbf{A} = \begin{bmatrix} a & b \\ c & d \end{bmatrix}$ is ± 1 the transformation is an *isometry*, and the positive sign indicates that the isometry is proper while the negative shows that it is improper and that the object is handed as a result of the transformation. In general, the absolute value of the determinant $(ad - bc)$ measures the change of scale resulting from the transformation as can be seen from the *stretches* and the *enlargement*. The matrices for *one-way stretches* are of the form

$$\begin{bmatrix} m & 0 \\ 0 & 1 \end{bmatrix} \text{ or } \begin{bmatrix} 1 & 0 \\ 0 & n \end{bmatrix}$$

with areal changes in the image proportional to $|m| : 1$ or $|n| : 1$ with respect to the original. A *two-way stretch* is given by a matrix of the form

$$\begin{bmatrix} m & 0 \\ 0 & n \end{bmatrix}$$

with an areal change of $|mn| : 1$. While an *enlargement* (or *dilation*) is effectively straight *scalar multiplication* since

$$\begin{bmatrix} m & 0 \\ 0 & m \end{bmatrix} = m\mathbf{I}.$$

Here the areal enlargement is $m^2 : 1$. The *shear* matrix has just one zero and can take the form

$$\begin{bmatrix} 0 & b \\ c & d \end{bmatrix}, \begin{bmatrix} a & b \\ c & 0 \end{bmatrix}, \text{or} \begin{bmatrix} a & 0 \\ c & d \end{bmatrix}, \begin{bmatrix} a & b \\ 0 & d \end{bmatrix}$$

with areal changes of $| bc | : 1$ in the first two, and $| ad | : 1$ in the second.

These matrix operations are employed in computer-aided design where graphic presentation requires the ability to manipulate forms – to change location, orientation, sense, reflection, dilation or enlargement, or to present an axonometric (affine) view. To develop perspectives it is necessary to define points in terms of coordinates called *homogeneous* or *barycentric*. Here instead of just two coordinates to define a point in the plane, we require three since absolute length is no longer as important as the *ratio* of lengths. The transformation matrix for a *perspectivity* in two dimensions is a square matrix of order 3. In three dimensions our affine coordinates have three components, and the affine transformation matrices are square of order 3; while we require four coordinates in projective space for perspectivities accompanied by a square transformation matrix of order 4.

Dilation and enlargement are everyday *affinities* in an architect's office as plans are produced at various scales for a variety of purposes, but there are two instances in actual building projects which are of some interest. The first occurs in an early project by Frank Lloyd Wright for a house for the wealthy Chicago industrialist Harold McCormick. This is an example of dilation. At one end of the house, by a pool, is a half-scale playhouse for children where all the detailing of the adults' house are reproduced in miniature (Figure 4.15). The other example is reported by Le Corbusier and concerns his project for the Governor's Palace at Chandigarh, the new capital of the Punjab. Le Corbusier designed many of his later buildings on a proportional scale of dimensions (see Chapter 9) which he named *le Modulor*. There are two interlocking scales giving, for example, on one scale a door-height of 2260 mm and

Figure 4.15
View of the McCormick House by Frank Lloyd Wright showing the half-scale playhouse

[3] From Le Corbusier, *Le Corbusier 1910–1960*, Zurich, Editions Girsberger, 1960, p. 206.

on the other 2959 mm. Le Corbusier describes this case of erroneous enlargement in his inimitable style (and we love him for it):

> 'Le Palais du Gouverneur couronne le Capitol. Son plan, sa silhouette sont le produit des strictes données du problème. Au cours de trois années, 1951–1953, le projet développé a pris corps. 1953: Crise! Le coût est infiniment trop élevé! Que s'est-il produit? Les plans étant acceptés, on avait revu les hauteurs et les largeurs de toutes choses...et l'on avait glissé (puisque c'était pour le Gouverneur!) du côté des cotes les plus fortes du Modulor. Le volume s'avère double du précédent! Et l'échelle du Palais démesurée! On avait bâti à l'échelle des géants!
>
> Tout fut reconsidéré. Le choix de valeurs suffisantes plus basses du Modulor fit baisser de moitié le cube de la bâtisse et nous réinstalla a l'échelle des hommes. Les plans d'exécution achevés demontrèrent qu'ainsi nous avions replacé le Gouverneur dans une maison d'homme.'[3]

Finally, there is a special set of zero-one matrices which form a multiplicative group called *permutation* matrices. These are defined as

$$\mathbf{P}_{j_1 j_2 j_3 \cdots j_n} = [\mathbf{e}_{j_1} \quad \mathbf{e}_{j_2} \quad \mathbf{e}_{j_3} \ \ldots \ \mathbf{e}_{j_n}]$$

where $j_1, j_2, j_3, \ldots, j_n$ are a permutation or rearrangement of the numbers $1, 2, 3, \ldots n$, and $\mathbf{e}_{j_1}, \mathbf{e}_{j_2}, \mathbf{e}_{j_3}, \ldots, \mathbf{e}_{j_n}$ are elementary $n \times 1$ column vectors with the $j_1, j_2, j_3, \ldots, j_n$ elements respectively equal to 1 and all the others zero. For example, with $n = 3$

$$\mathbf{e}_1 = \begin{bmatrix} 1 \\ 0 \\ 0 \end{bmatrix}, \quad \mathbf{e}_2 = \begin{bmatrix} 0 \\ 1 \\ 0 \end{bmatrix}, \quad \mathbf{e}_3 = \begin{bmatrix} 0 \\ 0 \\ 1 \end{bmatrix},$$

so that $$\mathbf{P}_{123} = \begin{bmatrix} 1 & 0 & 0 \\ 0 & 1 & 0 \\ 0 & 0 & 1 \end{bmatrix} = [\mathbf{e}_1 \quad \mathbf{e}_2 \quad \mathbf{e}_3] = \mathbf{I}$$

and $$\mathbf{P}_{231} = \begin{bmatrix} 0 & 0 & 1 \\ 1 & 0 & 0 \\ 0 & 1 & 0 \end{bmatrix} = [\mathbf{e}_2 \quad \mathbf{e}_3 \quad \mathbf{e}_1]$$

[4] See Bruce Martin, 'The Smallest Building; the Genesis of the Mark 8 Telephone Box' in *RIBA Journal*, August 1969, pp. 320–5.

and so on. Pre-multiplication by a permutation matrix interchanges the rows of a matrix, while post-multiplication permutes the columns. Later we shall make use of these matrices in describing shape.

There is, however, an interesting example of the use of permutation in design in Bruce Martin's new telephone boxes[4] for the General Post Office (Figures 4.16–4.17). The plan form is square. The elements consist of a panel containing all the services on one side of the square, *s*, an entrance to the booth, *e*, and two window walls, *w* and *w*. In cyclic order the plan of the telephone box might be represented in vector form, $\mathbf{b} = [s \quad e \quad w \quad w]^T$, or any of the 12 distinguishable permutations of these four elements. These are comprehensively described by **Pb** where **P** is a 4 × 4 permutation matrix. Next, Martin has designed the box so that the door may be hung in two handed positions – hinged to the right or to the left. The door in each of its three permissible

Figure 4.16
The permutations of the new Mark 8 telephone box designed for the General Post Office by Bruce Martin

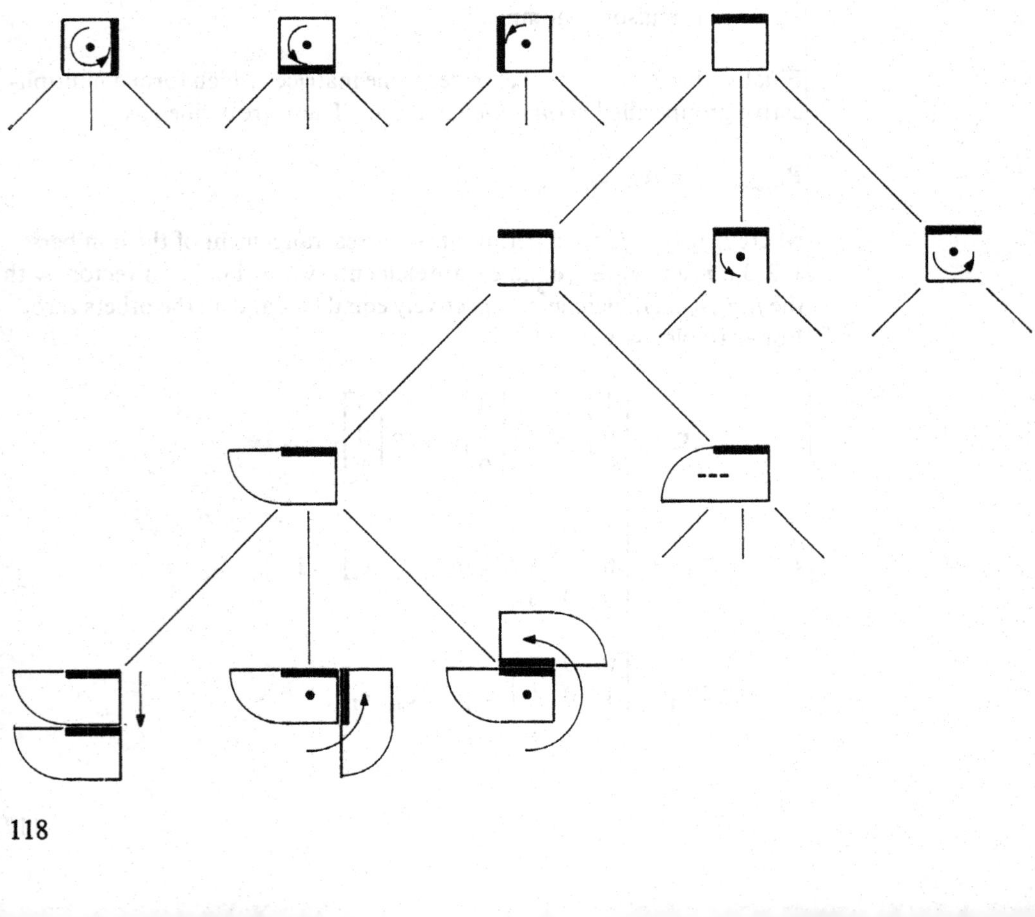

Figure 4.17 **Tableau showing the seventy-two combinations of pairs of Mark 8 telephone boxes**

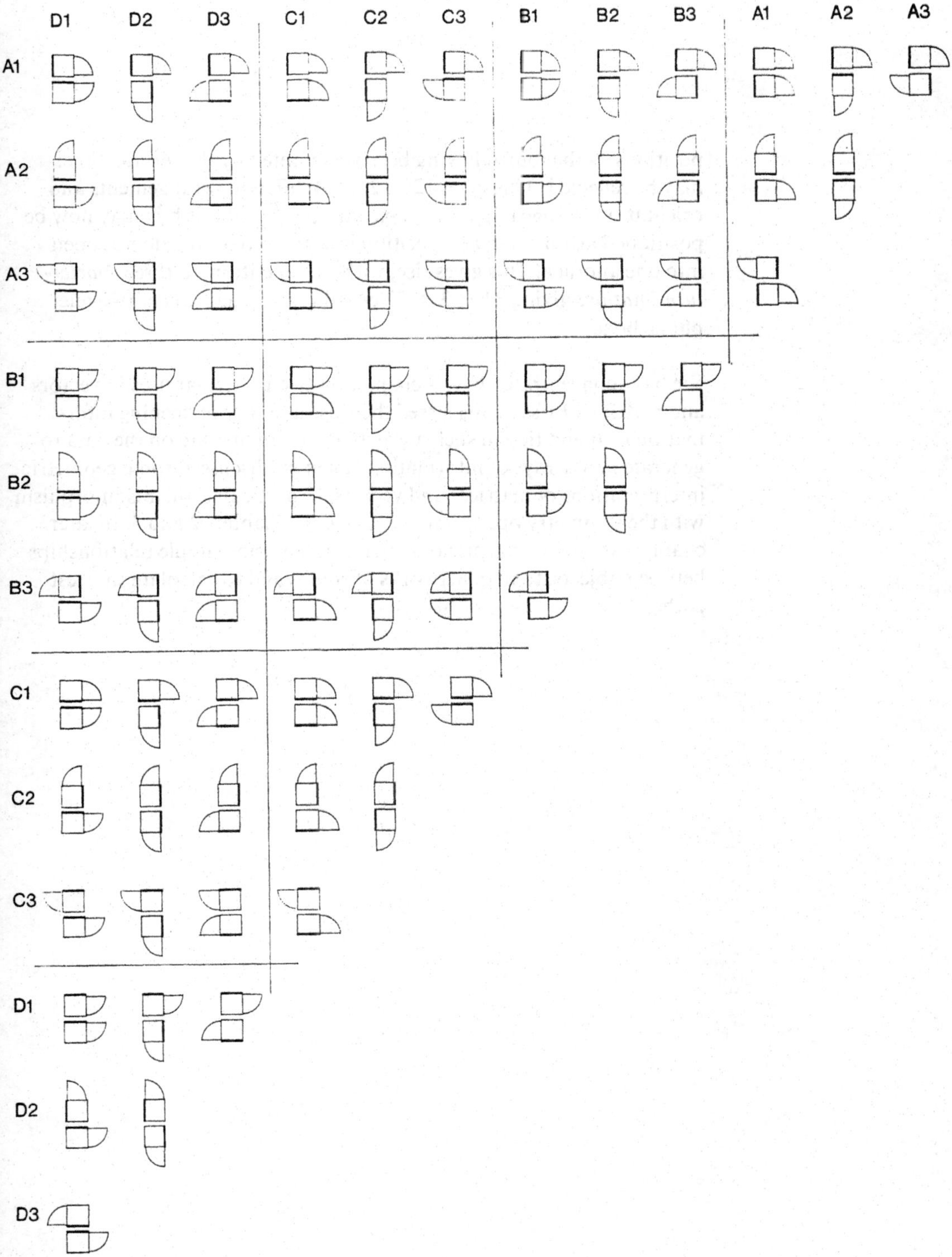

positions — the fourth having been pre-empted by the service panel – may be reflected. This gives $12 \times 2 = 24$ different arrangements and orientations of one telephone box. An *isometry* of this box may now be positioned against it in any position in which two doors do not open into one another! This gives rise, without repetition, to three *quarter-turn rotations* giving $24 \times 3 = 72$ possible arrangements of two telephone boxes.

We have concentrated in this chapter on two particular uses of vectors and matrices. First we discussed their use as a way of keeping information neat and tidy in such a way that we can operate on the data to generate new arrays of information. Then we discussed some geometric interpretations of matrices and vectors, in particular, their isomorphism with the symmetry operations discussed in Chapters 2 and 3. In later chapters we shall make use of matrices to describe simple relationships between objects, to map networks of routes and to calculate shortest paths.

5 Point sets and modular spaces

In any quantitative studies of building or environmental geometry it is clearly desirable to be able to describe the forms with some mathematical precision. This is a task which, in general, lacks the elegance and simplicity we usually expect of mathematics. It may be this absence of aesthetic content, in the mathematical sense, that has resulted in the comparative neglect of the study of shape and form in the literature. Historically, mathematicians have tended to interest themselves in regular and semi-regular figures. Theory here has ancient roots going back to Egyptian and Greek artists and philosophers. Recently there has been a revival of interest in the Archimedean and Platonic space-filling solids in architectural design, notably through the work of Buckminster Fuller and a well-illustrated exposition by Keith Critchlow,[1] but on the whole it seems reasonable to suppose that three-dimensional symmetries are more relevant in the micro-worlds of molecular biology and chemistry where the force of gravity does not dominate, giving preference to one dimension – the vertical – over the others.[2] It is surprising that the analytical tools required for the systematic study of these figures were not fully available until the nineteenth century. The subject is particularly fascinating because of its association with physical and chemical structures and its relationship to the symmetry and patterns of many artistic and architectural products of man.

During this century there has been a growing interest in irregular forms. This has arisen in part through problems associated with convexity, and the importance of convexity in linear programming and related extremum problems. Also, developments in set theory and group theory since the nineteenth century make it possible to describe complex shapes more concisely.

In this book no attempt is made to derive a unique or universal method of describing form. It would seem sensible to be pragmatic about this matter, and to choose the most appropriate method in a given situation or for a particular purpose. A system which can only cope with rectangular forms will be more economic in describing such forms than a

[1] Keith Critchlow. *Order in Space; a Design Source Book*, London, Thames & Hudson, 1969.

[2] For example, R. Buckminster Fuller's work is acknowledged in passing by A. Klug and J. T. Finch, 'Structure of Viruses of the Papilloma – Polyoma Type, 1. Human Warts' in the *Journal of Molecular Biology*, vol. 2, 1965, pp. 403–23. Also see H. H. Jaffe and Milton Orchin, *Symmetry in Chemistry*, New York, Wiley, 1965.

[3] Albert Farwell Bemis and John Burchard. *The Evolving House*, Cambridge, Mass., The Technology Press, MIT, 1933–36. This work is in three volumes. See particularly, vol. 3, *Rational Design*.

more general system capable of handling non-rectangular forms. There is no great virtue in having a universal system when it is not required. If the forms are rectilinear, but non-rectangular, then we shall be forced to use a more generalized method of description, but it is likely to be that much more cumbersome to manipulate. If the forms are curvilinear, then the description becomes even more difficult. To use a curvilinear method of description for rectangular forms would burden us with considerable irrelevancy and redundancy.

We shall start with a simple modular 'building block' description of shape and form. In two dimensions this is a method used in ecology, geography, statistics and urban studies in the form of a square grid or system of quadrats. In three dimensions the method was used for the first time by Abbé Haüy at the end of the eighteenth century to describe crystal forms by means of identical *molécules intégrantes*. In the thirties of this century the American designer, Albert Farwell Bemis, introduced the method systematically into modern architecture through his pioneering work on modular coordination.[3] And recently many computer-aided design methods have adopted, often for parsimonious reasons, modular space elements. The method has the important merits of simplicity, economy and wide application. The modular description serves as a useful introduction to some basic ideas in the theory of sets, in particular, sets of points.

In Chapter 6 we relax the constraint of modularity while retaining rectangularity. We introduce further set-theoretic notions such as the cartesian product of two sets and the power set. We start with the description of one-dimensional components, and then proceed to describe two-dimensional panels and three-dimensional blocks. We show, conceptually, how a building plan, or form, is stored in a computer, and discuss some of the combinatorial problems of packing, stacking and nesting rectangular forms together. In Chapter 7, we look at non-rectangular two-dimensional forms. The method suggested should be of value in describing sites and complex floor plans. Basically we employ matrices and vectors to denote these shapes, and use theorems on convexity to manipulate them. We give formulae for the calculation of areas, perimeters and centroids for plane irregular figures whether these are discontinuous or multiply connected. Briefly we shall look at some of the problems of measuring the 'shapeiness' of shape as opposed merely to describing it. The geographer William Bunge has made a

[4] Lionel March and Michael Trace. *The Land Use Performance of Selected Arrays of Built Forms*, Working Paper 2, University of Cambridge, Land Use and Built Form Studies, 1968.

[5] This approach was suggested by illustrations in James F. Gray, *Sets, Relations and Functions*, New York, Holt, Rinehart and Winton, 1962, which provides an easy introduction to set theory. For fully programmed self-instruction see Myra McFadden, *Sets, Relations and Functions*, New York, McGraw-Hill, 1963. A more rigorous account is given by J. Donald Monk, *Introduction to Set Theory*, New York, McGraw-Hill, 1969.

contribution here, and so too have microbiologists concerned with chromosome patterns. That measures of shape are desirable in architectural studies would seem to be self-evident. For example, we are used to the idea of density control in urban planning, that in a given district no more than so much floor space may be built on each hectare. But what does this mean when the sites may vary a great deal in *shape*? Are there more precise ways of measuring shape than the verbal 'long and thin', 'roughly triangular', 'squarish'? March and Trace[4] have suggested that *shape* of site may be as important an influence on development and land use performance as its *area*, but it is the latter that is used in administering legislation and not the former, for the practical reason that up until now the one has been quantifiable and the other has not. Finally, we shall look at two architectural problems related to sunlighting and overshadowing, and to the obstruction of view through a window of a room caused by external buildings. Both problems may be handled in set-theoretic terms by making use of some fundamental ideas of convexity, in particular a remarkable theorem by Carathéodory which enables us to test whether points lie inside or outside a convex set.

Perhaps the easiest configurations to describe are those that are modular and rectangular, that is to say, shapes which can be compounded of a small rectangular element such as a square *quadrat* in two dimensions and a small cube, a *cubelet*, in three. Effectively, what we are looking for is a simple way of naming the elements and a means of saying how they are combined. This gives us an opportunity to discuss some fundamental ideas in *set theory*.[5]

Let us look first at a simpler problem; that of naming and assembling *units* of a fixed length along a line. Imagine a line, extended indefinitely in two directions, marked off in unit lengths. Take one of the marks and give it the name 0. Call each mark to the right 1, 2, 3,... in order, and each to the left $-1, -2, -3$,.... In this way we map the set of integers, $Z = \{\ldots -3, -2, -1, 0, 1, 2, 3, \ldots\}$ onto points of a line.

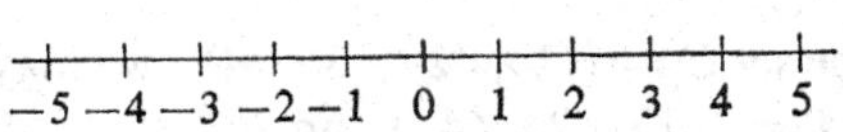

Figure 5.1

Let us name the unit which lies between i and $i + 1$, $[i, i + 1]$ or $\langle i \rangle$ for short. Using the notation for specifying sets we define a unit element $\langle i \rangle$ as

$$[i, i + 1] = \{x \in R \mid i \leqslant x \leqslant i + 1, i \in Z\}.$$

This mathematical sentence says that the unit $\langle i \rangle$ consists of *all* those points at a distance x from 0 which lie between, or are equal to, i and $i + 1$ where, x is a real number ($x \in R$) and i is an integer ($i \in Z$). Thus $\langle -4 \rangle$, $\langle 0 \rangle$, $\langle 1 \rangle$, and $\langle 4 \rangle$ are the names of the units illustrated.

Figure 5.2

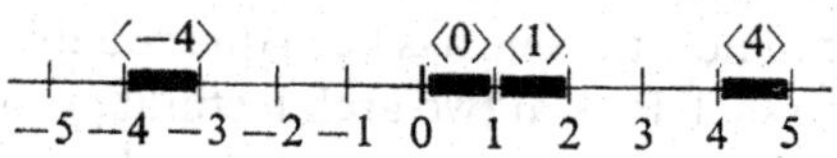

Figure 5.3

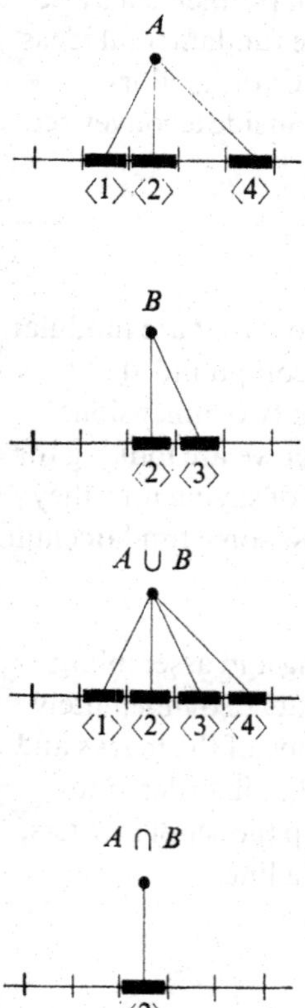

From a formal point of view it is useful to describe a combination of units by employing the set-theoretic operations of *union* and *intersection*. The *union* of two sets A and B is a new set, denoted by $A \cup B$ (pronounced 'A cup B'), consisting of elements that are in either A or B or both.

$$A \cup B = \{x \mid x \in A \text{ or } x \in B\}.$$

For example, if $A = \{\langle 1 \rangle, \langle 2 \rangle, \langle 4 \rangle\}$ is a set of units as described above,
and $B = \{\langle 2 \rangle, \langle 3 \rangle\}$ is another set,
then $A \cup B = \{\langle 1 \rangle, \langle 2 \rangle, \langle 3 \rangle, \langle 4 \rangle\}$.

Also the *intersection* of two sets A and B is a new set, denoted by $A \cap B$ (pronounced 'A cap B') consisting of all elements which belong to both A and B

$$A \cap B = \{x \mid x \in A \text{ and } x \in B\}.$$

Thus taking the two sets above $A \cap B = \{\langle 2 \rangle\}$ since $\langle 2 \rangle$ is the only element appearing in both sets. Note that in our examples $A \cup B = B \cup A$ and $A \cap B = B \cap A$: in fact, this *commutative* property of these operations is always true (Figure 5.3).

Out of the set $C = \{\langle 1 \rangle, \langle 2 \rangle, \langle 3 \rangle, \langle 4 \rangle\}$ we may select elements to form the sets $A = \{\langle 1 \rangle, \langle 2 \rangle, \langle 4 \rangle\}$ and $B = \{\langle 2 \rangle, \langle 3 \rangle\}$ (Figure 5.4). A and B

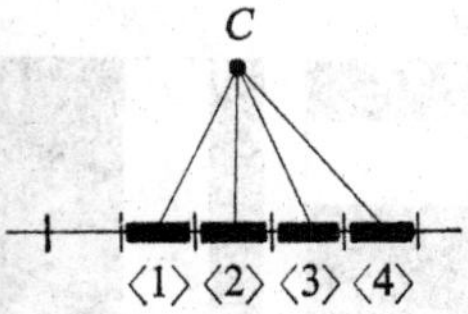

Figure 5.4

Figure 5.5

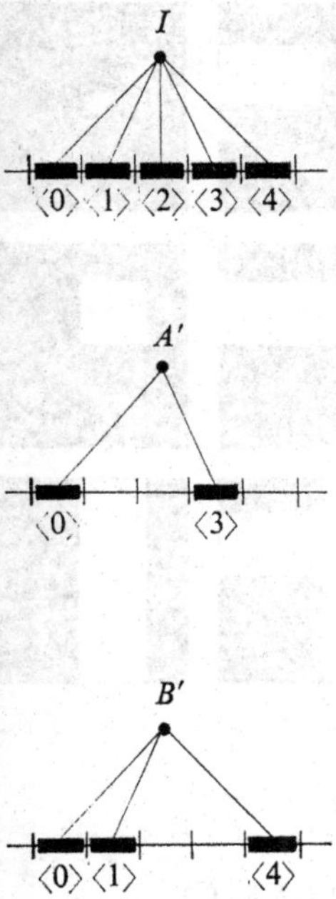

are called *sub-sets* of C, and both A and B are said to be *contained* in C. Employing a notation, $\subset$, analogous to the more familiar inequality 'less than', $<$, we write $A \subset C$, $B \subset C$. The large containing set from which all subsets are subsequently taken for the purposes of discussion is known as the *universe of discourse*, I. The set composed of all elements of I *not* belonging to A is called the *complement* of A and is denoted by A'.

$$A' = \{x \in I \mid x \notin A\}.$$

Hence for A and B,
if $I = \{\langle 0\rangle, \langle 1\rangle, \langle 2\rangle, \langle 3\rangle, \langle 4\rangle\}$
then $A' = \{\langle 0\rangle, \langle 3\rangle\}$
and $B' = \{\langle 0\rangle, \langle 1\rangle, \langle 4\rangle\}$ (Figure 5.5).

Clearly $A \cup A' = I$, but $A \cap A'$ is a set without elements since A and A', by definition, do not share elements in common. This empty set is usually denoted by $\emptyset$. Thus $A \cap A' = \emptyset$. Two sets A and B for which $A \cap B = \emptyset$ are said to be *disjoint*, and they clearly hold no elements in common. Note, however,that $A \cup A = A \cap A = A$: this is known as the *idempotent* relationship. The basic laws of the algebra of sets are shown below and illustrated in Figure 5.6.

Inclusion $\subset$:

Reflective:	$A \subset A$
Anti-symmetric:	$A \subset B, B \subset A$ implies $A = A$
Transitive:	$A \subset B, B \subset C$ implies $A \subset C$
Universal bounds:	$\emptyset \subset A \subset I$

Union $\cup$;
Intersection $\cap$:

Commutative:	$A \cup B = B \cup A$ $A \cap B = B \cap A$
Associative:	$(A \cup B) \cup C = A \cup (B \cup C)$ $(A \cap B) \cap C = A \cap (B \cap C)$

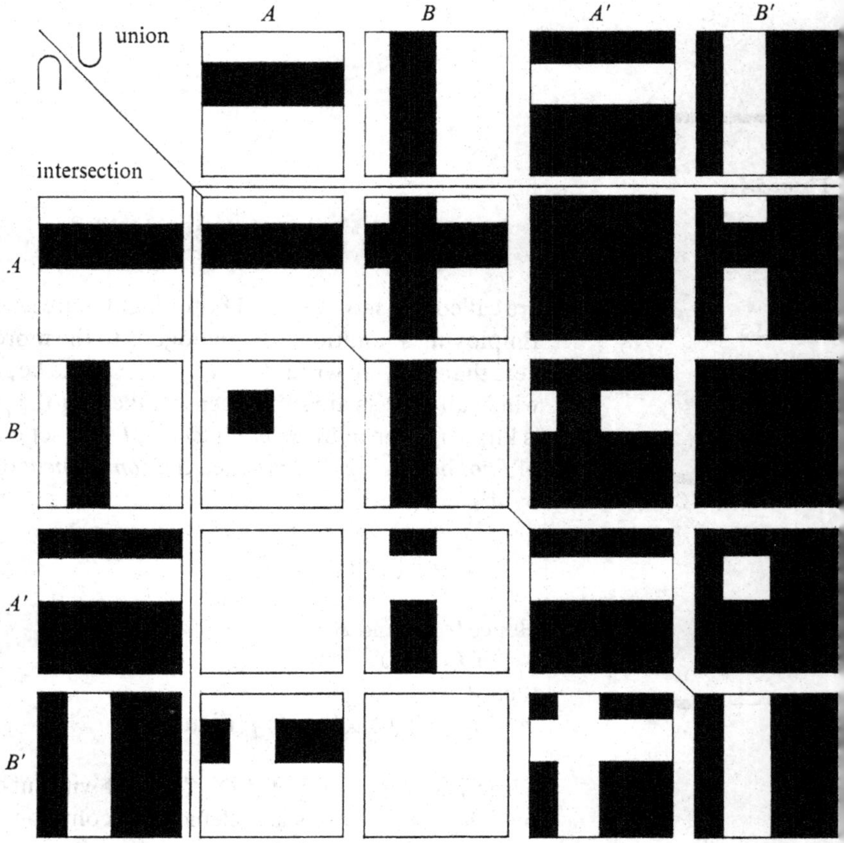

Figure 5.6
Tableau illustrating the operations of union and intersection on two sets A and B and their complements A' and B'. The two solid squares represent the universe of discourse given by $A \cup A' = I$, $B \cup B' = I$; while the two empty squares represent the null sets given by $A \cap A'$ and $B \cap B'$

Idempotent:

$$A \cup A = A$$
$$A \cap A = A$$

Distributive:

$$A \cup (B \cap C) = (A \cup B) \cap (A \cup C)$$
$$A \cap (B \cup C) = (A \cap B) \cup (A \cap C)$$

Consistency:

$A \subset B$ is equivalent to $A \cup B = B$
$A \subset B$ is equivalent to $A \cap B = A$

Properties of $\emptyset$:

$$\emptyset \cup A = A$$
$$\emptyset \cap A = \emptyset$$

Properties of I:

$$I \cup A = I$$
$$I \cap A = A$$

Complementarity:

$$A \cup A' = I$$
$$A \cap A' = \emptyset$$

Duality:

$$(A \cup B)' = A' \cap B'$$
$$(A \cap B)' = A' \cup B'$$

Involution:

$$(A')' = A$$

A number of units combined end-to-end make up a *component*. Thus [0, 1] and [1, 2] make a component two units in length. Let us call this component [0, 2] so that

$$[0, 2] = [0, 1] \cup [1, 2].$$

But a component such as [1, 5] is only clumsily written out in full

$$[1, 5] = [1, 2] \cup [2, 3] \cup [3, 4] \cup [4, 5]$$

and the contraction

$$[1, 5] = \bigcup_{i=1}^{4} [i, i + 1]$$

may be used. This says that the component [1, 5] is made up of (is the union of) all elements $\langle i \rangle$ where i takes the values 0 to 4, that is, 0, 1, 2, 3, 4. Thus, in general,

$$[a_1, a_2] = \bigcup_{i=a_1}^{a_2 - 1} \langle i \rangle.$$

The length, or dimension, of the component $[a_1, a_2]$ is $a_2 - a_1$. We write

$$\mathrm{d}\,[a_1, a_2] = [a_2 - a_1].$$

Its centre of mass is the point at a length $\dfrac{a_1 + a_2}{2}$ from O,

$$\mathrm{cm}\ [a_1, a_2] = \left[\frac{a_1 + a_2}{2}\right].$$

So far our element exists along a line in one dimension, but we may extend the idea into two and three dimensions without difficulty. Consider, for example, a square element whose sides are of unit length. We shall call such an element a *quadrat*.

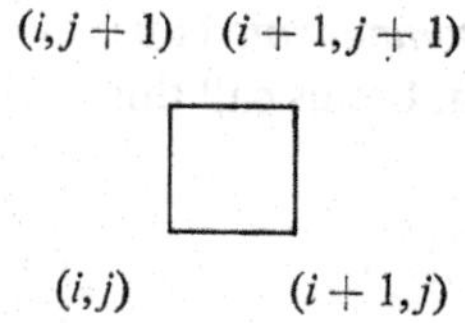

Figure 5.7

Quadrats, in a plane marked by a *modular grid*, may be identified by four numbers $\begin{bmatrix} i, i+1 \\ j, j+1 \end{bmatrix}$ which, taken two at a time, one from the top row and one from the bottom, represent the four vertices of the quadrat (Figure 5.7): (i, j), $(i, j+1)$, $(i+1, j+1)$ and $(i+1, j)$. To save space, it will often be more convenient to express these numbers as $[i, i+1; j, j+1]$ where we use a semi-colon to mark where the next row starts. We shall also contract the name of a quadrat further, as we have done for the unit element above, to $\langle i; j \rangle$. We use each name as appropriate, just as the name Candida may be contracted to Candy, or yet more to Can.

Figure 5.8

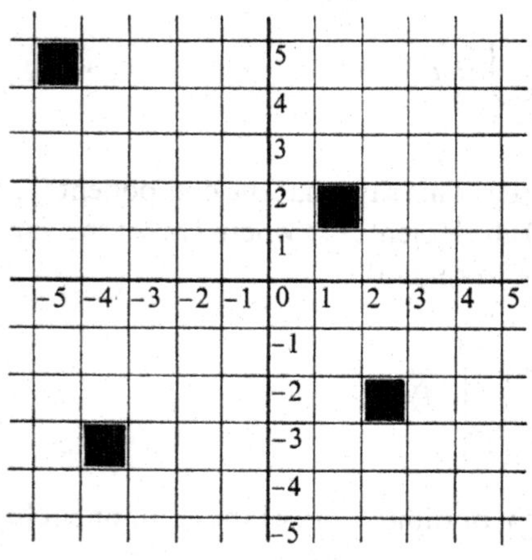

We shall define the quadrat $\langle i; j \rangle$ by

$$[i, i+1; j, j+1] = \{(x, y) \in R^2 \mid i \leqslant x \leqslant i+1, j \leqslant y \leqslant j+1, (i, j) \in Z^2\}.$$

This statement tells us that the quadrat which we name $\langle i; j \rangle$ consists of all those points (x, y) in two-dimensional real space, R^2, such that x lies between i and $i + 1$, and y lies between j and $j + 1$ where i and j are integers. Some typical quadrats are illustrated in Figure 5.8: the reader is invited to identify

$$\begin{bmatrix} 1, 2 \\ 1, 2 \end{bmatrix}, \langle 2; -3 \rangle, [1, 2; 1, 2], \langle -4; -4 \rangle, \langle 1; 1 \rangle \text{ and } \langle -5; 4 \rangle.$$

Simple rectangular shapes, or components, consisting of a number of quadrats are described by the operation of union. Thus the component which we shall call [1, 4; 1, 2] is given in full by

$$\begin{bmatrix}1,4\\1,2\end{bmatrix}=\begin{bmatrix}1,2\\1,2\end{bmatrix}\cup\begin{bmatrix}2,3\\1,2\end{bmatrix}\cup\begin{bmatrix}3,4\\1,2\end{bmatrix}$$

$$=\langle 1;1\rangle\cup\langle 2;1\rangle\cup\langle 3;1\rangle$$

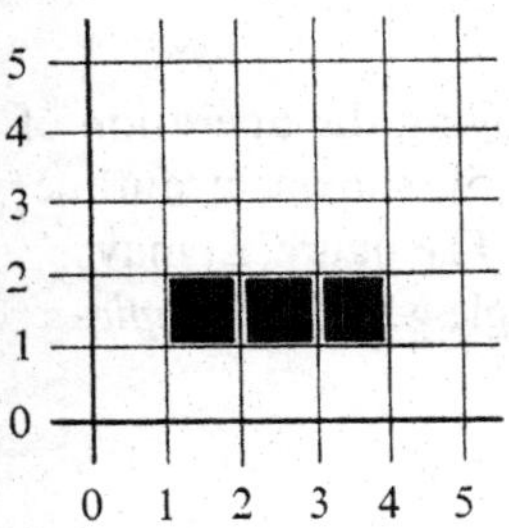

Figure 5.9

which may be further contracted as before to

$$\begin{bmatrix}1,4\\1,2\end{bmatrix}=\bigcup_{i=1}^{3}\langle i;1\rangle$$

which simply says that the rectangular component is the union of all quadrats $\langle i;1\rangle$ where i successively takes the values of 1 to 3, namely, 1, 2, 3 (Figure 5.9).

Similarly the descriptions of [1, 2; 1, 5] and [1, 4; 1, 5] are given by

$$\begin{bmatrix}1,2\\1,5\end{bmatrix}=\bigcup_{j=1}^{4}\langle 1;j\rangle \text{ and } \begin{bmatrix}1,4\\1,5\end{bmatrix}=\bigcup_{i=1}^{3}\bigcup_{j=1}^{4}\langle i;j\rangle$$

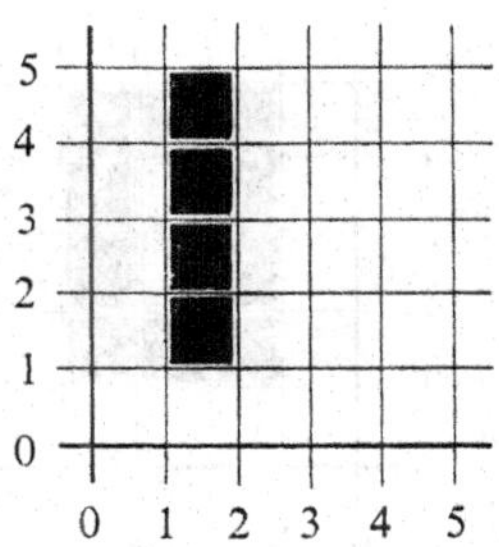

Figure 5.10

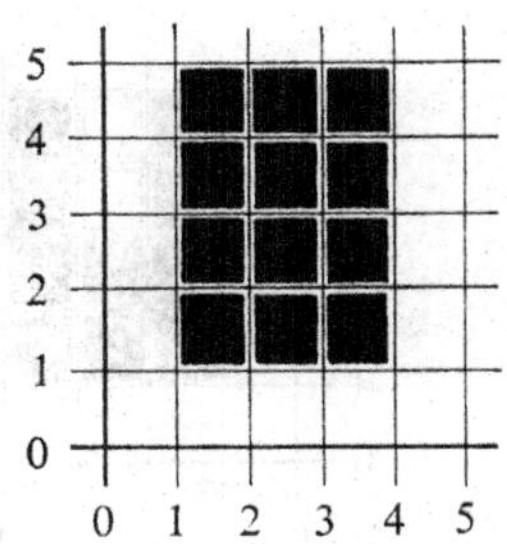

Figure 5.11

where the latter requires a double union of both i and j (Figures 5.10 and 5.11).

[6] Dean Hawkes and Richard Stibbs. *The Environmental Evaluation of Buildings. 1. A Mathematical Model*, Working Paper 15, University of Cambridge, Land Use and Built Form Studies, 1969, pp. 12–22.

Figures 5.12 and 5.13

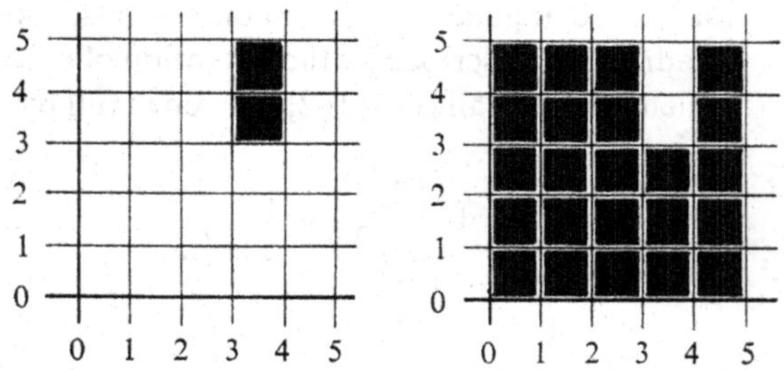

More complex shapes may be described by employing the operation of intersection. For example, by removing [3, 4; 3, 5], shown in outline, from [1, 4; 1, 5] we obtain a black *L*-shaped figure. The figure, *L*, may be expressed as the *intersection* of the larger rectangle with the *complement* of the smaller [3, 4; 3, 5]′, that is,

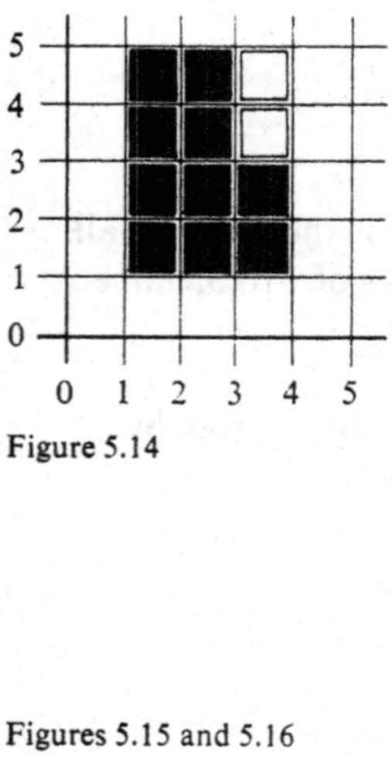

Figure 5.14

$$L = \begin{bmatrix} 1, 4 \\ 1, 5 \end{bmatrix} \cap \begin{bmatrix} 3, 4 \\ 3, 5 \end{bmatrix}'.$$

In the same way the *U*-shaped and *O*-shaped figures may be described by suitably selecting the appropriate 'subtracted area' (Figures 5.12–5.16). Thus,

$$U = \begin{bmatrix} 1, 4 \\ 1, 5 \end{bmatrix} \cap \begin{bmatrix} 2, 3 \\ 3, 5 \end{bmatrix}' \text{ and } O = \begin{bmatrix} 1, 4 \\ 1, 5 \end{bmatrix} \cap \begin{bmatrix} 2, 3 \\ 2, 4 \end{bmatrix}'.$$

Figures 5.15 and 5.16

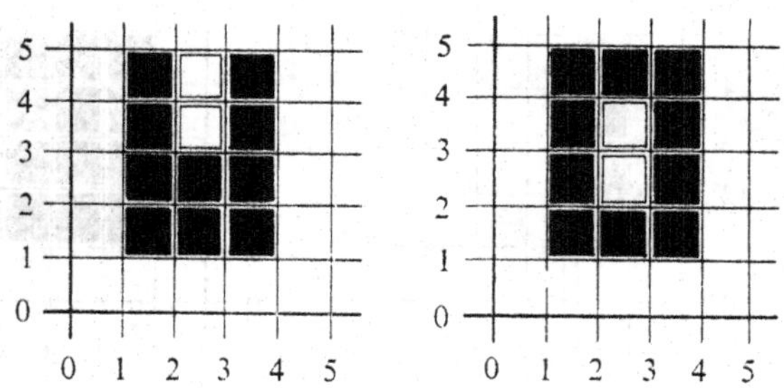

These expressions serve to illustrate in formal terms the method of 'subtracted areas' used by Hawkes and Stibbs[6] for putting the plans of a room into a computer. These authors do not, however, restrict themselves to modular spaces and this is something we shall consider later.

Figure 5.17

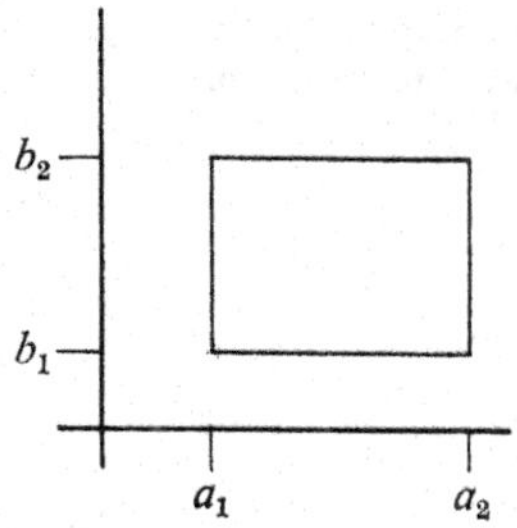

In general,

$$\begin{bmatrix} a_1, a_2 \\ b_1, b_2 \end{bmatrix} = \bigcup_{i=a_1}^{a_2-1} \bigcup_{j=b_1}^{b_2-1} \langle i;j \rangle$$

represents a rectangle with one corner located at (a_1, b_1) and its diagonally opposite one at (a_2, b_2) as illustrated in Figure 5.17. The rectangle has dimensions of length $a_2 - a_1 = a$, say, and $b_2 - b_1 = b$

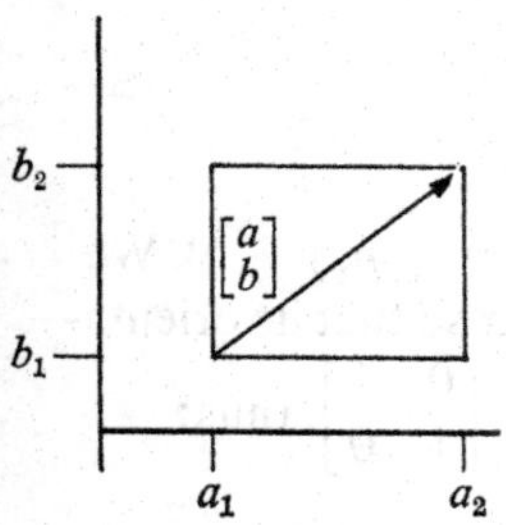

Figure 5.18

$$\mathrm{d} \begin{bmatrix} a_1, a_2 \\ b_1, b_2 \end{bmatrix} = \begin{bmatrix} a_2 - a_1 \\ b_2 - b_1 \end{bmatrix} = \begin{bmatrix} a \\ b \end{bmatrix}.$$

Note that $[a \quad b]^T$ is the vector which represents the diagonal of the rectangle (Figure 5.18). The area of the rectangle is clearly ab, and its centre of mass is given by the position vector (Figure 5.19)

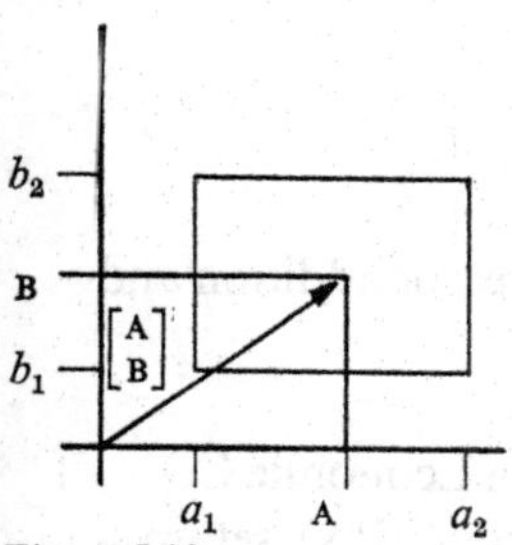

Figure 5.19

$$\mathrm{cm}\,[a_1, a_2; b_1, b_2] = \begin{bmatrix} \dfrac{a_1 + a_2}{2} & \dfrac{b_1 + b_2}{2} \end{bmatrix}^T = [\text{A} \quad \text{B}]^T.$$

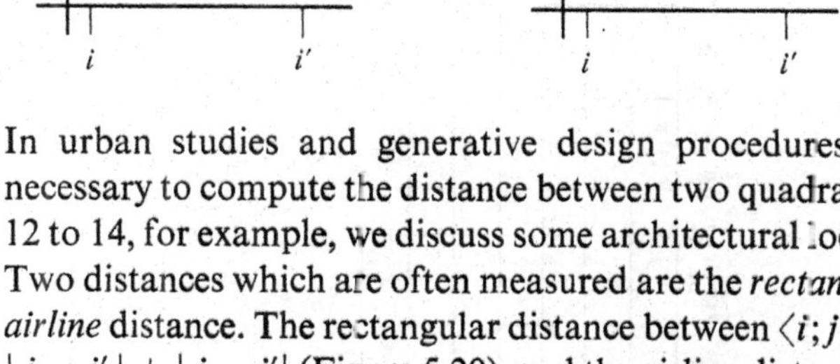

Figures 5.20 and 5.21

In urban studies and generative design procedures, it is frequently necessary to compute the distance between two quadrats – in Chapters 12 to 14, for example, we discuss some architectural location problems. Two distances which are often measured are the *rectangular* and the *airline* distance. The rectangular distance between $\langle i;j \rangle$ and $\langle i';j' \rangle$ is $|i - i'| + |j - j'|$ (Figure 5.20), and the airline distance, by Pythagoras' theorem, is $(|i - i'|^2 + |j - j'|^2)^{\frac{1}{2}}$ (Figure 5.21); while the two distances between the centroids (A, B) and (A′, B′) of two component spaces are $|\text{A} - \text{A}'| + |\text{B} - \text{B}'|$ and $(|\text{A} - \text{A}'|^2 + |\text{B} - \text{B}'|^2)^{\frac{1}{2}}$ respectively.

[7] See the Second Report of European Productivity Agency Project 174, *Modular Co-ordination*, Paris, OEEC Publications, 1961, pp. 103–9.

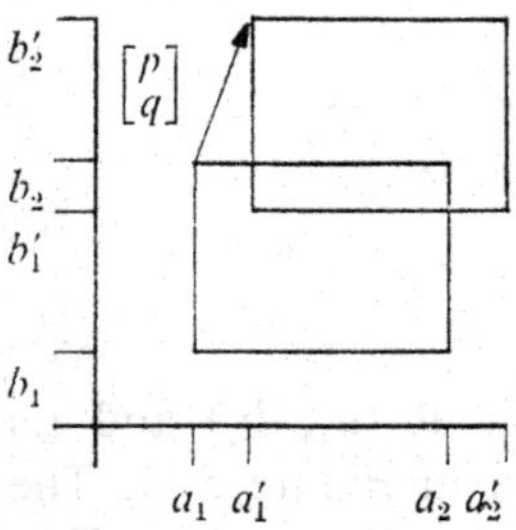

Figure 5.22

A rectangular space $[a_1, a_2; b_1, b_2]$ is fixed in location and dimensions, but we may *translate* (Figure 5.22) the rectangle to a new position by a transformation

$$\begin{bmatrix} a_1, a_2 \\ b_1, b_2 \end{bmatrix} + \begin{bmatrix} p \\ q \end{bmatrix} = \begin{bmatrix} a_1 + p, a_2 + p \\ b_1 + q, b_2 + q \end{bmatrix} = \begin{bmatrix} a_1', a_2' \\ b_1', b_2' \end{bmatrix}$$

where the vector $[p \quad q]^T$ carries all points (x, y) to $(x + p, y + q)$. We may also *reflect* (Figure 5.23) the rectangle in $y = x$ so that its orientation is changed, for this we use the zero-one matrix $\begin{bmatrix} 0 & 1 \\ 1 & 0 \end{bmatrix}$, thus:

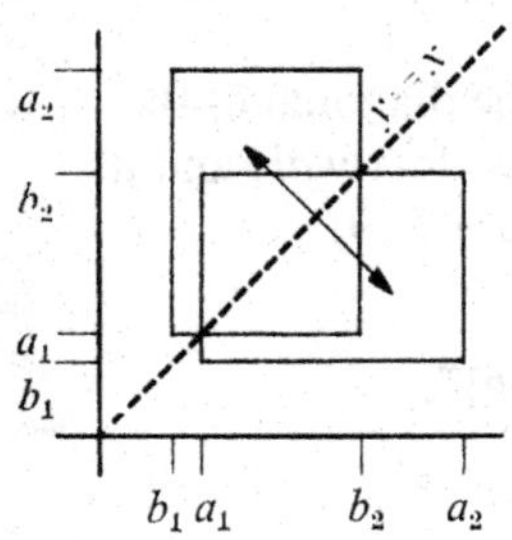

Figure 5.23

$$\begin{bmatrix} 0 & 1 \\ 1 & 0 \end{bmatrix} \begin{bmatrix} a_1, a_2 \\ b_1, b_2 \end{bmatrix} = \begin{bmatrix} b_1, b_2 \\ a_1, a_2 \end{bmatrix}.$$

Observe that each column $\begin{bmatrix} a_1 \\ b_1 \end{bmatrix}$ and $\begin{bmatrix} a_2 \\ b_2 \end{bmatrix}$ of $\begin{bmatrix} a_1, a_2 \\ b_1, b_2 \end{bmatrix}$ is treated as a separate *vector* for the purposes of vector addition and matrix multiplication.

In 1954 a branch of the Organization for European Economic Co-operation, called the European Productivity Agency (EPA), set up a project on modular coordination[7] – a subject which we look at from a mathematical point of view in more detail in Chapters 8 and 9. Part of the exercise involved various countries in producing test buildings. One

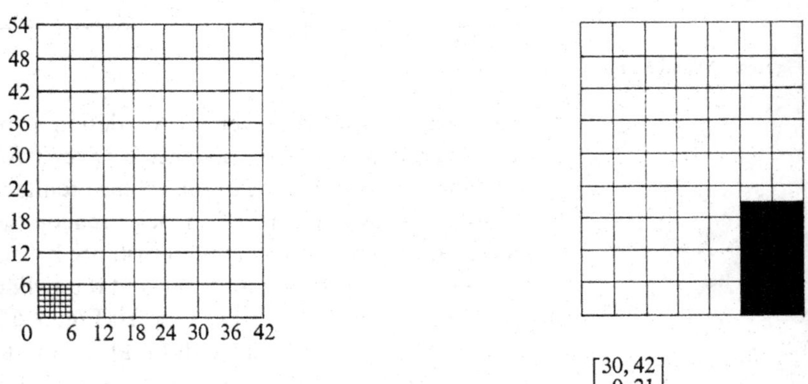

Figure 5.24 Blocking-in the plan of the modular EPA test building at Hemel Hempstead, Hertfordshire

of the United Kingdom's test buildings was a research centre consisting of three separate single-storey buildings designed to provide accommodation for general office work, administration and laboratory work.

The building was designed on a 4-inch (100 mm) module, and planned on a grid based on a multiple of eight times this module, that is, 32 in. Taking this planning module as a unit we may describe the plans of the various parts of the design in relation to the reference grid. The simple block plan of the test building (Figure 5.24) is given by

$$\begin{bmatrix} 30, 42 \\ 0, 21 \end{bmatrix} \cup \begin{bmatrix} 30, 42 \\ 26, 54 \end{bmatrix} \cup \begin{bmatrix} 0, 24 \\ 0, 12 \end{bmatrix} \cup \begin{bmatrix} 24, 30 \\ 6, \ 8 \end{bmatrix} \cup \begin{bmatrix} 35, 37 \\ 21, 26 \end{bmatrix}.$$

From this we immediately have the dimensions of the component plans: they are, in order, (42 − 30) by (21 − 0), or 12 × 21, and by similar calculation, 12 × 28, 24 × 12, 6 × 2, and 2 × 5. From their plan description it is possible to tell that two of the buildings are aligned in the y-direction. They do so because they share the same x-component for the units making up their width, namely [30, 42]. We can also see that these two buildings are separated. Their y-components [0, 21] and [26, 54] are not contiguous, showing a gap [21, 26] between them. However, this is the y-component of [35, 37; 21, 26] which can link the two buildings together if it falls within their alignment. This it does since [35, 37] lies within [30, 42], that is 30,..., 35, 36, 37,..., 42. Thus the two buildings are linked. In the same way we know that the remaining building is linked.

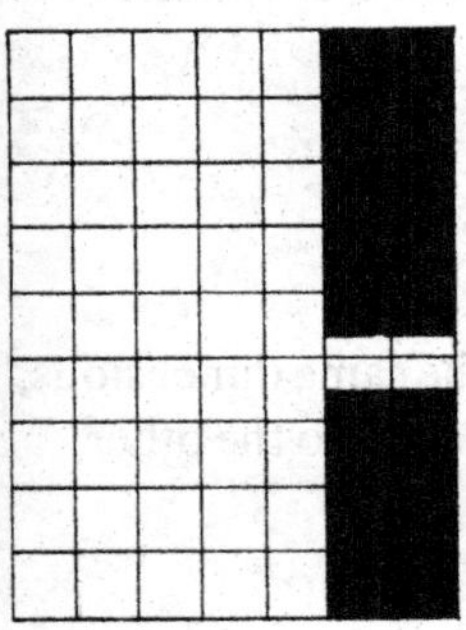

$\cup \begin{bmatrix} 30, 42 \\ 26, 54 \end{bmatrix}$

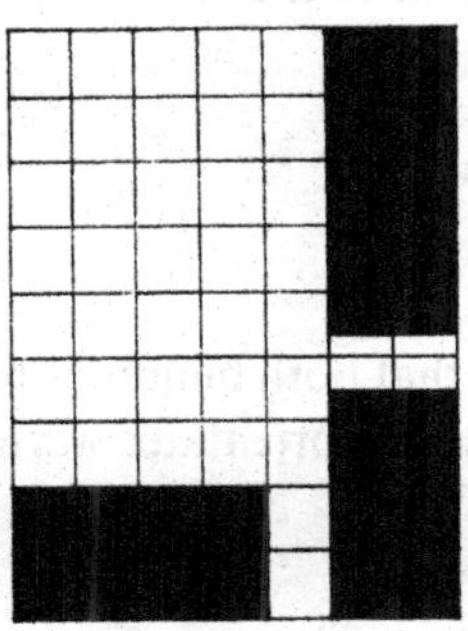

$\cup \begin{bmatrix} 0, 24 \\ 0, 12 \end{bmatrix}$

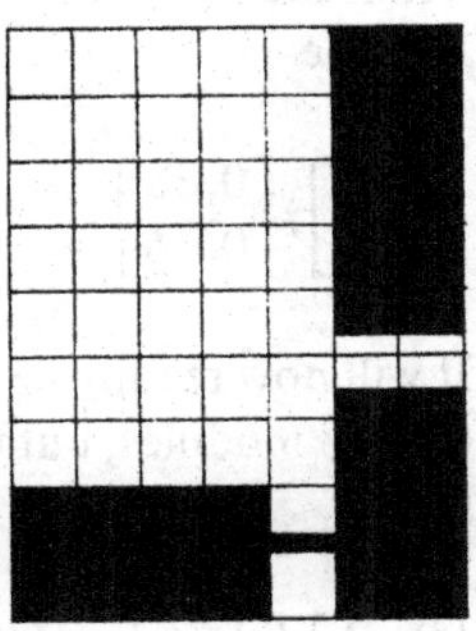

$\cup \begin{bmatrix} 24, 30 \\ 6, \ 8 \end{bmatrix}$

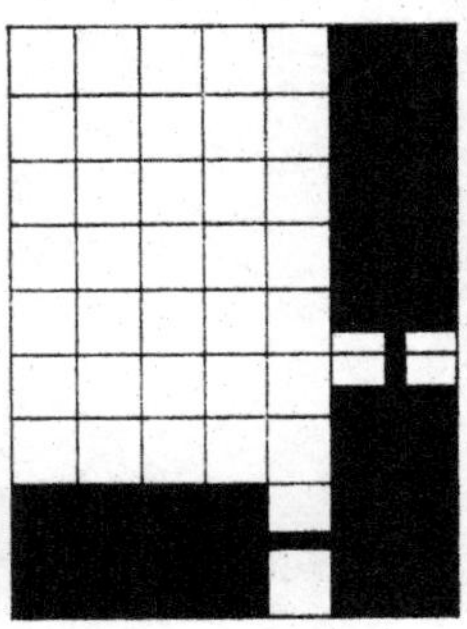

$\cup \begin{bmatrix} 35, 37 \\ 21, 26 \end{bmatrix}$

[8] An introduction to this architect's work is given by Arthur Drexler, *Ludwig Mies van der Rohe*, New York, George Braziller Inc., 1960.

An architect who based many of his buildings on a rectangular and modular grid was Mies van der Rohe,[8] the former director of the Bauhaus at Dessau and an outstanding pioneer of modern architecture. On coming to the United States in the late thirties, Mies designed the master plan for the new campus at the Illinois Institute of Technology in Chicago. The whole block plan for the new buildings was set out on a 12 ft by 12 ft grid. We illustrate the original 1940 plan together with the numerical description of each building (Figure 5.25). The reader is invited to identify each building on the plan from its description.

One interesting property of the numerical description is that it tells us, among other things, which buildings are aligned along the same frontage. For example, buildings in 1, 4, 8 and 16 are aligned. Their respective descriptions are

$$\begin{bmatrix} 1, & 9 \\ 31, & 53 \end{bmatrix}, \begin{bmatrix} 15, & 49 \\ 31, & 39 \end{bmatrix}, \begin{bmatrix} 57, & 85 \\ 31, & 47 \end{bmatrix}, \begin{bmatrix} 133, & 159 \\ 31, & 47 \end{bmatrix}.$$

Each description shares one term, 31, in common. All these buildings 'stand' on the line $y = 31$, but buildings 8 and 16 also have the term 47 in common, which shows that these buildings 'hang' from $y = 47$. Now take building 11 and another building from 1 whose descriptions are

$$\begin{bmatrix} 1, & 7 \\ 7, & 25 \end{bmatrix}, \begin{bmatrix} 87, & 105 \\ 20, & 26 \end{bmatrix}.$$

Translate both building plans to the origin (0, 0). Their descriptions become

$$\begin{bmatrix} 0, & 6 \\ 0, & 18 \end{bmatrix}, \begin{bmatrix} 0, & 18 \\ 0, & 6 \end{bmatrix}.$$

It will now readily be seen that both buildings have the same dimensions, 6 by 18 modules, but that one is orientated at right-angles to the other, since

$$\begin{bmatrix} 0 & 1 \\ 1 & 0 \end{bmatrix} \begin{bmatrix} 0, & 6 \\ 0, & 18 \end{bmatrix} = \begin{bmatrix} 0, & 18 \\ 0, & 6 \end{bmatrix}.$$

Figure 5.25
Master plan of the Illinois Institute of Technology, Chicago, as designed by Mies van der Rohe in 1940

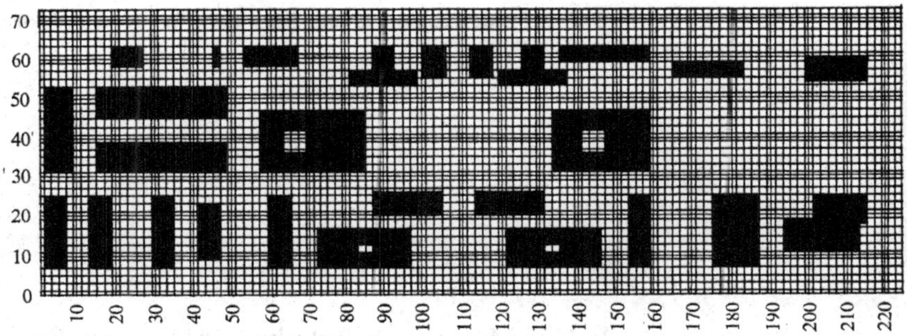

1 Armour Research Foundation Laboratories	[1, 7; 7, 25], [1, 9; 31, 53] [13, 19; 7, 25]
2 Boiler plant	[19, 27; 58, 63]
3 Central vault	[45, 47; 58, 63]
4 ARF Engineering Research Buildings	[15, 49; 31, 39], [15, 49; 45, 53]
5 Institute of Gas Technology Laboratory	[29, 35; 7, 25]
6 Institute of Gas Technology Building	[41, 47; 9, 23]
7 School of Architecture and Design	[59, 65; 7, 25]
8 Student Union and Auditorium	[57, 85; 31, 47] ∩ [63, 69; 36, 42]′
9 Minerals and Metals Research Building	[53, 67; 58, 63]
10 Electrical Engineering and Physics Building	[81, 99; 53, 57] ∪ [87, 93; 57, 63], [100, 106; 55, 63]
11 Lewis Institute	[87, 105; 20, 26]
12 Mechanical Engineering Building	[72, 97; 7, 17] ∩ [83, 87; 11, 13]′
13 Chemical Engineering and Metallurgy Building	[121, 146; 7, 17] ∩ [131, 135; 11, 13]′
14 Chemistry Building	[113, 131; 20, 26]
15 Alumni Memorial Hall	[153, 159; 7, 25]
16 Library and Administrative Building	[133, 159; 31, 47] ∩ [141, 147; 36, 42]′
17 Civil Engineering and Mechanics Building	[112, 118; 55, 63], [119, 137; 53, 57] ∪ [125, 131; 57, 63], [135, 159; 59, 63]
18 Association of Railroads Building	[165, 183; 55, 59]
19 Association of Railroads Laboratory	[199, 215; 54, 60]
20 Field House	[175, 187; 7, 25]
21 Gymnasium and Pool	[193, 213; 11, 19] ∪ [201, 215; 19, 25]

[9] Albert Farwell Bemis and John Burchard. *The Evolving House*, vol. 3, *Rational Design*, pp. 69–70.

In 1958 Mies van der Rohe built the Seagram Building at 375 Park Avenue, New York. This is probably the most widely admired high-rise office building in the world, and one which very much influenced commercial architecture during the sixties. The block plan, based on the structural bay grid, of the Seagram is

$$\begin{bmatrix}1,6\\3,6\end{bmatrix} \cup \begin{bmatrix}2,5\\6,7\end{bmatrix} \cup \begin{bmatrix}0, & 7\\7, & 10\end{bmatrix}.$$

The reader is invited to complete the plan on the site grid (Figure 5.26).

Figure 5.26

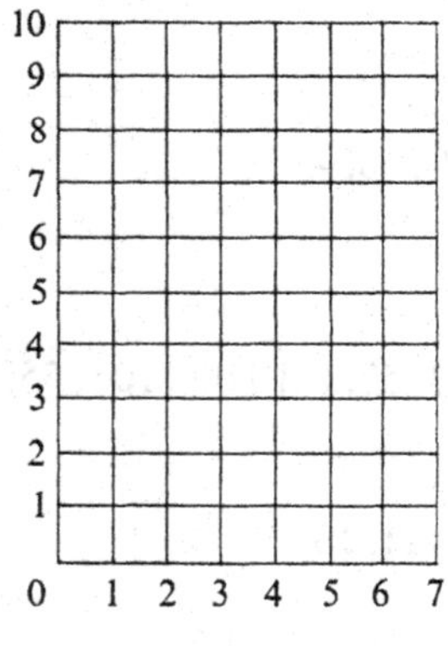

This method of describing shape may be extended into three dimensions. Here the two-dimensional quadrat becomes a three-dimensional cubelet. In 1784 the 'father of crystallography', Abbé Haüy, published a work entitled *Essai d'une Théorie sur la Structure des Crystaux Appliquée à Plusieurs Genres de Substances Crystallisées* in which he put forward the idea that crystals could be dissected along cleavage lines into a smallest possible unit, a *molécule intégrante*, by the repetition of which the whole crystal could be reconstituted. The shape of the fundamental unit was chosen according to the particular system of symmetry to which a substance belonged, and our illustrations show his ideas in the cubic system using small unit cubelets (Figure 5.27). Irregular three-dimensional forms can be approximated in a similar way.

In the 1930s Albert Farwell Bemis proposed that a cube might be used as a module in building design and component standardization.[9] The idea was not new, although the context of industrial mass production

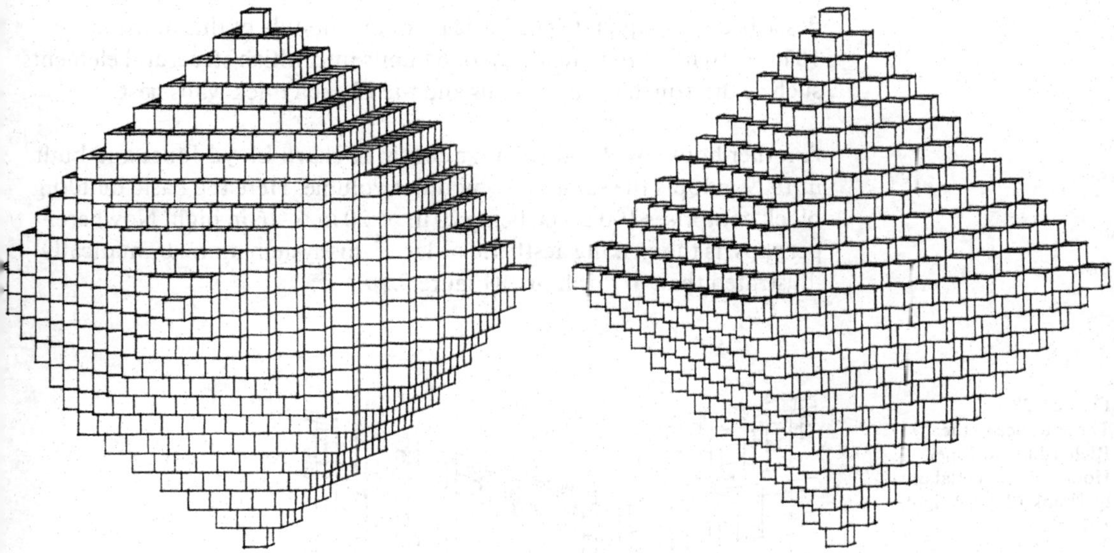

Figure 5.27
Two crystal forms built from Haüy's cubelets:
a, the completed rhombic dodecahedron
b, an octahedron

gave it fresh force. Indeed J. N. L. Durand, Haüy's contemporary, frequently employs a cubic *molécule intégrante* in his *Leçons d'Architecture* of 1819 as a powerful way of modulating space, while Viollet-le-Duc, as we have seen, accepted the approach as the natural *principe* of architectural organization, observing its use in many Gothic buildings. For Bemis the building is designed within 'a total matrix of cubes', a rectangular outline of space, large enough to include all the physical parts of the building. 'The delineation of the structure within the total matrix may be visualized by first removing from within the matrix all the space cubes not comprised in the building volume. The entire exterior surface thus defined coincides with cube surfaces but not necessarily with the surfaces of the grand matrix. The voids that constitute rooms, doors and windows can then be defined by the elimination within the house volume of the cubes filling these spaces. The complete and exact form of the structure is now defined. It is divided into units of volume, cubes of the same size, and all measurements may be expressed as multiples of the module.' Essentially this is analogous to the two dimensional system we have just described. Bemis chose a 4 in × 4 in × 4 in cubelet as his fundamental modular volume; a unit employed by Frank Lloyd Wright ten years before in his Californian concrete block houses such as the Mrs George Madison Millard House described in Chapter 3. The basic planning module of this house is 16 in × 16 in × 16 in made up of 64 units into which structural elements such as the 8-inch wide columns and the 12-inch thick walls nest.

Another house by Wright, for his cousin Richard Lloyd Jones and built in 1929, adopts the same 4 in^3 modular volume. Here the basic building block comprises 100 units, being 20 in × 20 in × 16 in high. Nowhere, perhaps, is the striking aesthetic order of this system more dramatically illustrated than in this house (Figure 5.28).

Figure 5.28
The entrance to the Richard Lloyd Jones House, Tulsa, Oklahoma, by Frank Lloyd Wright, 1929

In his *Encyclopaedia of Architecture*, Joseph Gwilt shows the entire nave of the thirteenth-century cathedral at Amiens contained within a cube made up of 216 smaller cubes, 23 ft 6 in along each edge (Figure 5.29).

He writes:

> '... there can be little doubt that this simple figure (the cube) served as a means of measuring the quantities, of either solid or void, in every period of the constructive arts; certainly none presents to the architect a better means of comprehending or of

[10] Joseph Gwilt (revised by Wyatt Papworth). *An Encyclopaedia of Architecture*, London, Longmans, Green, 1881, p. 1016.

measuring quantity, and none is more readily subdivided, or rendered subservient to the taste of the designer, whatever may be the architecture he is anxious to imitate.'[10]

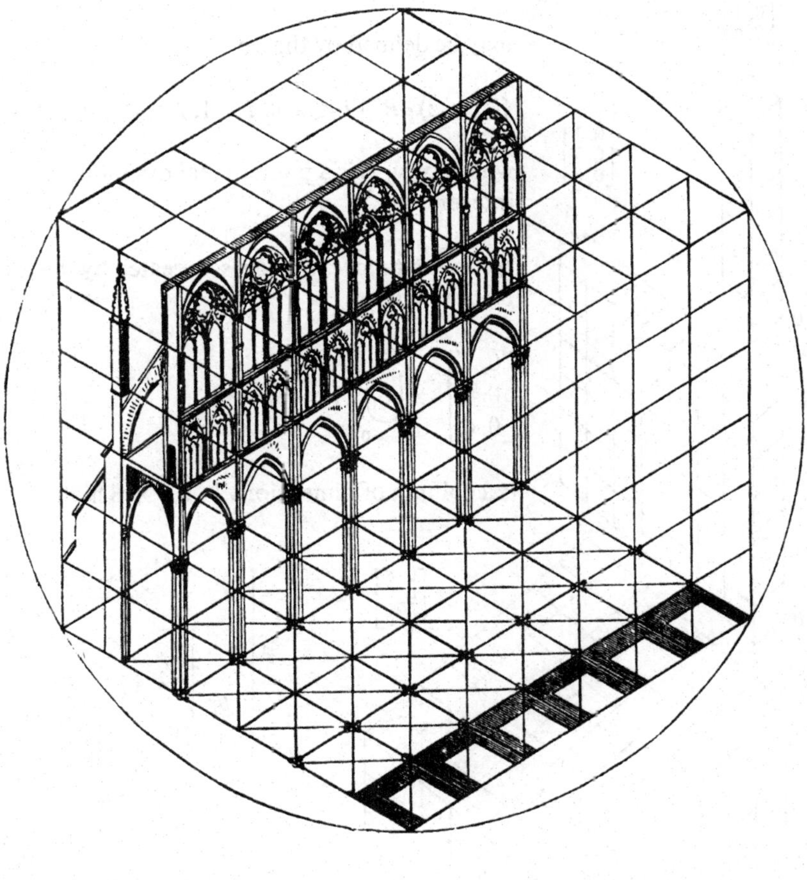

Figure 5.29
Joseph Gwilt's diagram of the nave of Amiens Cathedral

But there was very little imitation in Sir Joseph Paxton's 1851 Crystal Palace, the Building for the Exhibition of the Industry of all Nations, and yet, as Gwilt points out, this remarkable building was also based on a cubic module. Paxton's units, being 24 ft on each side, were almost identical to those that rule the nave at Amiens, and they also correspond to the two-storey-high cubic modules of Mies van der Rohe's Seagram Building.

In three dimensions, the elemental cubelet

$$\begin{bmatrix} i, i & +1 \\ j, j & +1 \\ k, k+1 \end{bmatrix} \text{ alias } [i, i+1; j, j+1; k, k+1] \text{ alias } \langle i; j; k \rangle$$

may be defined by the set

$$\{(x, y, z) \in R^3 \mid i \leqslant x \leqslant i+1, j \leqslant y \leqslant j+1, k \leqslant z \leqslant k+1, (i, j, k) \in Z^3\}$$

where (x, y, z) is a point in real three-dimensional space, and i, j, k are integers.

A modular *block* is a new set created by the union of these cubelets. For example,

$$\begin{bmatrix} 0, 1 \\ 0, 1 \\ 0, 6 \end{bmatrix} = \bigcup_{k=0}^{5} \langle 1; 1; k \rangle$$

is a column of dimensions $1 \times 1 \times 6$ high,

Figure 5.30

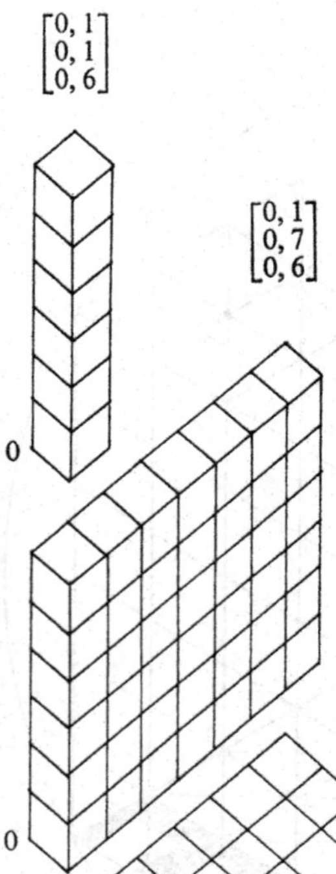

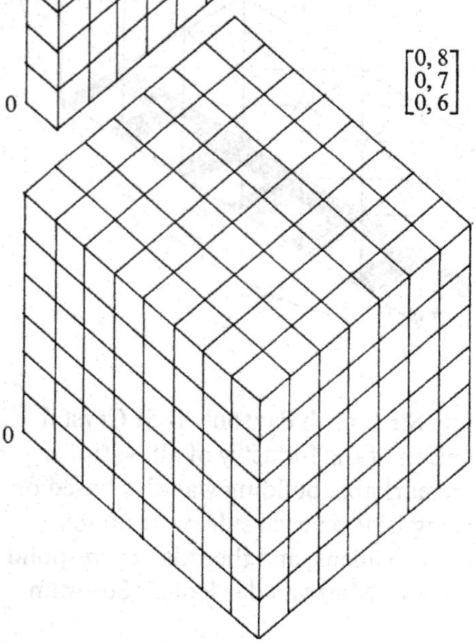

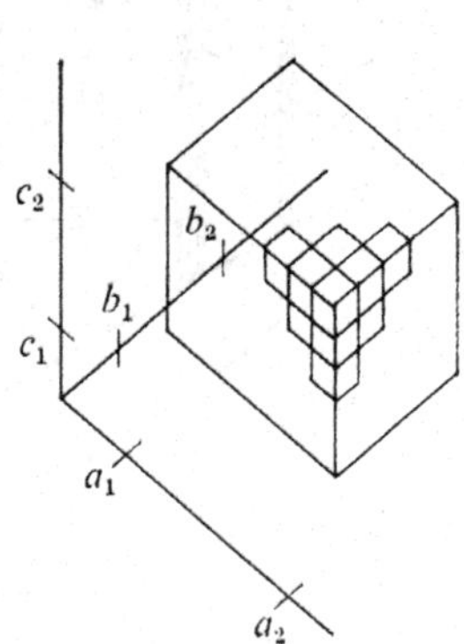

block:

$$\begin{bmatrix} a_1, a_2 \\ b_1, b_2 \\ c_1, c_2 \end{bmatrix}$$

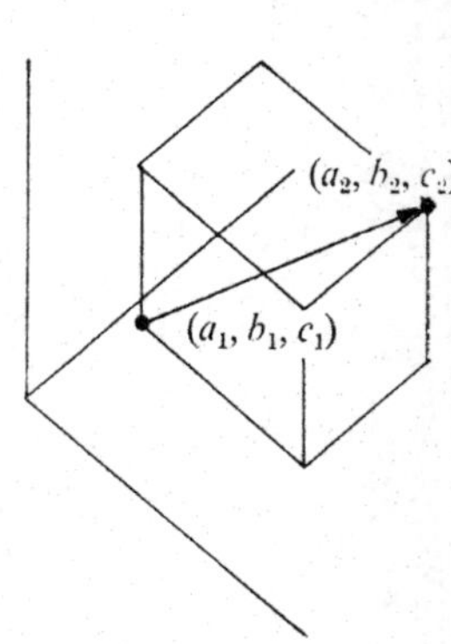

vector:

$$\mathrm{d} \begin{bmatrix} a_1, a_2 \\ b_1, b_2 \\ c_1, c_2 \end{bmatrix} = \begin{bmatrix} a_2 - a_1 \\ b_2 - b_1 \\ c_2 - c_1 \end{bmatrix}$$

$$\begin{bmatrix} 0, 1 \\ 0, 7 \\ 0, 6 \end{bmatrix} = \bigcup_{j=0}^{6} \bigcup_{k=0}^{5} \langle 1; j; k \rangle$$

is a slab, 1 × 7 × 6 high, and

$$\begin{bmatrix} 0, 8 \\ 0, 7 \\ 0, 6 \end{bmatrix} = \bigcup_{i=0}^{7} \bigcup_{j=0}^{6} \bigcup_{k=0}^{5} \langle i; j; k \rangle$$

is a block, 8 × 7 × 6 high (Figure 5.30).

In general,

$$\begin{bmatrix} a_1, a_2 \\ b_1, b_2 \\ c_1, c_2 \end{bmatrix} = \bigcup_{i=a_1}^{a_2-1} \bigcup_{j=b_1}^{b_2-1} \bigcup_{k=c_1}^{c_2-1} \langle i; j; k \rangle$$

defines a block with one corner located at (a_1, b_1, c_1) and its diagonally opposite corner located at (a_2, b_2, c_2). This block has dimensions given by

$$\mathrm{d} \begin{bmatrix} a_1, a_2 \\ b_1, b_2 \\ c_1, c_2 \end{bmatrix} = \begin{bmatrix} a_2 - a_1 \\ b_2 - b_1 \\ c_2 - c_1 \end{bmatrix}$$

where the dimensions are seen as the components of the vector representing the diagonal of the block (Figure 5.30).

The block form of Mies van der Rohe's Seagram Building is given by

$$\begin{bmatrix} 0, 14 \\ 0, \ 6 \\ 0, \ 5 \end{bmatrix} \cup \begin{bmatrix} 4, 10 \\ 0, \ 6 \\ 5, 11 \end{bmatrix} \cup \begin{bmatrix} 4, 10 \\ 6, \ 8 \\ 0, 42 \end{bmatrix} \cup \begin{bmatrix} 2, 12 \\ 8, 14 \\ 0, 42 \end{bmatrix}.$$

Of course, this is just one way of describing the block form. Equally as well, we could have employed the method of subtracted volumes to carve the building out of its treble-cube envelope (Figure 5.31).

Figure 5.31
The development of the block form of the Seagram Building (architect: Mies van der Rohe), New York, by the 'subtraction of volumes' from a notional building envelope, I, made up of three cubes

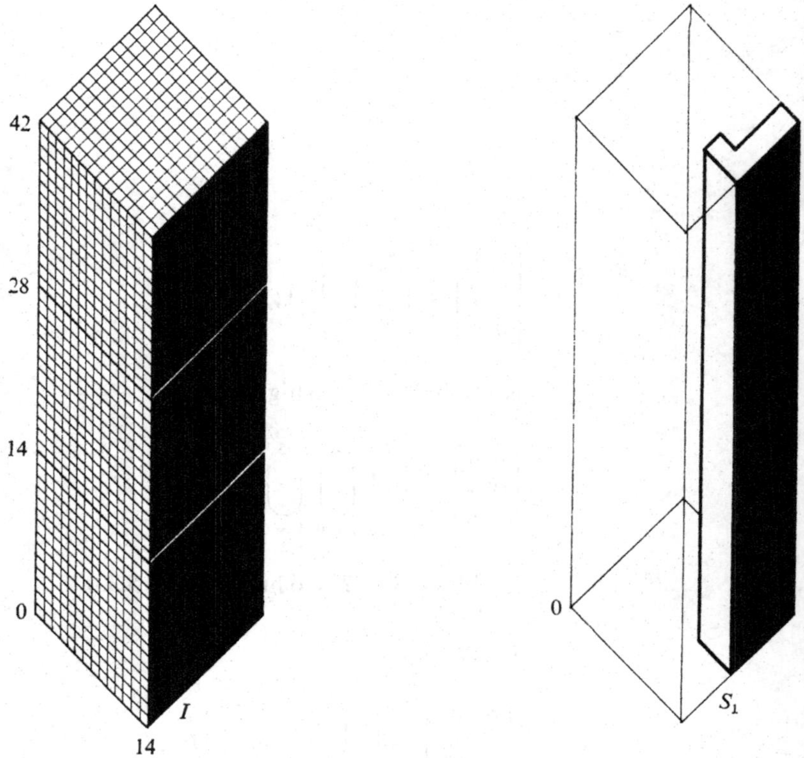

$$S_1 = \begin{bmatrix} 12, 14 \\ 6, 14 \\ 0, 42 \end{bmatrix} \cup \begin{bmatrix} 10, 12 \\ 6, 8 \\ 0, 42 \end{bmatrix}$$

$$S_2 = \begin{bmatrix} 0, 2 \\ 6, 14 \\ 0, 42 \end{bmatrix} \cup \begin{bmatrix} 2, 4 \\ 6, 8 \\ 0, 42 \end{bmatrix}$$

$$S_3 = \begin{bmatrix} 10, 14 \\ 0, 6 \\ 5, 42 \end{bmatrix}$$

$$S_4 = \begin{bmatrix} 0, 4 \\ 0, 6 \\ 5, 42 \end{bmatrix}$$

$$S_5 = \begin{bmatrix} 4, 10 \\ 0, 6 \\ 11, 42 \end{bmatrix}$$

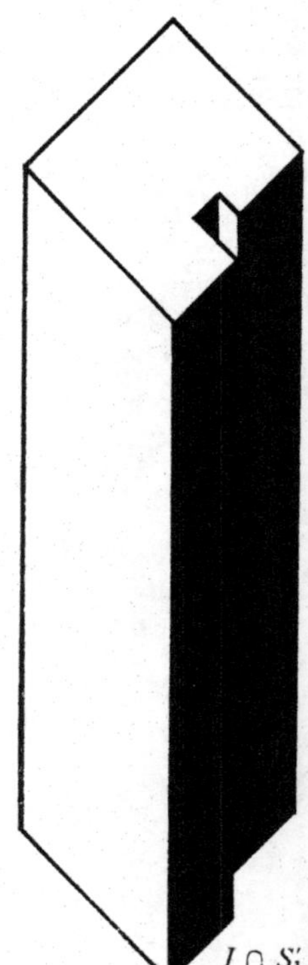

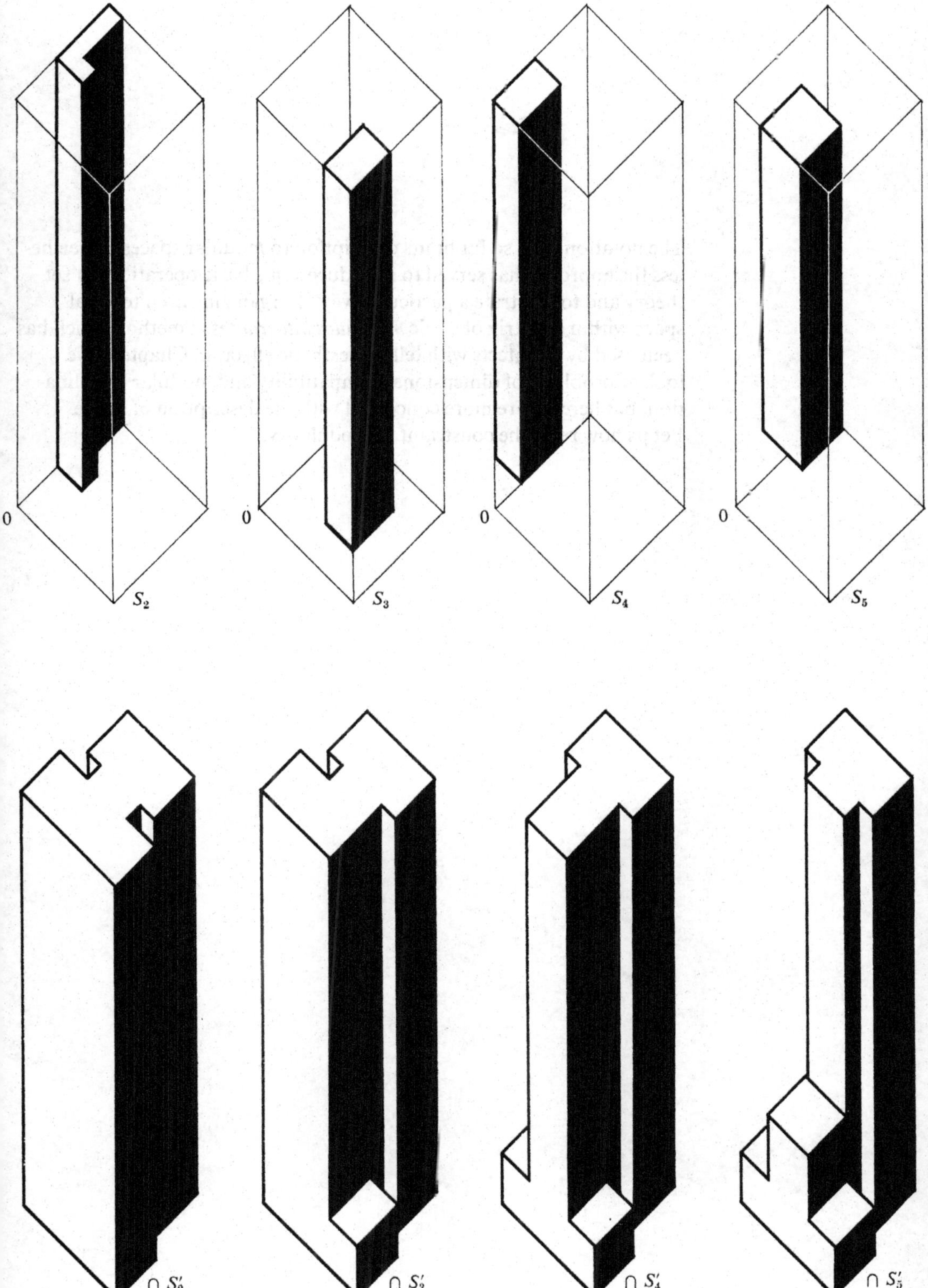
0
0
0
0
S_2
S_3
S_4
S_5
$\cap\ S'_2$
$\cap\ S'_3$
$\cap\ S'_4$
$\cap\ S'_5$

The notation used so far limits description to modular spaces, nevertheless the approach has served to introduce some basic operations of set theory and to illustrate a particular way of organizing architectural space within a matrix of cubic *molécules intégrantes*, a method which has been used by architects with telling aesthetic effect. In Chapter 8 we look at problems of dimensional compatibility and modular coordination, but here we are more concerned with the description of shape. Let us now relax the constraint of modularity.

6 Stacking, nesting and fitting

Consider the set of points, H_1, such that if $x \in H_1$, then x is a real number greater than or equal to a number a_1, that is,

$$H_1 = \{x \in R \mid a_1 \leqslant x\}.$$

This statement may be shown (Figure 6.1) as a set of points along a line. Effectively, the statement divides the line representing the real number system, R, into two halves, one half being the set of points in H_1 and

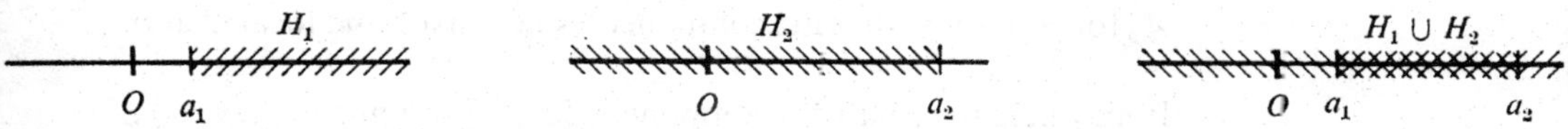

Figure 6.1

the other comprising points representing elements in its complement, H_1', where $H_1 \cup H_1' = R$. For this reason H_1 is called a *halfline*. Consider now the halfline H_2 defined by

$$H_2 = \{x \in R \mid x \leqslant a_2\}.$$

Suppose that $a_1 \leqslant a_2$, then the intersection of the two halflines H_1 and H_2 defines an *interval* of points A given by

$$A = H_1 \cap H_2 = \{x \in R \mid a_1 \leqslant x \leqslant a_2\}.$$

Such an interval is usually abbreviated $[a_1, a_2]$ and this is the notation we adopt here. We shall also speak of the *component* $[a_1, a_2]$ which fills the interval of that name. From the context, it should be clear whether we are talking about the interval or component. The points a_1 and a_2 are called the *extremum points*. The length of the interval is $a_2 - a_1$, and this is the element of the 1×1 vector $\mathrm{d}\,[a_1, a_2] = [a_2 - a_1]$. We shall denote the length of A by $\mid A \mid$.

Two intervals $A = [a_1, a_2]$ and $B = [b_1, b_2]$ are said to be *identical* if $a_1 = b_1$ and $a_2 = b_2$. Then we write $A \equiv B$: such identical intervals obviously have equal lengths, but there are intervals with equal lengths which are not identical. For example, two intervals $A_1 = [x_1, a + x_1]$ and $A_2 = [x_2, a + x_2]$ have equal lengths, a, but their positions differ if x_1 is not the same as x_2. Two components (or intervals), A and B, are *equal* if $\mid A \mid = \mid B \mid$ whether or not a_1 is equal to b_1, and a_2 to b_2. For equality we write $A = B$. In the example, $\mathrm{d}\,A_1 = \mathrm{d}\,A_2 = [a]$ and this is a vector of length a in R. Components of the same length, but of no fixed addresses, are said to be *free*. There is a clear analogy between free components and vectors: both have the property of having length without having a definite location. We denote all free components by $\underline{A}$ if they are *equal* to the fixed component A. Thus

$$\underline{A} = [x, |A| + x] \text{ for all } x \in R.$$

We may fix $\underline{A}$ by giving x a definite value such as p, say. We then write A_p for $\underline{A}$ to show that the component is now fixed and located at p.

It may help to visualize a component, $[a_1, a_2]$ as a fine hollow rod of length $a_2 - a_1$ which is threaded onto a line of real numbers and positioned at a_1 at one end and at a_2 at the other. Each rod has slightly elastic sides so that any one rod may be slipped over or inside another.

In general, two components $A = [a_1, a_2]$ and $B = [b_1, b_2]$ are either disjoint, that is separate, or they overlap. There are five possibilities (Figure 6.2), and there are seven distinct ways in which components may be joined so that their ends abut (Figure 6.3). Components may be separate, may partially overlap, may completely overlap so that one is contained within the other, or they may be identical. Components may be joined externally one to the other, in which case they are said to *stack*, or internally in such a way as to *nest* or, perhaps, to *fit* when they are identical. If a set of components $A_1, A_2 \ldots A_n$ stack to fill an interval B they are said to *pack* the interval: a situation analogous, for example, to the problem of assembling building components of given widths to make up a wall section.

Formally, $A \cap B \neq \emptyset$, if $a_2 \geqslant b_1$ and $b_2 \geqslant a_1$, and then

$$A \cup B = [\min\{a_1, b_1\}, \max\{a_2, b_2\}]$$
$$A \cap B = [\max\{a_1, b_1\}, \min\{a_2, b_2\}]$$

where

$$\min\{a, b\} = \begin{cases} a \text{ if } a < b \\ a = b \text{ if } a = b \\ b \text{ if } a > b \end{cases}$$

$$\max\{a, b\} = \begin{cases} a \text{ if } a > b \\ a = b \text{ if } a = b \\ b \text{ if } a < b. \end{cases}$$

Figure 6.2

Condition	Type of overlap	Graph
1. if $A \cap B = \emptyset$	none; disjunctive	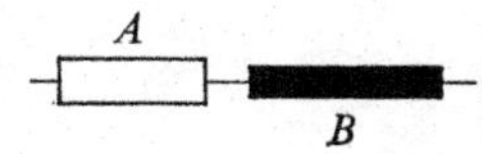
but, if $A \cap B \neq \emptyset$:		
2. $A \not\subset B, B \not\subset A$	partial	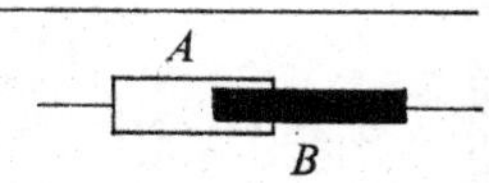
3,(4). $A \not\subset B, B \subset A$ $(A \subset B, B \not\subset A)$	inclusive	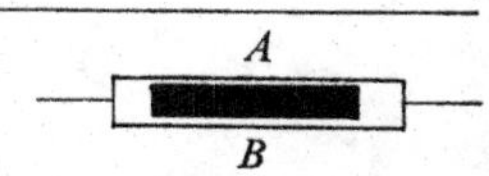
5. $A \subset B, B \subset A$	identical	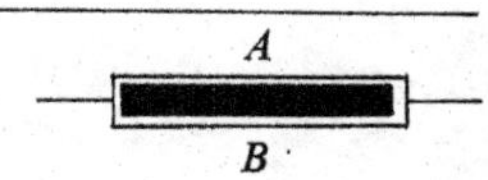

Figure 6.3

Condition	Type of conjunction	Graph
$A \not\subset B, B \not\subset A$ 1. $a_2 = b_1$ 2. $a_1 = b_2$	stacking	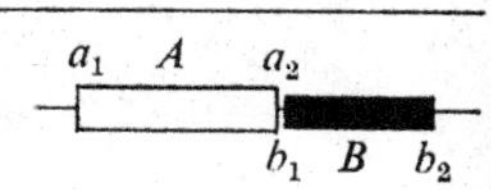 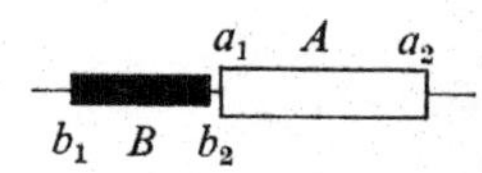
$A \not\subset B, B \subset A$ $(A \subset B, B \not\subset A)$ 3,(5). $a_1 = b_1$ 4,(6). $a_2 = b_2$	nesting	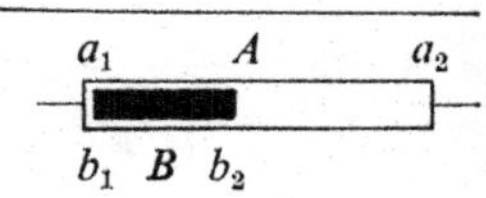 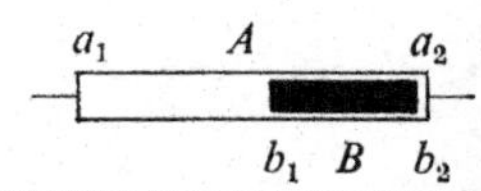
$A \subset B, B \subset A$ 7. $\begin{cases} a_1 = b_1 \\ a_2 = b_2 \end{cases}$	fitting	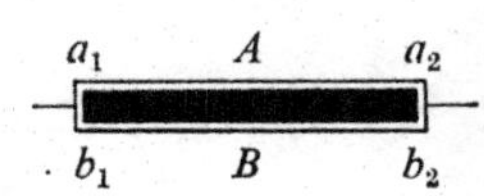

Packing in one dimension amounts to the task of selecting from a set of free components particular combinations which can be assembled to make up the length of the interval to be packed. If the set of permitted lengths is unrestrained within the real number system then packing accords with the everyday laws of arithmetic, but if, as we shall discuss in Chapters 8 and 9, the set is restricted to multiples of a module or to a set of numbers belonging to some proportional series then packing becomes more difficult.

Packing, even of simple rectangular spaces, is far more complicated in two and three dimensions, yet this is an essential task of architectural design: how to arrange rooms within a given plan form, or their volumes inside a building shell or envelope. The problem is not confined to architecture and is to be found in designing modules for electronic equipment and the economic cutting of sheet materials into smaller, but various, rectangles. While graphic representation and descriptive geometry have served the designer well and have much to recommend them, systematic and computer-aided design methods often require quantitative, numerical descriptions of two- and three-dimensional spaces. To move into these high-order spaces we need to introduce the idea of a *product set*.

A *pair* is a set containing two elements. Usually order is of no importance in a set, but sometimes we want to take order into account. Recall the example of putting on socks and shoes in Chapter 2: as a pair {socks, shoes} = {shoes, socks}, but as a direction for dressing (socks, shoes) put in *that* order is much sounder advice than the suggestion (shoes, socks). Such a pair is said to be ordered, and in general when a set has a specific order it is written with brackets rather than braces. An *ordered pair* (a, b) consists of two elements a and b so that a is the *first coordinate* and b is the *second coordinate*.

The *product* of two sets A and B is a new set consisting of all ordered pairs (a, b) where $a \in A$ and $b \in B$. It is denoted by AB. That is,

$$AB = \{(a, b) \mid a \in A, b \in B\}.$$

Thus, taking the sets $A = \{1,2,4\}$ and $B = \{2,3\}$ we have

$AB = \{(1,2), (1,3), (2,2), (2,3), (4,2), (4,3)\}$.

The generation of ordered pairs becomes obvious in tabular form

B \ A	1	2	4
3	(1,3)	(2,3)	(4,3)
2	(1,2)	(2,2)	(4,2)

and this suggests that products may be mapped onto points in two-dimensional space using cartesian coordinates. For this reason the operation is known as the *cartesian product*. Note that the product is not usually commutative, for example,

$BA = \{(2,1), (2,2), (2,4), (3,1), (3,2), (3,4)\}$

so that $AB \neq BA$.

Given two components A and B it is clear that we need to inspect and compare the values of the extremum points a_1, a_2, b_1, b_2 before we can tell whether the components are disjoint, or whether they nest, stack or fit. It is a matter of looking at the pairs in the product set

$$\{a_1,a_2\}\ \{b_1,b_2\} = \{(a_1,b_1), (a_1,b_2), (a_2,b_1), (a_2,b_2)\}$$
$$= \{(a_i,b_j) \mid i = 1,2 \text{ and } j = 1,2\}.$$

A neat way of doing this is to consider a set of 2×2 matrices $E = \{[e_{ij}]\}$ whose elements take the values 0 and 1 according to the rule

$$e_{ij} = \begin{cases} 1, \text{ if } a_i - b_j = 0 \\ 0, \text{ if } a_i - b_j \neq 0. \end{cases}$$

The numbers 1 and 0 are binary values corresponding to the statements 'it is *true* that components A and B conjoin' (1), or 'it is *false*' (0). Assuming A and B are not degenerate, there are just six possible zero-one *conjunction matrices* in E. These matrices are like codes which tell us the type of conjunction, if any, existing between the two components.

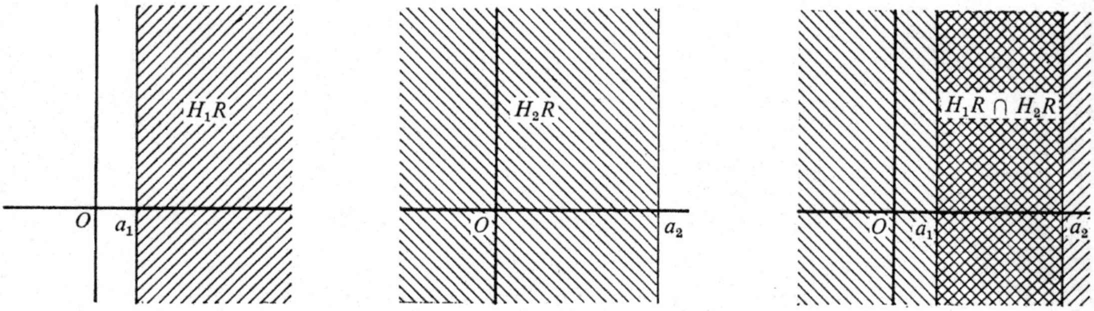

Figure 6.4

The set, $E = \{\mathbf{O}, \mathbf{E}_{11}, \mathbf{E}_{22}, \mathbf{E}_{12}, \mathbf{E}_{21}, \mathbf{I}\}$, of conjunction matrices is shown below.

Number of ones	**Conjunction matrix**	$\mathbf{E} \in E$	**Conjunction type**
none	$\begin{bmatrix} 0 & 0 \\ 0 & 0 \end{bmatrix}$	$\mathbf{O}$	none, disjoint
one	$\begin{bmatrix} 1 & 0 \\ 0 & 0 \end{bmatrix}, \begin{bmatrix} 0 & 0 \\ 0 & 1 \end{bmatrix}$	$\mathbf{E}_{11}, \mathbf{E}_{22}$	nesting
one	$\begin{bmatrix} 0 & 1 \\ 0 & 0 \end{bmatrix}, \begin{bmatrix} 0 & 0 \\ 1 & 0 \end{bmatrix}$	$\mathbf{E}_{12}, \mathbf{E}_{21}$	stacking
two	$\begin{bmatrix} 1 & 0 \\ 0 & 1 \end{bmatrix}$	$\mathbf{I}$	identity, fitting

The cartesian product RR, or R^2, is the set of all points (x, y) in two-dimensional space with real coordinates x and y. Similarly RRR, or R^3, is the set of points (x, y, z) in real three-dimensional space. This is why we have used the shorthand R, R^2 and R^3 for one-, two- and three-dimensional space previously. It should be observed in passing that R^3, say, does not of itself imply a rectangular coordinate system, but this is the only system we use here.

Earlier we looked at the halflines H_1 and H_2. What meaning can be given to the cartesian product H_1R? By definition

$$H_1R = \{(x, y) \in R^2 \mid x \in H_1, y \in R\} \text{ and } H_1 = \{x \in R \mid a_1 \leqslant x\}$$

so that x may take any value greater than or equal to a_1, while y is unrestricted within the set of real numbers. H_1R, then, defines what we call a halfplane. Similarly H_2R is a halfplane (Figure 6.4). The intersection of these two halfplanes

$$H_1R \cap H_2R = (H_1 \cap H_2)\, R = AR$$

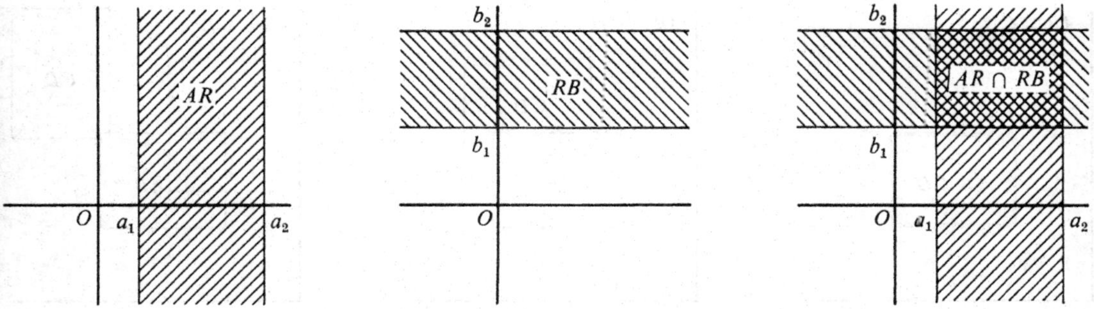

Figure 6.5

since $H_1 \cap H_2$ is the interval $A = [a_1, a_2]$. The intersection is thus a band of width $a_2 - a_1$ in the x-direction and of infinite length in the y-direction.

The product RB where $B = [b_1, b_2]$ may be represented by an infinite band running parallel to the x-axis. The intersection of the band AR and the band RB is a rectangular area which we shall call a *panel* (Figure 6.5). More briefly, the panel may be defined directly by the cartesian product of its two components A and B

$$AB = \{(x, y) \in R^2 \mid x \in A, y \in B\}.$$

To be specific we may write out AB more fully as

$$\begin{bmatrix} a_1, a_2 \\ b_1, b_2 \end{bmatrix} \text{ or } [a_1, a_2; b_1, b_2]$$

as we have done already for modular rectangles. Whereas in the modular case we only permitted integer values of a_1, a_2, b_1, b_2, now we allow them to take any real value.

Earlier we defined a free component $\underline{A}$ as one which is equal to component A, that is to say $\underline{A}$ is a component having the same length, $|A|$, as A while remaining free in regard to its position. Consider the fixed panel AB and the free components $\underline{A}$ and $\underline{B}$ which correspond to its sides. We shall say that the *free panel* $\underline{AB}$ is *equal* to AB and we see that it is a panel of equal dimensions and orientation, but of unspecified location. The free panel $\underline{AB}$ is a *translation* of AB,

$$\underline{AB} = \begin{bmatrix} a_1, a_2 \\ b_1, b_2 \end{bmatrix} + \begin{bmatrix} x \\ y \end{bmatrix} = \begin{bmatrix} a_1 + x, a_2 + x \\ b_1 + y, b_2 + y \end{bmatrix}, \text{ for all } (x, y) \in R^2.$$

On the other hand the free panel $\underline{BA}$ is an isometric transformation of AB which preserves length but not orientation and position. Thus

$$\underline{BA} = \begin{bmatrix} 0 & 1 \\ 1 & 0 \end{bmatrix} \begin{bmatrix} a_1, a_2 \\ b_1, b_2 \end{bmatrix} = \begin{bmatrix} b_1, b_2 \\ a_1, a_2 \end{bmatrix}$$

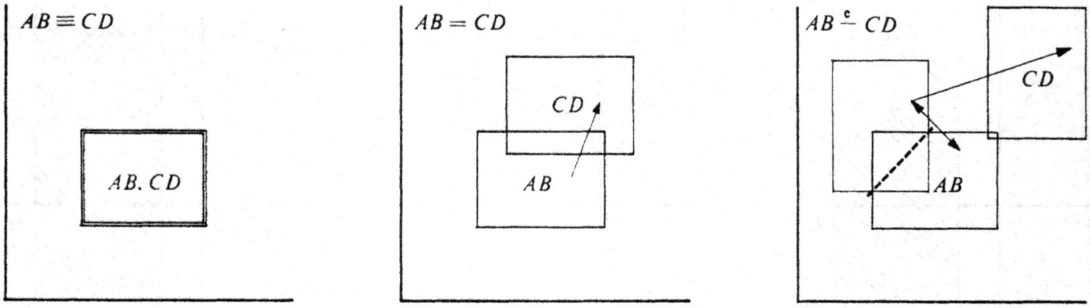

Figure 6.6

is the reflection in $y = x$ of AB, and $\underline{B}\underline{A}$ is then any translation of BA. In general, we shall say that two panels AB and CD are *equivalent* ($\overset{e}{-}$) if $\{|A|, |B|\} = \{|C|, |D|\}$. We may summarize these conditions as follows (Figure 6.6):

Identity: $AB \equiv CD$, if $(A, B) = (C, D)$
Equality: $AB = CD$, if $(|A|, |B|) = (|C|, |D|)$
Equivalence: $AB \doteq CD$, if $\{|A|, |B|\} = \{|C|, |D|\}$.

Two panels $A\underline{B}$ and $\underline{A}B$ are clearly equal to AB but they are not completely free. They are, in fact, constrained within the bands AR and RB respectively (Figure 6.7). Normally, two panels AB and CD cannot be combined to form another rectangular panel, thus $AB \cup CD$ is not usually a panel, although $AB \cap CD$ is (Figure 6.8). For testing the

Figure 6.7

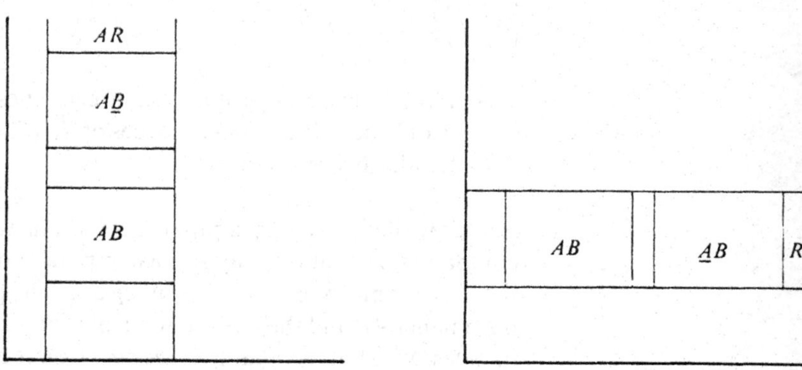

Figure 6.8

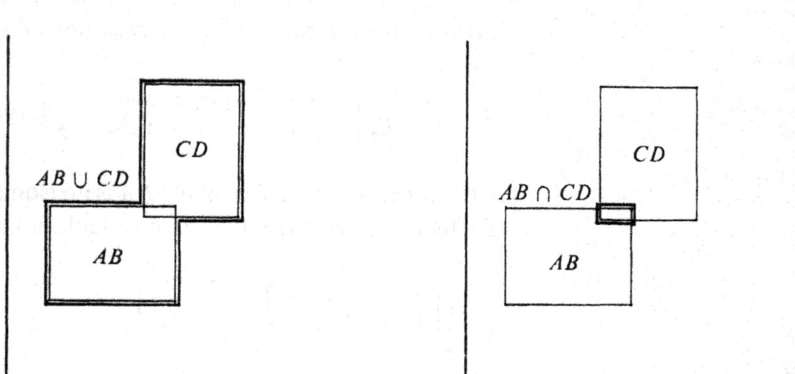

Figure 6.9
The thirty-six distinct ways in which two panels may be disjoint, aligned in one, two, or three directions, or fitted

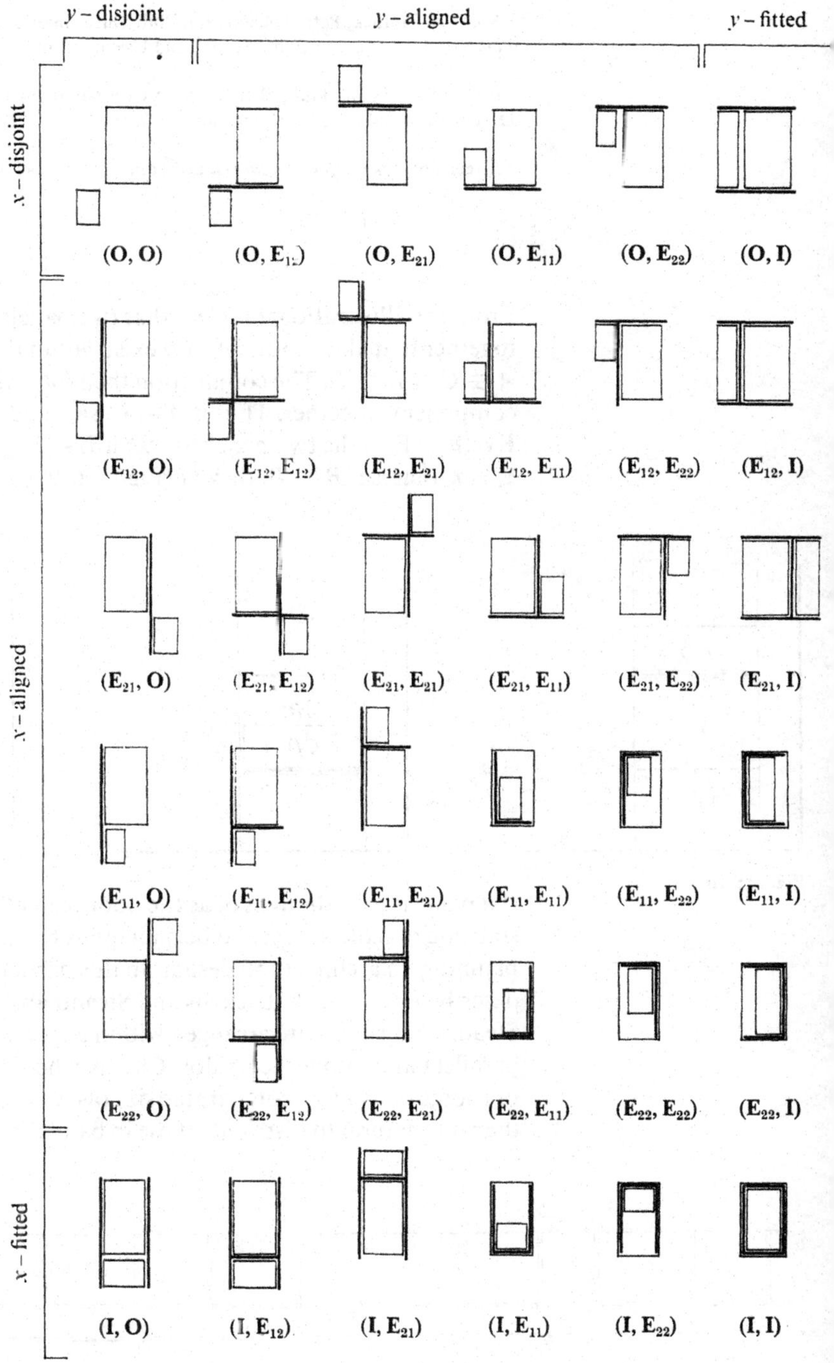

conjunction of panels we may use an ordered pair of conjunction matrices comprising a 2×2 matrix $\mathbf{E}$ from the set $E = \{\mathbf{O}, \mathbf{E}_{11}, \mathbf{E}_{22}, \mathbf{E}_{12}, \mathbf{E}_{21}, \mathbf{I}\}$ for the x-component and another matrix from E for the y-component. We shall call such an ordered pair the *conjoint* of panels AB and CD. The ways in which the panels may be disjoint, aligned or fitted are given by the thirty-six possible conjoints in E^2 (Figure 6.9).

[1] Nicholas Bullock, Peter Dickens and Philip Steadman. *A Theoretical Basis for Univers y Planning*, University of Cambridge, Land Use and Built Form Studies, 1968, pp. 117–34.

[2] Richard Stibbs and Philip Steadman. 'A Computer Aided System for Architectural Design and Analysis', *Cambridge Research*, vol. 2, no. 1, 1968, pp. 18–22.

[3] Jurgen Joedicke. *Office Buildings*, London, Crosby Lockwood, 1962.

From this illustration it is clear that two panels can only be joined together to make another if they exist within the same band, that is if $A \equiv C$ or $B \equiv D$. The conjunction then follows the rules of joining two components together. Thus if $A \equiv C$, say, and if, for B and D, $\mathbf{E} \in \{\mathbf{E}_{12}, \mathbf{E}_{21}\}$ the two panels *stack*; if $\mathbf{E} \in \{\mathbf{E}_{11}, \mathbf{E}_{22}\}$ they *nest*; and if $\mathbf{E} = \mathbf{I}$, that is if $B \equiv D$, they *fit* (Figure 6.10).

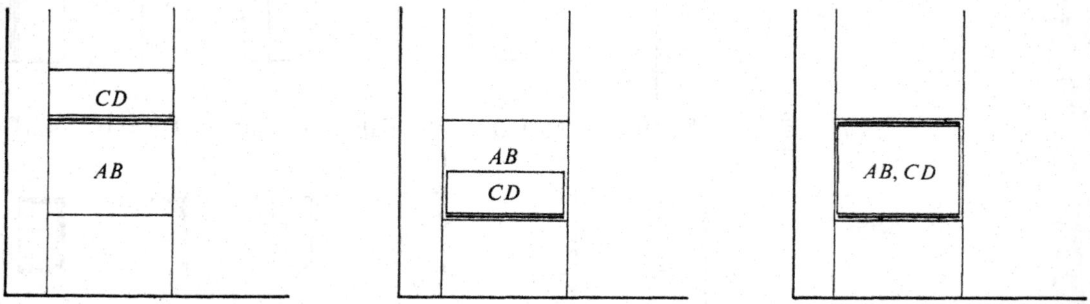

Figure 6.10

If a panel is now thought of as the floor plan of a rectangular room, the stacking of panels is seen to be analogous to the problem of 'band-planning' in architectural design. In this particular class of problem described by Bullock, Dickens and Steadman,[1] and by Stibbs and Steadman,[2] rooms are arranged within a rectangular building shell in parallel bands along a corridor. Clearly where there are a large number of rooms of roughly comparable size (as, typically, in an office building), then it is natural to distribute these in band arrangement: indeed,

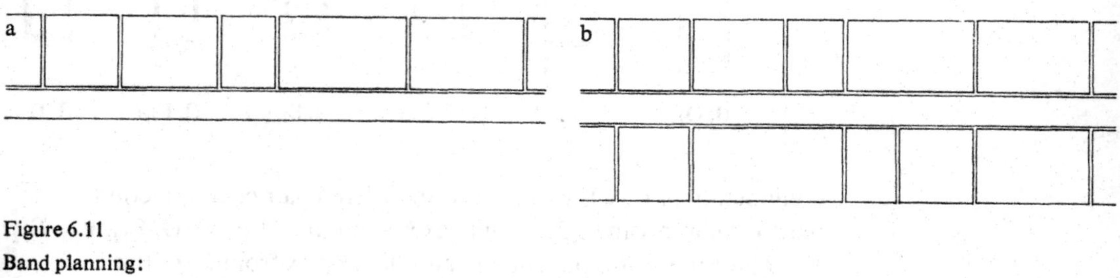

Figure 6.11
Band planning:

a, one-band with rooms on one side of corridor

b, two-band with rooms on both sides

c, three-band with two corridors serving three sets of rooms

d, four-band with two corridors serving four sets

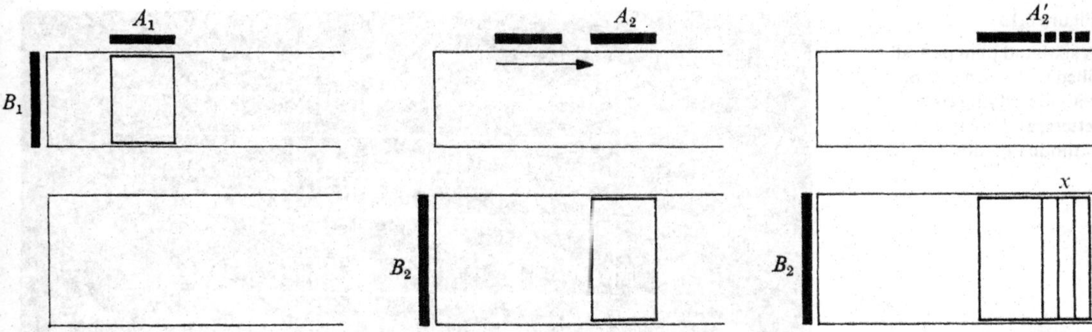

Figure 6.12

Joedicke[3] classifies some office buildings in terms of the number of bands, or zones, occurring in the plan (Figure 6.11).

The computer-aided method of band-planning proposed by Stibbs and Steadman involves a 'floating room' which has similarities with our free panel. The computer program allows the designer to fix a number of bands on the cathode ray tube by means of a light-pen. The floating room automatically appears in one of the bands. The room has an x-component $A_1 = [a_{11}, a_{12}]$ and takes up the y-component of the band: it is located in $B_1 = [b_{11}, b_{12}]$, say. The room is then moved by means of the light-pen to the band in which it is to be located. In our terms the component A_1 is translated to a new position A_2 where its product with a new band B_2, say, is formed to make the panel A_2B_2. The room may not be the right width and the designer uses the light-pen to widen or narrow the room on its right-hand side. This is equivalent to changing the component A_2 to

$$A_2' = [a_{21}, a_{22} \pm x]$$

where x is the length to be added or subtracted (Figure 6.12). Once one room is fixed a new floating room is created with the same x-component as the last fixed room. The procedure is then repeated. The program was

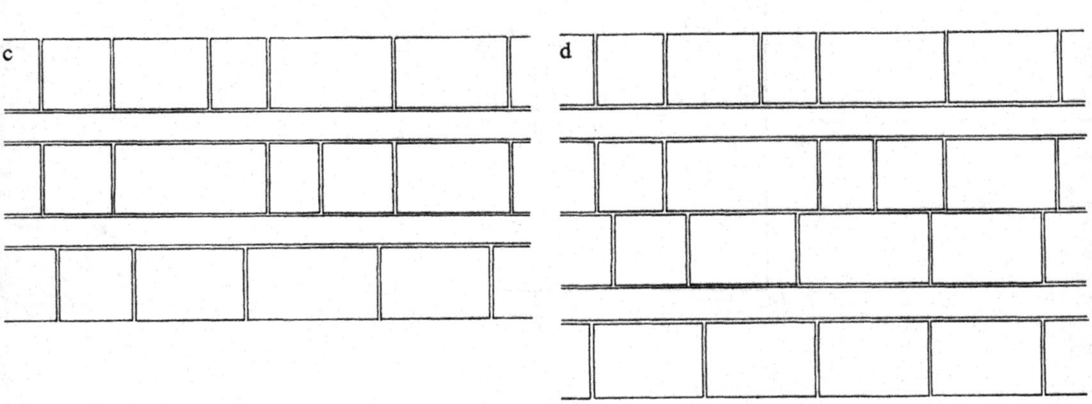

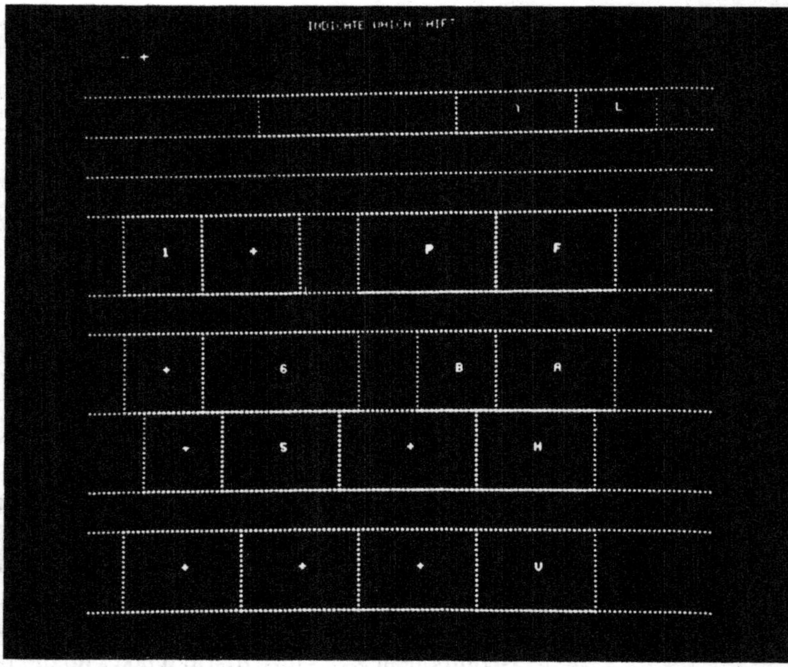

Figure 6.13
A six-band plan partially filled with rooms, some with identifying code letters, as displayed on a cathode ray tube

written as an interactive graphic method of putting a building plan into a computer where various aspects of its performance – circulation, lighting, heating and cooling loads – could be assessed (Figure 6.13).

Using the notation $[a_1, a_2; b_1, b_2]$ for the panel AB we are able to describe its *edges* and *vertices* in terms of products of the sets $\{a_1, a_2\}$ and $\{b_1, b_2\}$ and their subsets. The set of all subsets of a set is called its *power set*. The subsets of $\{a_1, a_2\}$ are $\{\emptyset\}$, $\{a_1\}$, $\{a_2\}$, $\{a_1, a_2\}$. Let $A_1 = \{a_1\}$ and $A_2 = \{a_2\}$. The power set, $P(A)$, of A is then the new set $\{\emptyset, A_1, A_2, A\}$. In the same way we may define the power set, $P(B)$, of B as $\{\emptyset, B_1, B_2, B\}$ where $B_1 = \{b_1\}$ and $B_2 = \{b_2\}$. Consider the cartesian product $P(A)P(B)$. This is the set

$$P(A)P(B) = \{\emptyset, AB, AB_1, AB_2, A_1B, A_2B, A_1B_1, A_1B_2, A_2B_1, A_2B_2\}$$

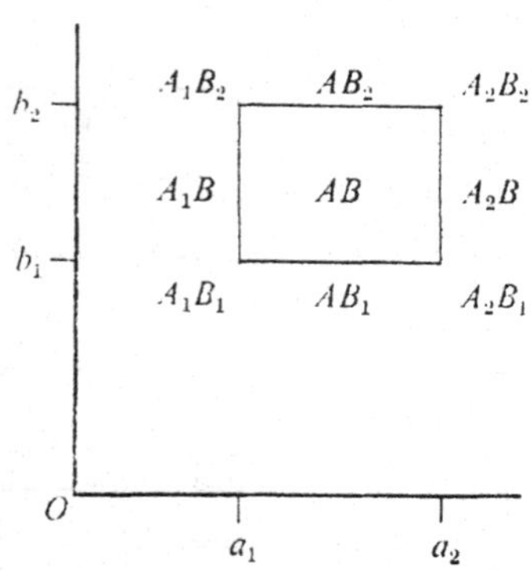

Figure 6.14

Note that $\emptyset X = X\emptyset = \emptyset$ no matter what X is. There are nine non-empty sets in $P(A)P(B)$ and these correspond to the panel, its four edges and its four vertices (Figure 6.14). We write these in full as

Panel	Edges	Vertices
$AB = \begin{bmatrix} a_1, a_2 \\ b_1, b_2 \end{bmatrix}$	$AB_1 = \begin{bmatrix} a_1, a_2 \\ b_1 \end{bmatrix}$	$A_1B_1 = \begin{bmatrix} a_1 \\ b_1 \end{bmatrix}$
	$AB_2 = \begin{bmatrix} a_1, a_2 \\ b_2 \end{bmatrix}$	$A_1B_2 = \begin{bmatrix} a_1 \\ b_2 \end{bmatrix}$
	$A_1B = \begin{bmatrix} a_1 \\ b_1, b_2 \end{bmatrix}$	$A_2B_1 = \begin{bmatrix} a_2 \\ b_1 \end{bmatrix}$
	$A_2B = \begin{bmatrix} a_2 \\ b_1, b_2 \end{bmatrix}$	$A_2B_2 = \begin{bmatrix} a_2 \\ b_2 \end{bmatrix}$

The vertices are located from the origin by the position vectors $[a_1 \quad b_1]^T, [a_1 \quad b_2]^T, [a_2 \quad b_1]^T, [a_2 \quad b_2]^T$. The edges are delineated by the *vector cones* which subtend them, and so is the panel itself (Figure 6.15).

Figure 6.15

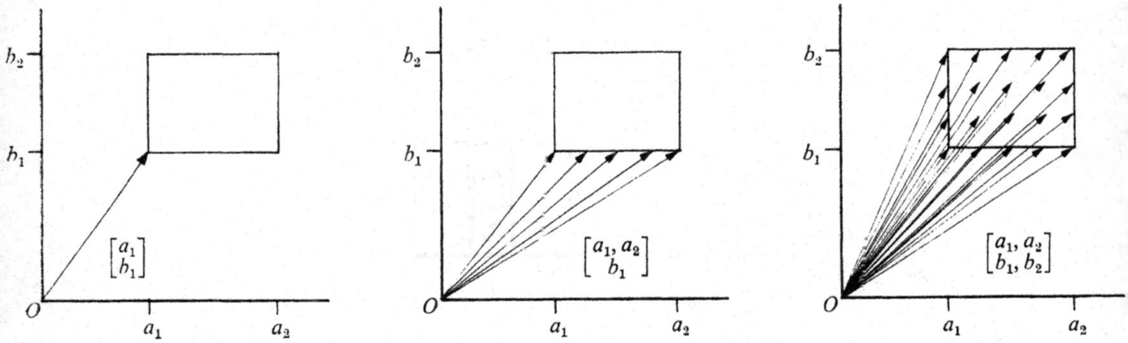

The schematic plan of the ground floor of one of Le Corbusier's Maisons Minimum can be written down in notation form. There are four rooms. The dimensions are given in units of 500 mm and measured roughly (for the purposes of the exercise) to the centre lines of the walls.

Figure 6.16
The assembly of a four-room plan

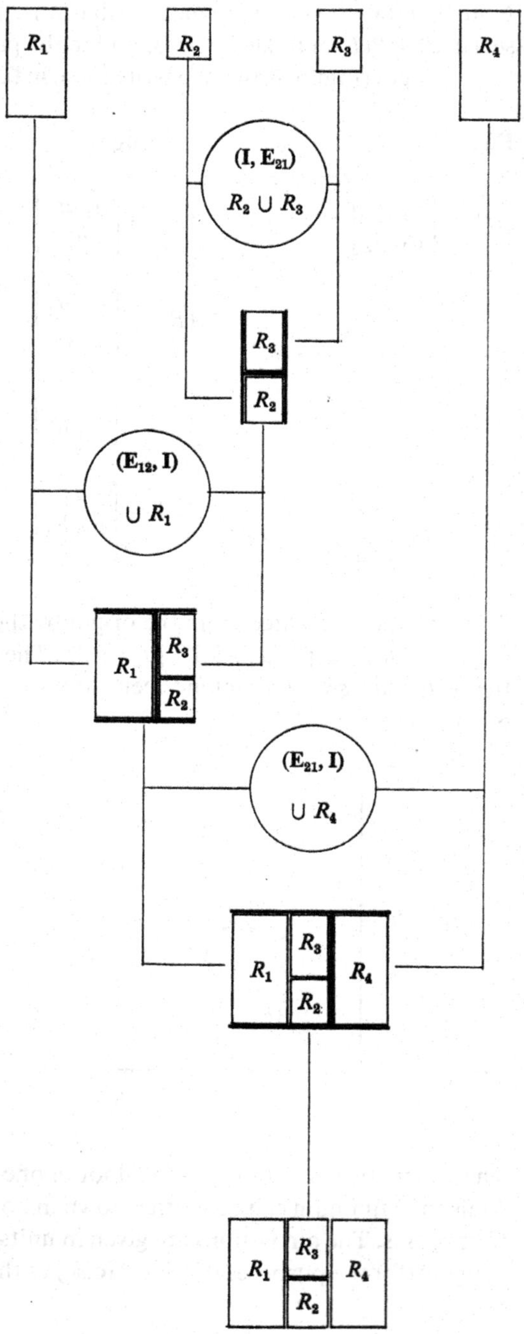

[4] See, for example, P. C. Gilmore and R. E. Gomory, 'Multistage Cutting Stock Problems in Two and More Dimensions', *Operational Research*, vol. 13, 1965, pp. 94–120.

$$R_1 = \begin{bmatrix} 0, & 6 \\ 0, & 11 \end{bmatrix}, R_2 = \begin{bmatrix} 6, & 10 \\ 0, & 5 \end{bmatrix}, R_3 = \begin{bmatrix} 6, & 10 \\ 5, & 11 \end{bmatrix}, R_4 = \begin{bmatrix} 10, & 16 \\ 0, & 11 \end{bmatrix}.$$

What can we say about the room arrangement? First we may note that $\mathrm{d}\,R_1 = \mathrm{d}\,R_4 = [6 \quad 11]^T$ so that the two rooms are equal. They also occupy the same band RB where $B = [0, 11]$.

They are, however, disjoint in the x-direction. The room R_2 shares the façade $y = 0$ with R_1 and R_4, while R_3 shares the other façade $y = 11$ along which R_1 and R_4 are aligned. We also see that R_2 and R_3 occupy the band AR where $A = [6, 10]$. Also, since their y-components [0, 5] and [5, 11] conjoin at $y = 5$ we see that together they make a new rectangular space

$$R_2 \cup R_3 = \begin{bmatrix} 6, & 10 \\ 0, & 11 \end{bmatrix}.$$

Now we see that this new space also occupies the band RB and that R_1 conjoins $R_2 \cup R_3$ at $x = 6$ and $R_2 \cup R_3$ conjoins R_4 at $x = 10$. All together the four rooms fill the rectangular plan $P = [0, 16; 0, 11]$.

The assembly of rooms may be illustrated in diagrammatic form (Figure 6.16). The procedure breaks down if the panels are arranged so the 'guillotine cuts' cannot be made. This is a cut which when started on one side of a rectangular assembly must traverse in a straight line to the other side. The term arises in connection with interesting problems related to the cutting of sheet materials in order to minimize wastage.[4] Consider the following assembly of panels,

Figure 6.17

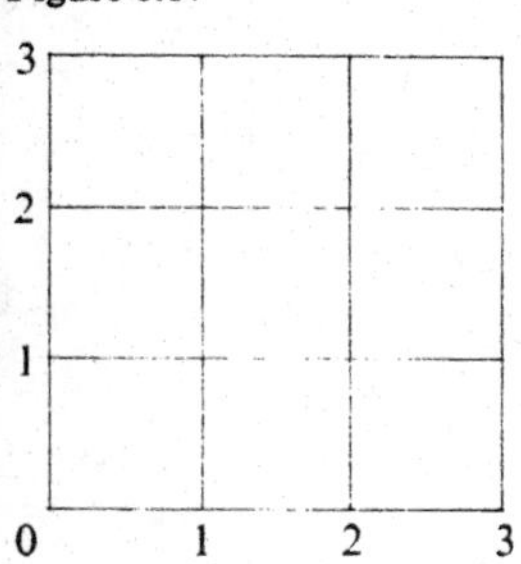

$$P_1 = \begin{bmatrix} 2, 3 \\ 0, 2 \end{bmatrix}, P_2 = \begin{bmatrix} 0, 2 \\ 0, 1 \end{bmatrix}, P_3 = \begin{bmatrix} 1, 2 \\ 1, 2 \end{bmatrix}, P_4 = \begin{bmatrix} 0, 1 \\ 1, 3 \end{bmatrix}, P_5 = \begin{bmatrix} 1, 3 \\ 2, 3 \end{bmatrix}.$$

Let us form a 5×5 array of conjoints for each pair P_i and P_j. (It is, of course, much easier to sketch out the assembly from this description and a 3×3 grid is provided (Figure 6.17) for the reader to do just this.) Nevertheless, we shall proceed, saving effort by computing only half the array of conjoints since this will suffice for our purpose:

Figure 6.18
The assembly of a five-room plan which cannot be 'guillotined'

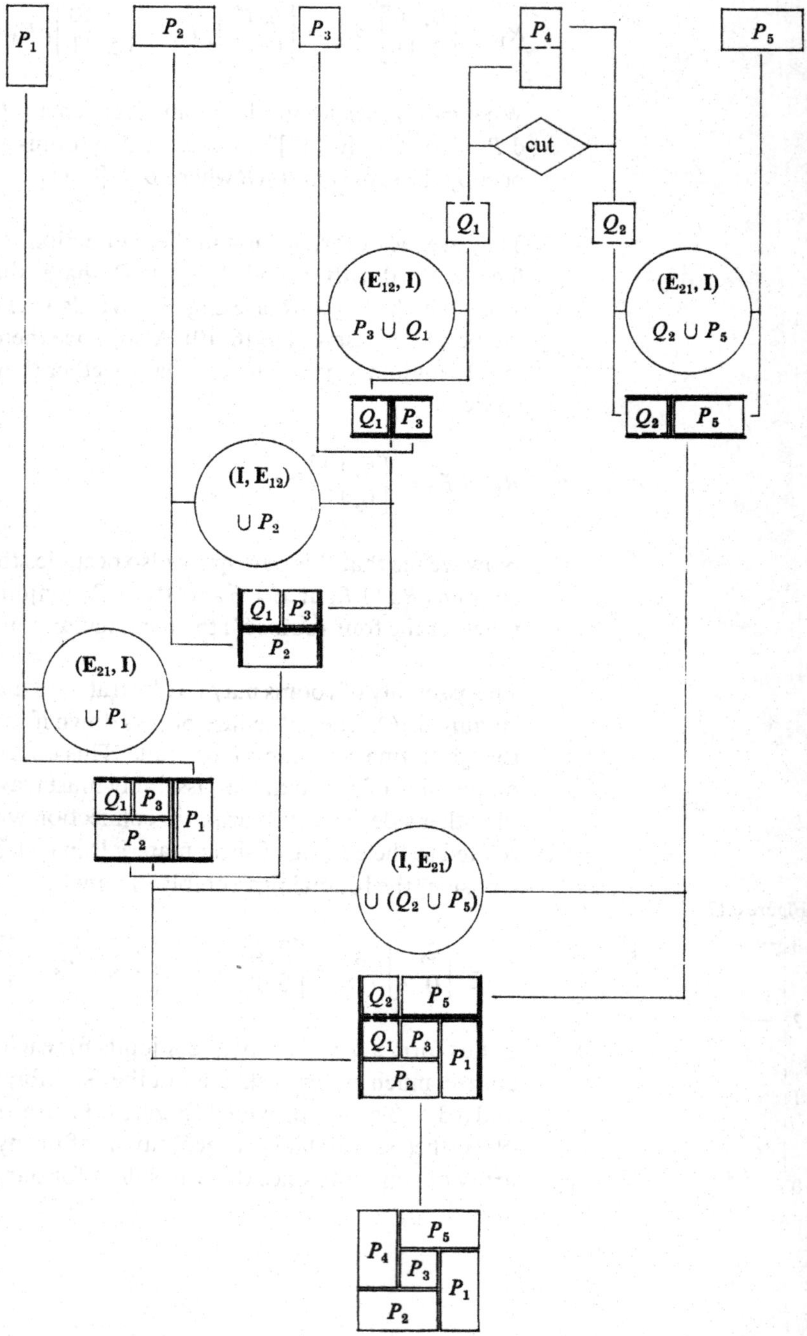

	P_1	P_2	P_3	P_4	P_5
P_1	.	$(\mathbf{E}_{12}, \mathbf{E}_{11})$	$(\mathbf{E}_{12}, \mathbf{E}_{22})$	$(\mathbf{O}, \mathbf{O})$	$(\mathbf{E}_{22}, \mathbf{E}_{21})$
P_2	.	.	$(\mathbf{E}_{22}, \mathbf{E}_{21})$	$(\mathbf{E}_{11}, \mathbf{E}_{21})$	$(\mathbf{O}, \mathbf{O})$
P_3	.	.	.	$(\mathbf{E}_{12}, \mathbf{E}_{11})$	$(\mathbf{E}_{11}, \mathbf{E}_{21})$
P_4	.	.	.	.	$(\mathbf{E}_{21}, \mathbf{E}_{22})$
P_5	.	.	.	.	.

We note that P_1 and P_4 are disjoint, as are P_2 and P_5, but that P_1 and P_2, P_1 and P_3, P_1 and P_4, P_2 and P_3, P_2 and P_4, and P_3 and P_4, P_3 and P_5, and P_4 and P_5 are all aligned in two directions. No pair, however, *fits* in any direction and we are unable to stack the panels. Panel P_3 is aligned with all four panels P_1, P_2, P_4, P_5. Let us take panel P_4 and nominally 'cut' it into two parts Q_1 and Q_2.

$$\begin{bmatrix} 0,1 \\ 1,3 \end{bmatrix} = \begin{bmatrix} 0,1 \\ 1,2 \end{bmatrix} \cup \begin{bmatrix} 0,1 \\ 2,3 \end{bmatrix} = Q_1 \cup Q_2.$$

Now Q_1 stacks with P_3, and Q_2 stacks with P_5. This 'cut' allows us to assemble in sequence all the remaining panels. The difficulty occurred because the original assembly could not be dissected with guillotine cuts (Figure 6.18).

In this chapter we are mainly concerned with the description of shape and not with problems of dissection or assembly. Some issues related to these problems are discussed in Chapters 10 and 11. One of the difficulties encountered in numerical description of shape for architectural design purposes is that it is *too* precise. An architect works within a range of tolerance permitted by the client's programme. Rarely will this say that a room must have *precisely* such and such dimension, or that its area or proportion must be *exactly* a given amount. Sometimes the brief will specify that one room must be next to another, but that is not the same thing as saying that the two rooms must be *aligned* in the way we have so far described. No, there is considerable looseness of fit about combinatorial problems in real architectural design, and man's judgement remains highly competitive against numeric, or even heuristic, computer methods. But for many design tasks, particularly where the architect, or engineer, wishes to test the performance of his proposal for one purpose or another, the computer is invaluable: and consequently the numeric description of the building geometry is virtually unavoidable.

Some degree of looseness can be expressed by the notation we have used. For example, the set of panels XY defined by

$$XY = \left\{ \begin{bmatrix} x_1, x_2 \\ y_1, y_2 \end{bmatrix} \middle| \, a \leqslant (x_2 - x_1) . (y_2 - y_1) \leqslant b, \text{for } a, b \geqslant 0 \right\}$$

is the infinite set of panels whose area is larger than a and less than b. In particular,

$$XY = \left\{ \begin{bmatrix} 0, x \\ 0, y \end{bmatrix} \middle| \, a \leqslant xy \leqslant b, \text{for } a, b \geqslant 0 \right\}$$

is the set of panels which share a corner at the origin (0, 0) whose area lies between a and b. Some of these panels can be extremely thin and excessively long. In practical terms we would probably want to constrain the proportions of the panels within useful limits. Consider the set

$$XY = \left\{ \begin{bmatrix} 0, x \\ 0, y \end{bmatrix} \middle| \, x \leqslant ay \leqslant a^2 x, \text{for } a \geqslant 1 \right\}.$$

This is a set of panels each of which lies within the proportion $1 : a$.

For example, let us suppose the client wants a room not smaller than 9 m^2 and not bigger than 16 m^2 (Figure 6.19 a–c). At the same time the room's length is not to exceed twice its width (Figure 6.19d–e). These requirements may be expressed formally as

$$XY = \left\{ \begin{bmatrix} 0, x \\ 0, y \end{bmatrix} \middle| \; 9 \leqslant xy \leqslant 16, x \leqslant 2y \leqslant 4x \right\}$$

and they can be shown graphically as a set consisting of all vertices (x, y) which satisfy the constraints. There may even be additional requirements related to orientation; that, for example, the dimension in the y-direction should not exceed 4·5 m, and in the x-direction should be less than 3 m (Figure 6.19 f–g). The intersection of all the sets of requirements provides the *solution set* within which an acceptable answer may be found. The client may wish to define the solution set yet more narrowly by adding the constraint that the perimeter, $2(x + y)$, should not exceed 14 m (Figure 6.19 h–i). Notice that some previous restrictions have been made redundant by these later requirements. It is

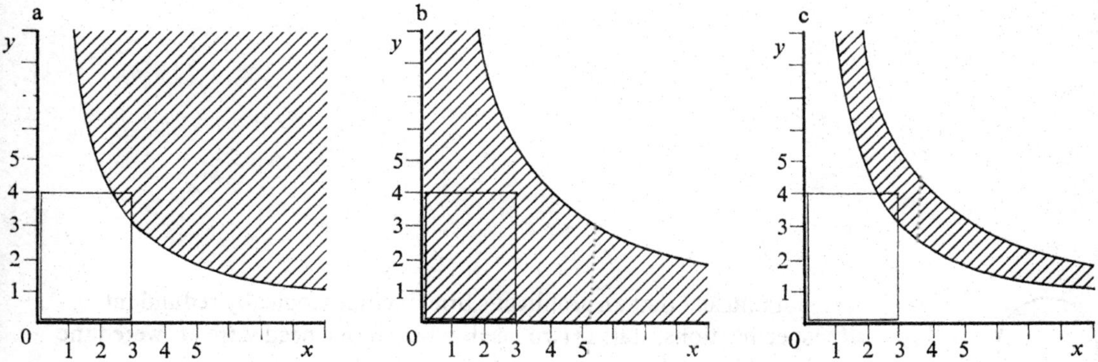

Figure 6.19
Constraints on room area and shape limit the permissible solution set (shaded):
a, b, area
d, proportion
f, dimensions
h, perimeter
c, e, g, i, the accumulation of previous constraints

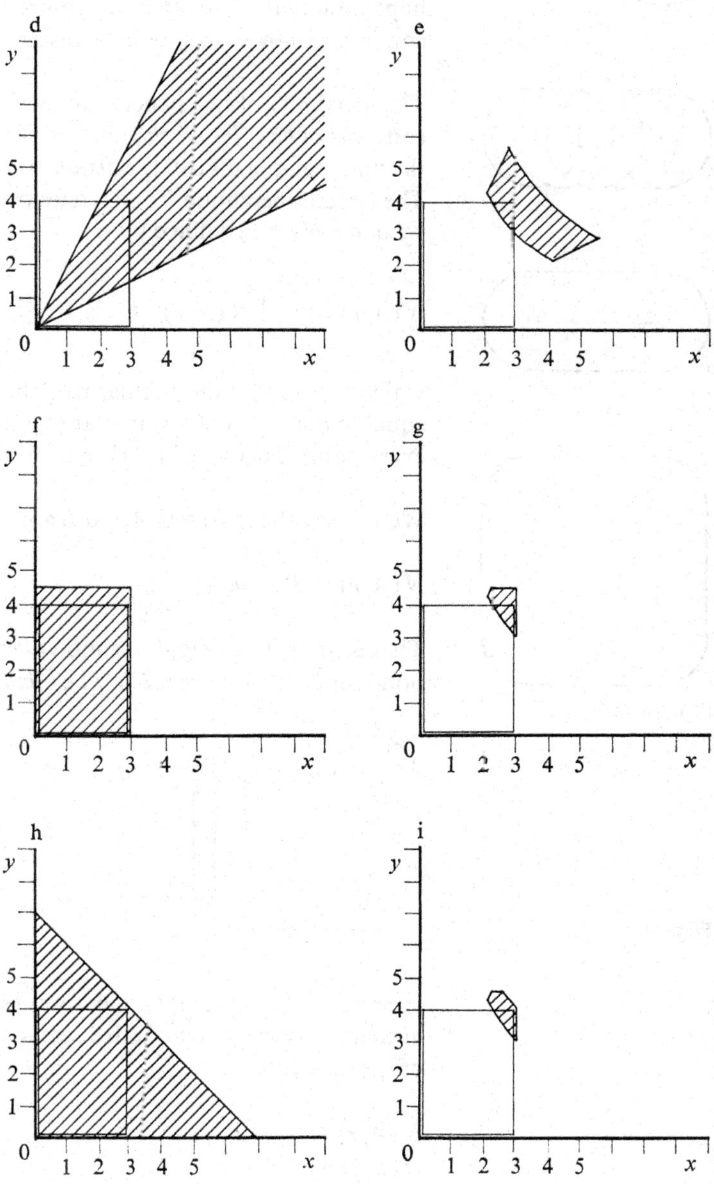

not unusual that an architect's brief includes logically redundant specifications, stating two needs where if one need were answered the second would be satisfied automatically. But such a situation is a happy one: far more often do requirements lead to contradiction and conflict, and these may only be resolved by practical reasonableness.

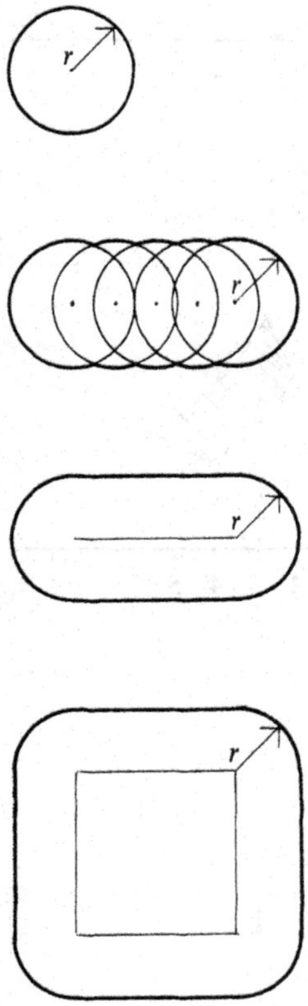

Figure 6.20

The proximity of two panels may be dealt with by introducing the concepts of *circular neighbourhood* of a plane set of points A and what the mathematicians call the *distance* between two bounded sets of points. The circular neighbourhood is a mapping of A onto another set which we name $N(A, r)$ defined by

$$N(A, r) = \bigcup_{X \in A} N(X, r)$$

which tells us that the circular neighbourhood N of A of radius r is equal to the union of all circular regions of radius r and centre X for every point X existing in A (Figure 6.20).

We will say that two sets A and B are *proximate* within a distance r if

$$N(A, r) \cap B \neq \emptyset.$$

That is to say, the neighbourhood of one intersects with the other in a non-empty set (Figure 6.21). The *minimum* value of r for which the

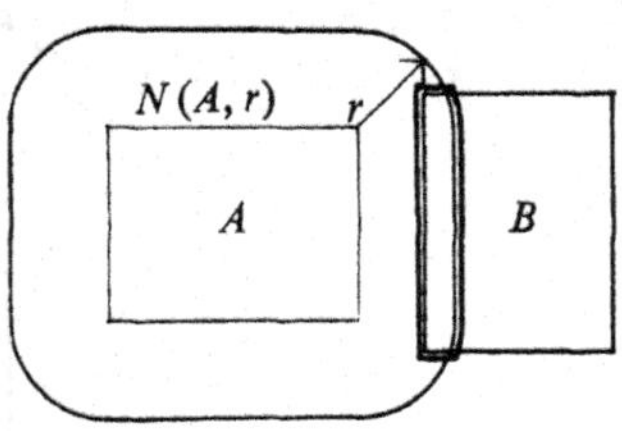

Figure 6.21

intersection is non-empty, measures the closeness of the two panels. The *distance* between two bounded sets A and B is defined as the *maximum* of r, s for which

$$N(A, r) \cap B = B,$$
$$N(B, s) \cap A = A.$$

Figure 6.22

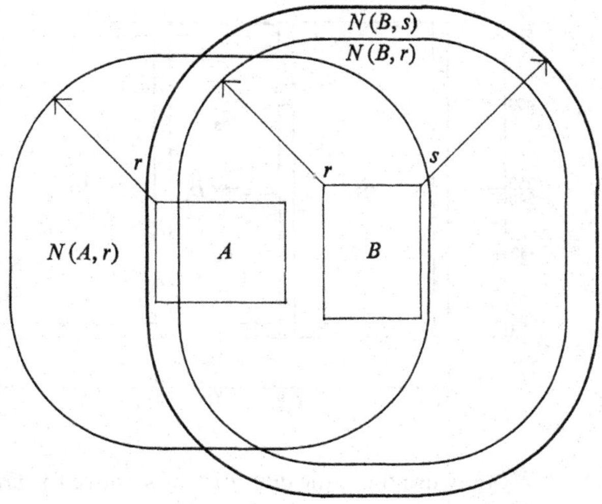

Effectively, the distance is the minimum radius of neighbourhood which ensures that both sets are contained by each other's neighbourhood (Figure 6.22).

The circular neighbourhood of a rectangular panel, or block (Figure 6.23), is a styled-up version with rounded corners. For our purposes we may modify the mathematician's neighbourhood concept in such a way

Figure 6.23
The circular neighbourhood of a block

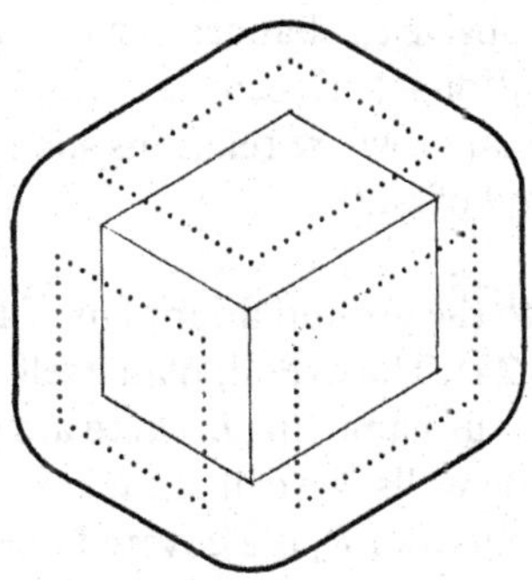

as to define a *rectangular tolerance zone* around a panel. We do this by mapping a panel AB onto a new set $T(AB, t)$ given by

$$T(AB, t) = \bigcup_{X \in AB} T(X, t)$$

where $T(X, t)$ is a square of semi-width t and centre X. This new set is a panel whose width and length have been increased by an amount t on all four sides, thus

$$T(AB, t) = \begin{bmatrix} a_1 - t, a_2 + t \\ b_1 - t, b_2 + t \end{bmatrix}, t \geqslant 0.$$

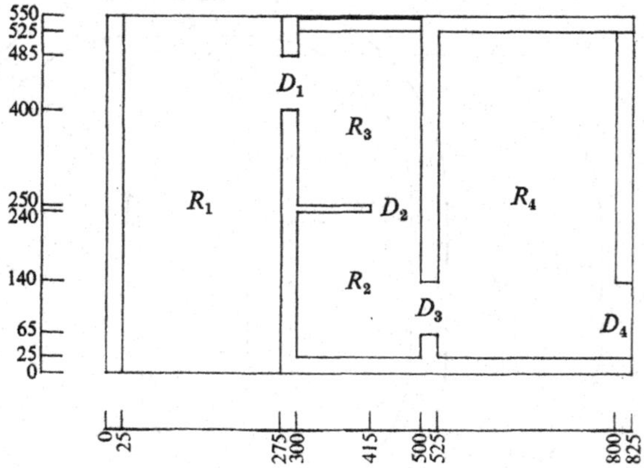

Figure 6.24
Dimensioned diagram of the ground plan of one of Le Corbusier's Maison Minimum (cm)

We may use the same definitions as above for *proximity* and *distance.* It will be clear that there is a close parallel between proximity and *positional tolerance*, where a manufactured component such as a door has to fit into a door opening. But proximity is also useful in building studies where some measure of which rooms are near to which is required. Such studies are often related to problems of allocating activities to rooms within a building in such a way as to reduce circulation. These 'looser' concepts of proximity and distance may be useful, for example, in tempering computer-aided design methods which rely on the exaggerated precision, yet oversimplification, of distance between the mean centres of individual spaces (see Chapter 14). The measures discussed above are more responsive to changes in room geometry – size, shape, orientation, and location – and seem to accord with our commonsense understanding of distance where relations such as 'quite near to' and 'in the neighbourhood of' suffice.

Let us return, now, to the ground floor plan of Le Corbusier's Maison Minimum (Figure 6.24). The overall plan itself measured to the outside wall is a rectangle: let us name this I. There are four 'rooms' which, measured to the inside walls, we call R_1, R_2, R_3, R_4. R_1 is a car port, R_2 and R_3 are utility rooms and R_4 is a covered family area. The main living-rooms and bedrooms are on the first floor and can be reached by means of an external staircase. There are four doorways and these we label D_1, D_2, D_3, D_4. The elements of the ground plan may be expressed numerically. The measurements are in centimetres.

Overall plan: $I \quad = [0, 825; 0, 550]$

Figure 6.25
The generation of the ground plan of the Maison Minimum using set-theoretic concepts

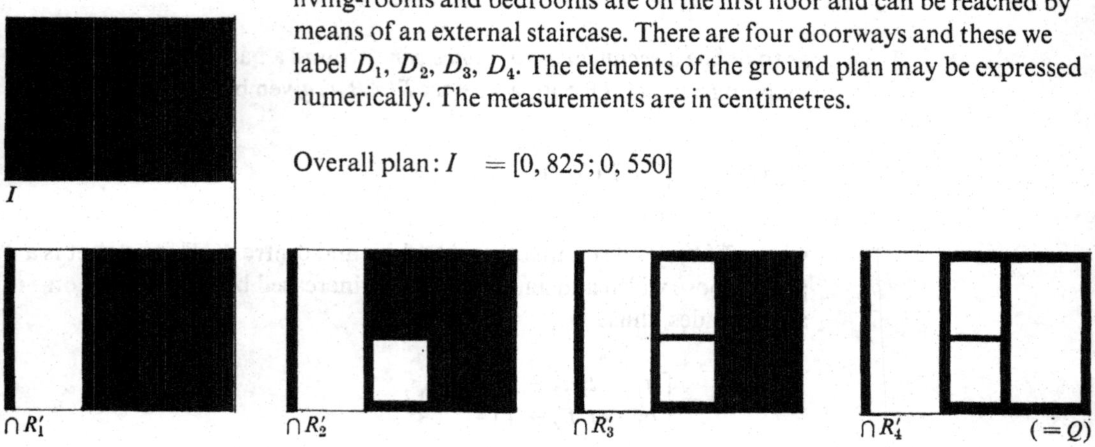

Rooms:	$R_1 = [25, 275; 0, 550]$
	$R_2 = [300, 500; 25, 240]$
	$R_3 = [300, 500; 250, 525]$
	$R_4 = [525, 800; 25, 525]$
Doorways:	$D_1 = [275, 300; 400, 485]$
	$D_2 = [415, 500; 240, 250]$
	$D_3 = [500, 525; 65, 140]$
	$D_4 = [800, 825; 25, 140]$

Previously, when we measured to the centre line of the wall, R_2 and R_3 conjoined. Now we see that they are still doubly aligned in the x-direction since they share the same x-component [300, 500] but there is a gap of 10 cm between the two y-components [25, 240] and [250, 525]. A *tolerance zone* of 10 cm around R_2 changes its y-component from [25, 240] to [15, 250] and we see that R_1 conjoins R_2 and R_3 along $y = 250$ within a tolerance of 10 cm (the wall thickness). The *distance* between the rooms R_1, R_2, R_3 and R_4, as we have defined it, is 500 cm.

The plan of this building may be generated (Figure 6.25), as it is within a computer by the method of 'subtracted areas', by intersecting the complement, R_i', of each room and the complement, D_i', of each doorway with the overall plan I. Thus the ground plan, Q, *without* doorways is

$$Q = I \cap R', \text{ where } R = \bigcup_{i=1}^{4} R_i$$

while the plan *with* doorways, P, is

$$P = Q \cap D', \text{ where } D = \bigcup_{i=1}^{4} D_i.$$

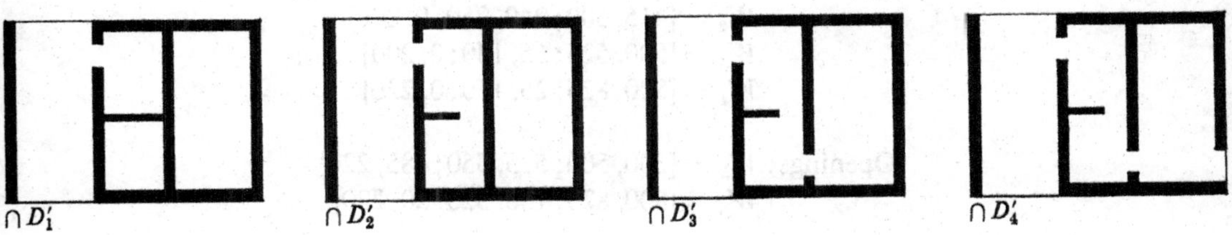

Figure 6.26

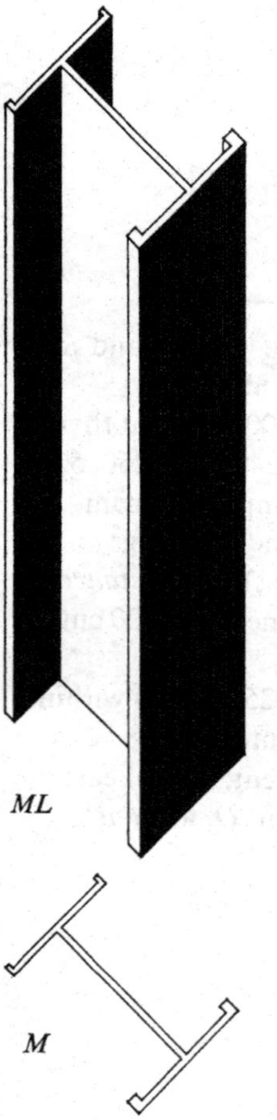

Before we do more, let us pause to consider the extension of cartesian products into three dimensions. The product, AB, of two components A and B is a two-dimensional panel. The product, ABC, of three components A, B and C is a three-dimensional *block*.

$$ABC = \{(x, y, z) \in R^3 \mid x \in A, y \in B, z \in C\}.$$

To be more specific we may write ABC out fully as

$$\begin{bmatrix} a_1, a_2 \\ b_1, b_2 \\ c_1, c_2 \end{bmatrix} \text{ or } [a_1, a_2; b_1, b_2; c_1, c_2]$$

following the notation we have adopted already in one and two dimensions.

We may think of any two-dimensional set as the cross-section of an extrusion in three dimensions. Thus a set M, representing the cross-section of an extruded bronze mullion from Mies van der Rohe's Seagram Building, may be projected into three dimensions by forming its cartesian product with a one-dimensional set L (Figure 6.26). In the same way we may raise the walls of Le Corbusier's ground plan, Q, into three dimensions by forming the cartesian product QS where S is the z-component, [0, 220], representing the storey-height of the ground floor.

Three of the door openings, D_1, D_2, D_3, are not so high; their z-component is [0, 200] leaving a 10 cm-deep lintel between the doorhead and the ceiling. There are two other openings in the design (Figure 6.27). One is a clerestory light in room R_3, and the other is a large void beside the 'doorway', D_4. These two openings and the doorways may be described as three-dimensional blocks.

$$\begin{aligned} \text{Doorways: } W_1 &= [275, 300; 400, 485; 0, 200] \\ W_2 &= [415, 500; 240, 250; 0, 200] \\ W_3 &= [500, 525; 65, 140; 0, 200] \\ W_4 &= [800, 825; 25, 150; 0, 220] \end{aligned}$$

$$\begin{aligned} \text{Openings: } W_5 &= [300, 500; 525, 550; 185, 220] \\ W_6 &= [800, 825; 150, 525; 30, 220]. \end{aligned}$$

The intersection of $T = QS$, the three-dimensional projection of the rooms without openings (Figure 6.28), with the complements, W_i', of the openings generates the three-dimensional representation of the design, U,

$$U = T \cap W', \text{ where } W = \bigcup_{i=1}^{6} W_i.$$

To illustrate set union in three-dimensional space we may complete the concrete form of the design by 'adding' the small 5 cm × 5 cm lintels over the openings W_5 and W_6, and over the entrances to the car port. The lintels may also be described as blocks.

Lintels: $L_1 = [820, 825; 25, 525; 215, 220]$
$L_2 = [300, 500; 545, 550; 215, 220]$
$L_3 = [25, 275; 0, 5; 215, 220]$
$L_4 = [25, 275; 545, 550; 215, 220]$.

The final concrete shell, V, is the union of T with all the lintels, L, that is

$$V = U \cup L, \text{ where } L = \bigcup_{i=1}^{4} L_i.$$

Figure 6.27

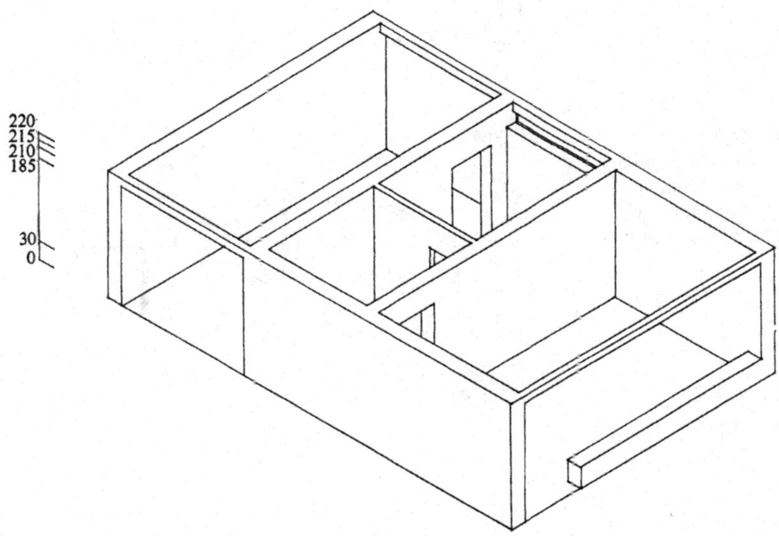

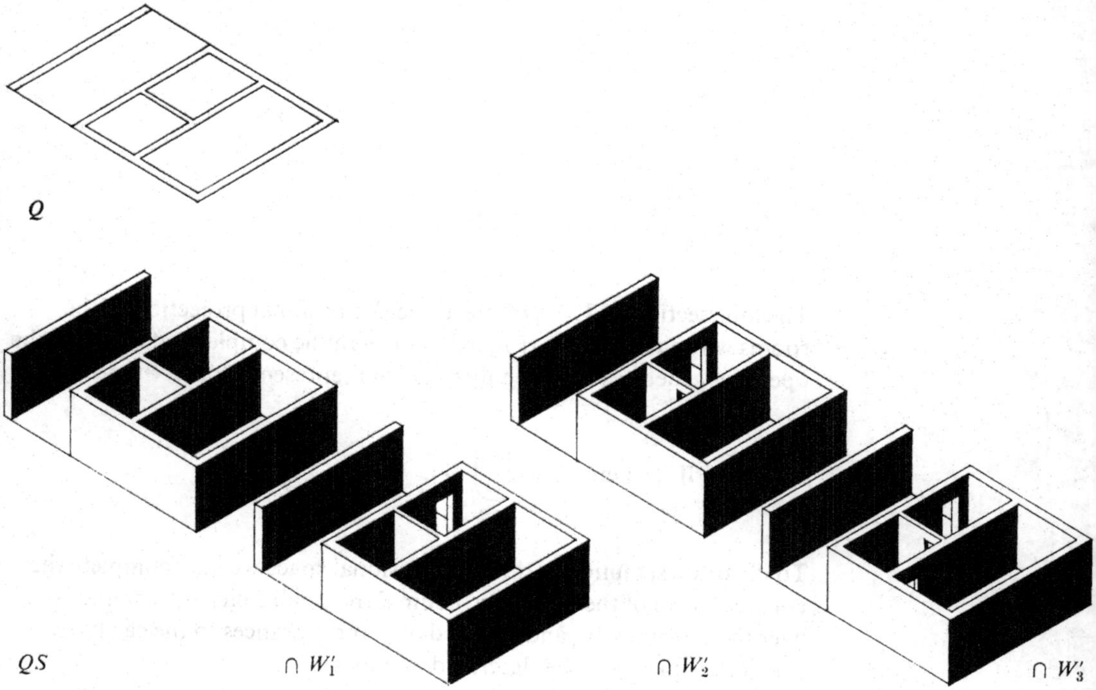

Figure 6.28

Architects often draw vertical sections through their buildings to show their structure. We have already seen how the product of a two-dimensional section with a component produces the solid extrusion in the example of Mies van der Rohe's bronze mullion. If we intersect the shell of Le Corbusier's house, V, with a plane we obtain a section (Figure 6.29). For example, planes A_1R^2 where A_1 is the degenerate x-component [450], and RA_2R where A_2 is the degenerate y-component [450] give us two cross-sections C_1 and C_2 through V on intersection:

$$C_1 = V \cap A_1R^2$$
$$C_2 = V \cap RA_2R.$$

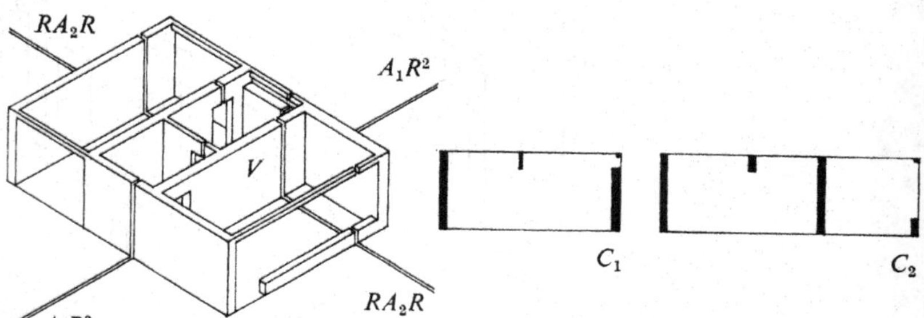

Figure 6.29
Cross-sections through the Maison Minimum generated by the intersections of sets of points

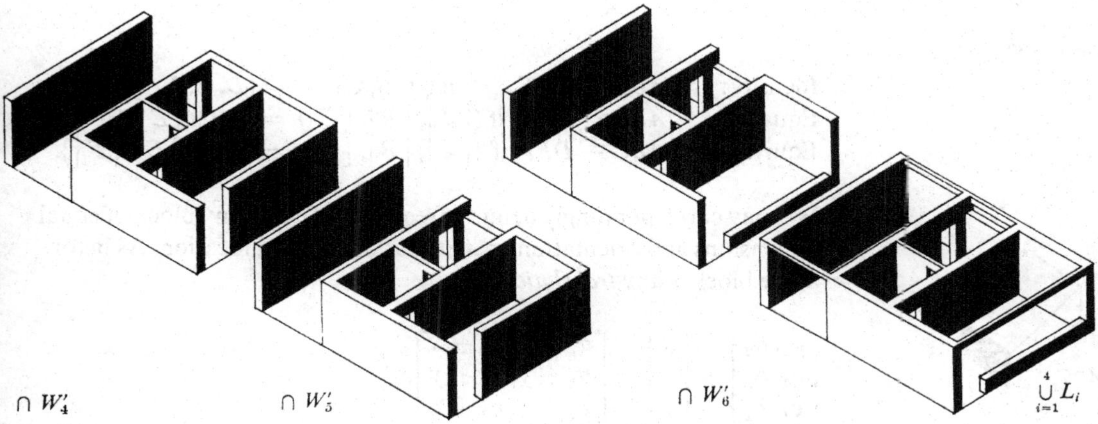

Let us reflect. To generalize, we have introduced a *component* $A = [a_1, a_2]$ in one dimension R, a *panel* $AB = [a_1, a_2; b_1\, b_2]$ in two dimensions R^2, and now a *block* $ABC = [a_1, a_2; b_1, b_2; c_1, c_2]$ in three, R^3. The pattern is clear. We originally considered a component as the intersection of two *halflines*, $H_1 \cap H_2 = A$; a *band* as the intersection of two *halfplanes*, $H_1R \cap H_2R = AR$, and a panel as the intersection of two bands $AR \cap RB = AB$. In three dimensions we have *halfspaces* – all the points which lie on or to one side of an infinite plane dividing space into two halves – and the intersection of two halfspaces, $H_1R^2 \cap H_2R^2 = AR^2$, is a *slice* of space bounded on either side by two planes. A block is the intersection of three such mutually perpendicular slices, that is,

$$ABC = AR^2 \cap RBR \cap R^2C.$$

Any set of points formed by the intersection of halfspace is said to be *convex*. We shall take a brief look at *convexity* when we consider non-rectangular forms, but here it is worth noting that components, panels and blocks are examples of *convex sets*.

Earlier we defined *identity*, *equality* and *equivalence* for a panel, and we may extend the definition to blocks ABC and DEF (Figure 6.30). Thus,

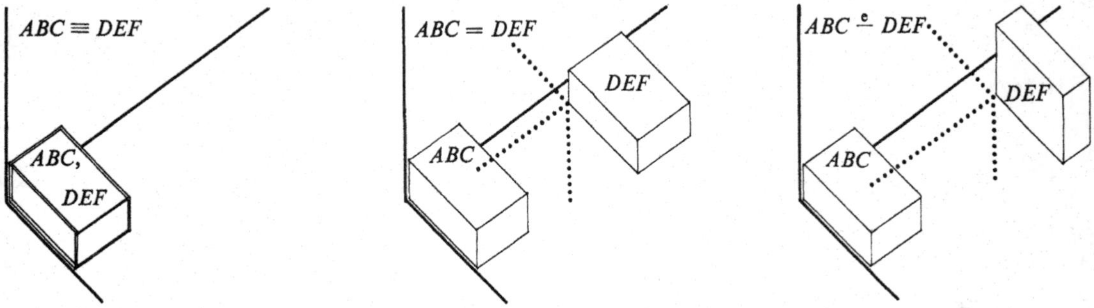

Figure 6.30

Identity: $ABC \equiv DEF$, if $(A, B, C) = (D, E, F)$
Equality: $ABC = DEF$, if $(|A|, |B|, |C|) = (|D|, |E|, |F|)$
Equivalence: $ABC \stackrel{e}{=} DEF$, if $\{|A|, |B|, |C|\} = \{|D|, |E|, |F|\}$.

We may correspondingly define a free block $\underline{ABC}$ as any block of equal dimensions and orientation as *ABC* but of different location. As before, a free block is any *translation*.

$$\begin{bmatrix} a_1, a_2 \\ b_1, b_2 \\ c_1, c_2 \end{bmatrix} + \begin{bmatrix} x \\ y \\ z \end{bmatrix} = \begin{bmatrix} a_1 + x, a_2 + x \\ b_1 + y, b_2 + y \\ c_1 + z, c_2 + z \end{bmatrix}.$$

Equivalent blocks, on the other hand, have equal dimensions but do not necessarily share orientation or location. Equivalence permutes *A*, *B* and *C*, so that there are six equivalent blocks including the identity: *ABC*, *ACB*, *CAB*, *CBA*, *BCA*, *BAC* (Figure 6.31). Note that the 3×3 zero-one permutation matrices which we discussed in Chapter 2 do indeed produce the equivalent blocks (remember each column of a block must be treated as a separate vector):

$$\mathbf{I} = [\mathbf{e}_1 \ \mathbf{e}_2 \ \mathbf{e}_3]: \quad \begin{bmatrix} 1 & 0 & 0 \\ 0 & 1 & 0 \\ 0 & 0 & 1 \end{bmatrix} \begin{bmatrix} a_1, a_2 \\ b_1, b_2 \\ c_1, c_2 \end{bmatrix} = \begin{bmatrix} a_1, a_2 \\ b_1, b_2 \\ c_1, c_2 \end{bmatrix} = ABC$$

$$\mathbf{P}_{132} = [\mathbf{e}_1 \ \mathbf{e}_3 \ \mathbf{e}_2]: \quad \begin{bmatrix} 1 & 0 & 0 \\ 0 & 0 & 1 \\ 0 & 1 & 0 \end{bmatrix} \begin{bmatrix} a_1, a_2 \\ b_1, b_2 \\ c_1, c_2 \end{bmatrix} = \begin{bmatrix} a_1, a_2 \\ c_1, c_2 \\ b_1, b_2 \end{bmatrix} = ACB$$

$$\mathbf{P}_{231} = [\mathbf{e}_2 \ \mathbf{e}_3 \ \mathbf{e}_1]: \quad \begin{bmatrix} 0 & 0 & 1 \\ 1 & 0 & 0 \\ 0 & 1 & 0 \end{bmatrix} \begin{bmatrix} a_1, a_2 \\ b_1, b_2 \\ c_1, c_2 \end{bmatrix} = \begin{bmatrix} c_1, c_2 \\ a_1, a_2 \\ b_1, b_2 \end{bmatrix} = CAB$$

and so on for the remaining permutation matrices $[\mathbf{e}_3 \ \mathbf{e}_2 \ \mathbf{e}_1]$, $[\mathbf{e}_3 \ \mathbf{e}_1 \ \mathbf{e}_2]$ and $[\mathbf{e}_2 \ \mathbf{e}_1 \ \mathbf{e}_3]$ to give *CBA*, *BCA* and *BAC*.

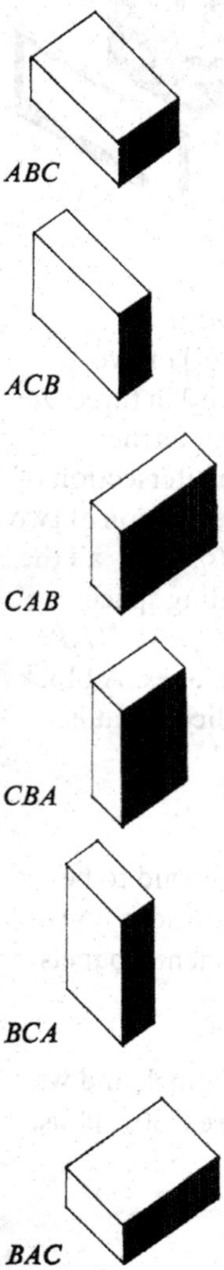

Figure 6.31

Figure 6.32

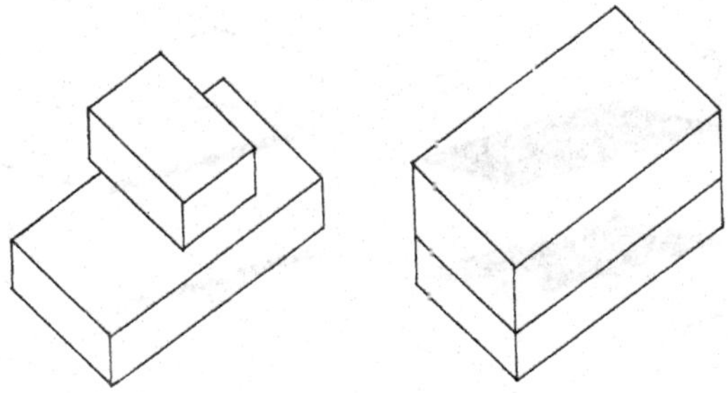

Like panels, the union of two boxes is not necessarily a third (Figure 6.32), but their intersection is. Indeed, one of the fundamental theorems of convex sets states that the intersection of any two convex sets is also convex. The conditions under which two blocks may be aligned or conjoined can be expressed by an ordered set of three 2×2 conjunction matrices from the set E shown on page 150. There are six elements in E, and the cartesian product E^3 gives us all the ordered sets of three, or *conjoints*, that are possible. We see that there are $6^3 = 216$ ways in which two blocks may be aligned or conjoined in space. The only situations in which two blocks may be joined to form a third is when they are in *extrusion* (Figure 6.33), that is to say, when at least two out of the three elements in their conjoint are the identity matrix **I**. There are three positions in the conjoint for the remaining element, and that can take one of four values – $\mathbf{E}_{11}$, $\mathbf{E}_{22}$, $\mathbf{E}_{12}$, $\mathbf{E}_{21}$ – if the blocks are to nest or stack. Altogether there are thus $3 \times 4 = 12$ ways in which two blocks may nest or stack and one way, of course, in which they fit, **(I, I, I)**. Assembling blocks to form others is no easier than stacking panels, while the same problem of guillotine cutting, which we discussed earlier, exists. The names of the faces, edges and vertices of the block ABC are given by forming the product of the power sets $P(A)$, $P(B)$ and $P(C)$ as we have done before for panels. Recall that $P(A) = \{\emptyset, A, A_1, A_2\}$ where $A_1 = [a_1]$ and $A_2 = [a_2]$ are degenerate components. $P(B)$ and $P(C)$ are similar. These elements are illustrated in Figure 6.34.

Figure 6.33

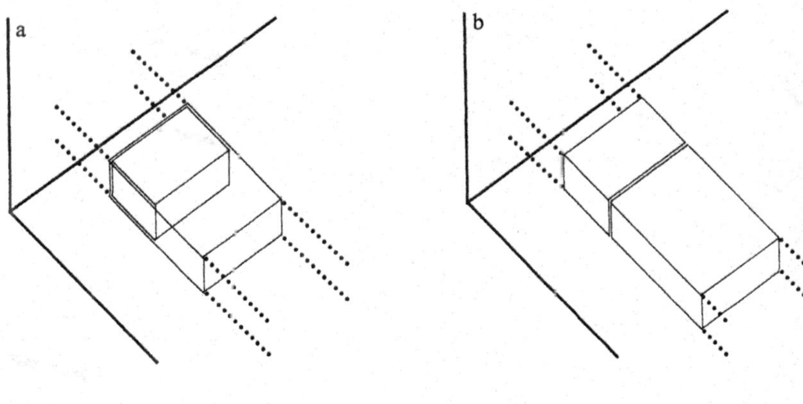

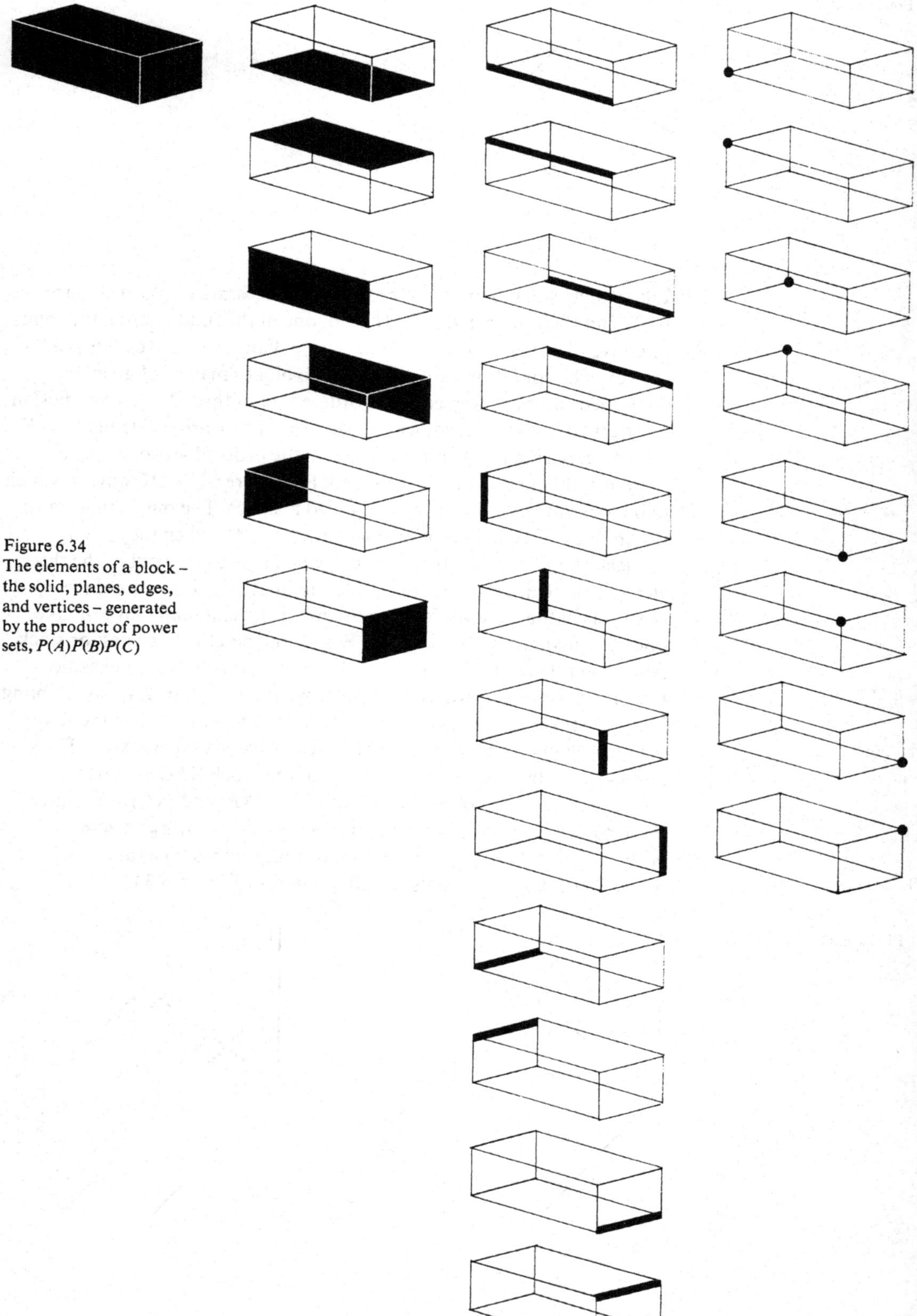

Figure 6.34
The elements of a block – the solid, planes, edges, and vertices – generated by the product of power sets, $P(A)P(B)P(C)$

	Block	Faces	Edges	Vertices
$P(A)P(B)P(C) - \{\emptyset,$	$ABC,$	$ABC_1,$	$AB_1C_1,$	$A_1B_1C_1,$
		$ABC_2,$	$AB_1C_2,$	$A_1B_1C_2,$
		$AB_1C,$	$AB_2C_1,$	$A_1B_2C_1,$
		$AB_2C,$	$AB_2C_2,$	$A_1B_2C_2,$
		$A_1BC,$	$A_1B_1C,$	$A_2B_1C_1,$
		$A_2BC,$	$A_1B_2C,$	$A_2B_1C_2,$
			$A_2B_1C,$	$A_2B_2C_1,$
			$A_2B_2C,$	$A_2B_2C_2\}.$
			$A_1BC_1,$	
			$A_1BC_2,$	
			$A_2BC_1,$	
			$A_2BC_2,$	

We may express a block, its faces, edges and vertices in full, typically, as

$$\text{Block: } K = \begin{bmatrix} x_1, x_2 \\ y_1, y_2 \\ z_1, z_2 \end{bmatrix}.$$

$$\text{Face: } K_f = \begin{bmatrix} x \\ y_1, y_2 \\ z_1, z_2 \end{bmatrix}, \text{ where } x \in \{x_1, x_2\}.$$

$$\text{Edge: } K_e = \begin{bmatrix} x \\ y \\ z_1, z_2 \end{bmatrix}, \text{ where } y \in \{y_1, y_2\}.$$

$$\text{Vertex: } K_v = \begin{bmatrix} x \\ y \\ z \end{bmatrix}, \text{ where } z \in \{z_1, z_2\}.$$

Once again they are seen to have a one-to-one correspondence with cones of vectors from the origin.

The description we have employed of rectangular forms shows a strong pattern: the kind of pattern incorporating symmetry, permutation and combinatorial logic which we hope demonstrates the strong affinities that *structure* in the arts has to mathematics, and mathematics has to the arts. The peculiar strength of mathematical structure allows mathematicians to contemplate geometries in unfamiliar realms and unseen dimensions. For example, we are now in a position to describe the four-dimensional boot-boxes that our quartermaster would require to transport the left-footed boots we referred to in Chapter 2. What do they 'look' like?

We have named a component A in one dimension, a panel AB in two dimensions, a block ABC in three. It seems reasonable to call a four-dimensional block – a *hyperblock* – $ABCD$. If a component in R^1 is defined by 2×1 halflines, a panel in R^2 by $2 \times 2 = 4$ halfplanes, a block in R^3 by $2 \times 3 = 6$ halfspaces, then we would expect a hyperblock in R^4 to be defined by $2 \times 4 = 8$ *half-hyperspaces*.

Table 6.1

	Point	**Component**	**Panel**	**Block**	**Hyperblock**
Space:	R^0	R^1	R^2	R^3	R^4
Vertices P_0:	$\binom{0}{0} 2^0 = 1$	$\binom{1}{1} 2^1 = 2$	$\binom{2}{2} 2^2 = 4$	$\binom{3}{3} 2^3 = 8$	$\binom{4}{4} 2^4 = 16$
Edges P_1:	·	$\binom{1}{0} 2^0 = 1$	$\binom{2}{1} 2^1 = 4$	$\binom{3}{2} 2^2 = 12$	$\binom{4}{3} 2^3 = 32$
Faces P_2:	·	·	$\binom{2}{0} 2^0 = 1$	$\binom{3}{1} 2^1 = 6$	$\binom{4}{2} 2^2 = 48$
Blocks P_3:	·	·	·	$\binom{3}{0} 2^0 = 1$	$\binom{4}{1} 2^1 = 8$
Hyperblocks P_4:	·	·	·	·	$\binom{4}{0} 2^0 = 1$
Total number of elements, $\Sigma_i P_i$:	$3^0 = 1$	$3^1 = 3$	$3^2 = 9$	$3^3 = 27$	$3^4 = 81$

Where $\binom{N}{n} 2^n = \frac{N!}{(N-n)!\,n!} 2^n$ is the number of ways n elements in an ordered arrangement of N terms may be selected and be replaced by one of two distinct elements.

[5] 'La construction rationnelle par cubes ne détruit pas l'initiative de chacun. Il n'y a qu'à en jouer suivant ses goûts.' Thus Le Corbusier describes his system of construction for the houses at Pessac, *Oeuvre Complète 1910–1929*, p. 69.

In R^1 there is $1! = 1$ way a component can be equivalent to A; in R^2 there are $2! = 2 \times 1 = 2$ ways a panel can be equivalent to AB; in R^3, $3! = 3 \times 2 \times 1 = 6$ ways a block can be equivalent to ABC; and, by induction, in R^4 there are $4! = 4 \times 3 \times 2 \times 1 = 24$ ways of equivalence with respect to $ABCD$. How many vertices, edges, faces and blocks (*sic*) does a hyperblock have? The mathematical pattern suggests the answer, and, as before, the elements of the cartesian product $P(A)P(B)P(C)P(D)$ will confirm it (Table 6.1).

This illustrates the algebraic system, and even if we do not concern ourselves with the detailed mathematics, nevertheless we can appreciate the abstract pattern of numbers. For the mathematician and the architect there can be immense aesthetic satisfaction in discovering an underlying pattern to outwardly disordered and unrelated phenomena. For both, the discovery may depend on the invention of a suitable pattern language to make the structure explicit. Good systems will not restrict freedom of thought or action, on the contrary they liberate, thus permitting speculations beyond the immediate circumstances of their invention.[5] Let us return from our escapade in four dimensions to the knobbly world in which we live. On earth – although many of the shapes and forms used in building tend to be rectangular, and the majority of architects accept the well-tried resilience and robustness of this 'cartesian' system – on earth most things are irregular and non-rectangular. We shall now take a look at some of the ways in which these forms may be described.

7 Irregular polygons and convexity

Suppose we want to describe the irregular two-dimensional area shown in Figure 7.1a. There are several ways this may be done, not least by representing it graphically, in the traditional way, as a *similarity*. But here we shall discuss methods of mapping the shape into numerical approximation. The first way we shall discuss depends on dividing the plane into quadrats or squares. If a square includes more than half of the original shape which it covers we let it represent that area; if not we ignore it. Thus in Figure 7.1b only one of the squares satisfies the requirement, and our quadrat approximation to the shape is very crude indeed. But as the quadrats decrease in size better and better approximations are possible (Figure 7.1c). We shall call the ratio of the areas Q of the quadrat and S of the shape the *resolution* of the approximation. It is clear as the ratio approaches zero, $Q/S \rightarrow 0$, that the resolution improves until approximations may be made as close to the original shape as desired. The process is analogous to rounding figures to a given number of decimal places, or mapping lengths into a set of modular dimensions.

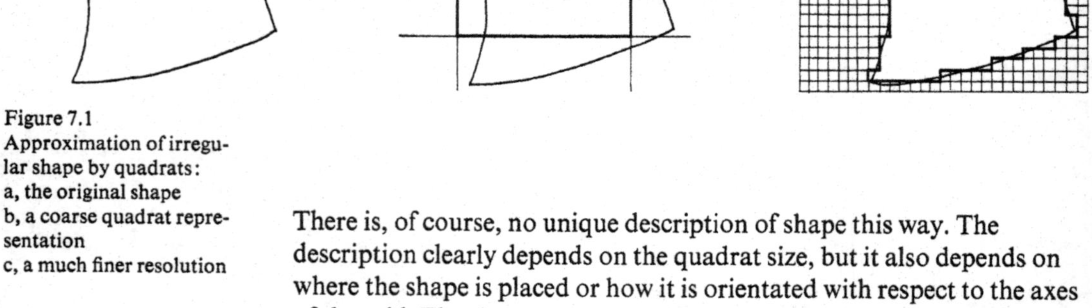

Figure 7.1
Approximation of irregular shape by quadrats:
a, the original shape
b, a coarse quadrat representation
c, a much finer resolution

There is, of course, no unique description of shape this way. The description clearly depends on the quadrat size, but it also depends on where the shape is placed or how it is orientated with respect to the axes of the grid. That is to say, translation and rotation of either the shape or the quadrat axes will change the approximation (Figure 7.2). It is sometimes argued that triangular or hexagonal grids are more neutral than the square, and while this is possibly true the increasing complexity of calculation and numerical description rarely make their use worthwhile.

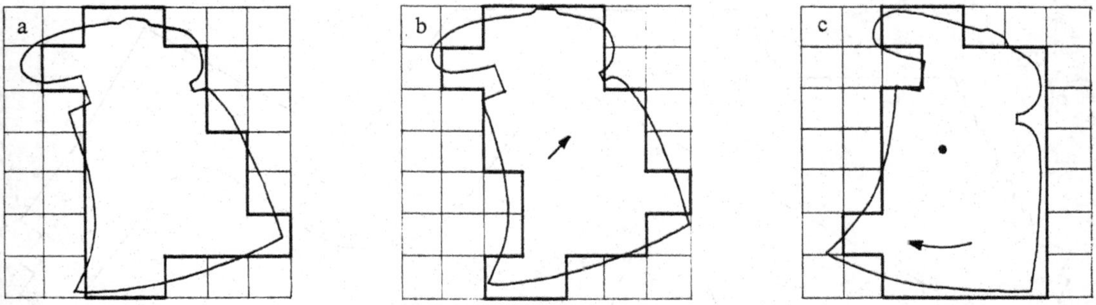

Figure 7.2
Quadrat approximation is not unique and depends on the placing of the grid:
a, original position
b, change of approximation after translation
c, change after rotation

Effectively, this is similar to the way in which a computer image of a shape is output in binary form by a line-printer. Figure 7.3 compares a coarse 7 × 7 zero-one matrix representation of the shape with a line-printer representation to a finer resolution (Figure 7.4). Clearly, as we have previously seen, quadrat representation is at its best when it is used to describe shapes which are both rectangular and modular and where the size and orientation of the grid is suitably chosen. Economic and concise notation may then be employed. With irregular shapes the description is bound to become more cumbersome and mapping into a zero-one matrix may become unwieldy, especially if the number of zeros exceeds the ones.

One way in which the numerical description may be made more terse is by scanning the shape in the manner of the cathode ray raster on a television screen. We move across the shape in a given number of intervals (twenty-five, say) marking the number of the interval at which we 'step' into the shape or out of it. On completing one line we return to the start, move down a line and repeat our search for the boundary points. Using a twenty-five line raster with twenty-five intervals we are able to define our test shape in fifty entries at a resolution twelve times better than given by the forty-nine entries of the 7 × 7 zero-one matrix (Figure 7.5). This raster description is in effect a vector representation. The reconstruction of the picture from the vector is a good example of the modulo arithmetic which we briefly touched on in Chapter 1.

Figure 7.3
Zero-one matrix representation of the original shape

$$\begin{bmatrix} 0 & 0 & 1 & 1 & 0 & 0 & 0 \\ 0 & 1 & 1 & 1 & 1 & 0 & 0 \\ 0 & 0 & 1 & 1 & 1 & 0 & 0 \\ 0 & 0 & 1 & 1 & 1 & 1 & 0 \\ 0 & 0 & 1 & 1 & 1 & 1 & 0 \\ 0 & 0 & 1 & 1 & 1 & 1 & 1 \\ 0 & 0 & 1 & 1 & 0 & 0 & 0 \end{bmatrix}$$

Figure 7.4
A 'line-printer' representation of the shape

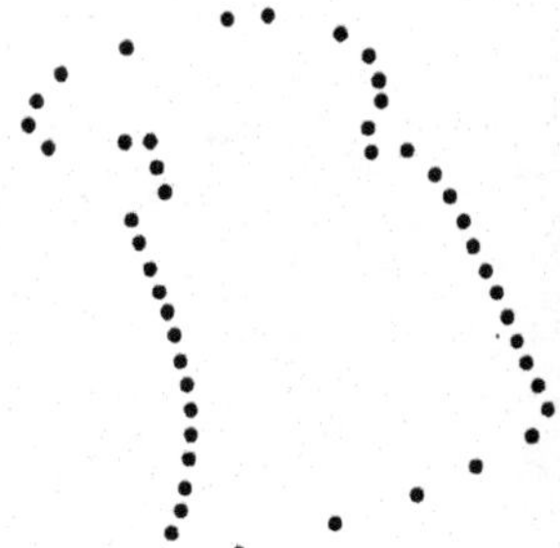

Figure 7.5
Raster representation of the shape

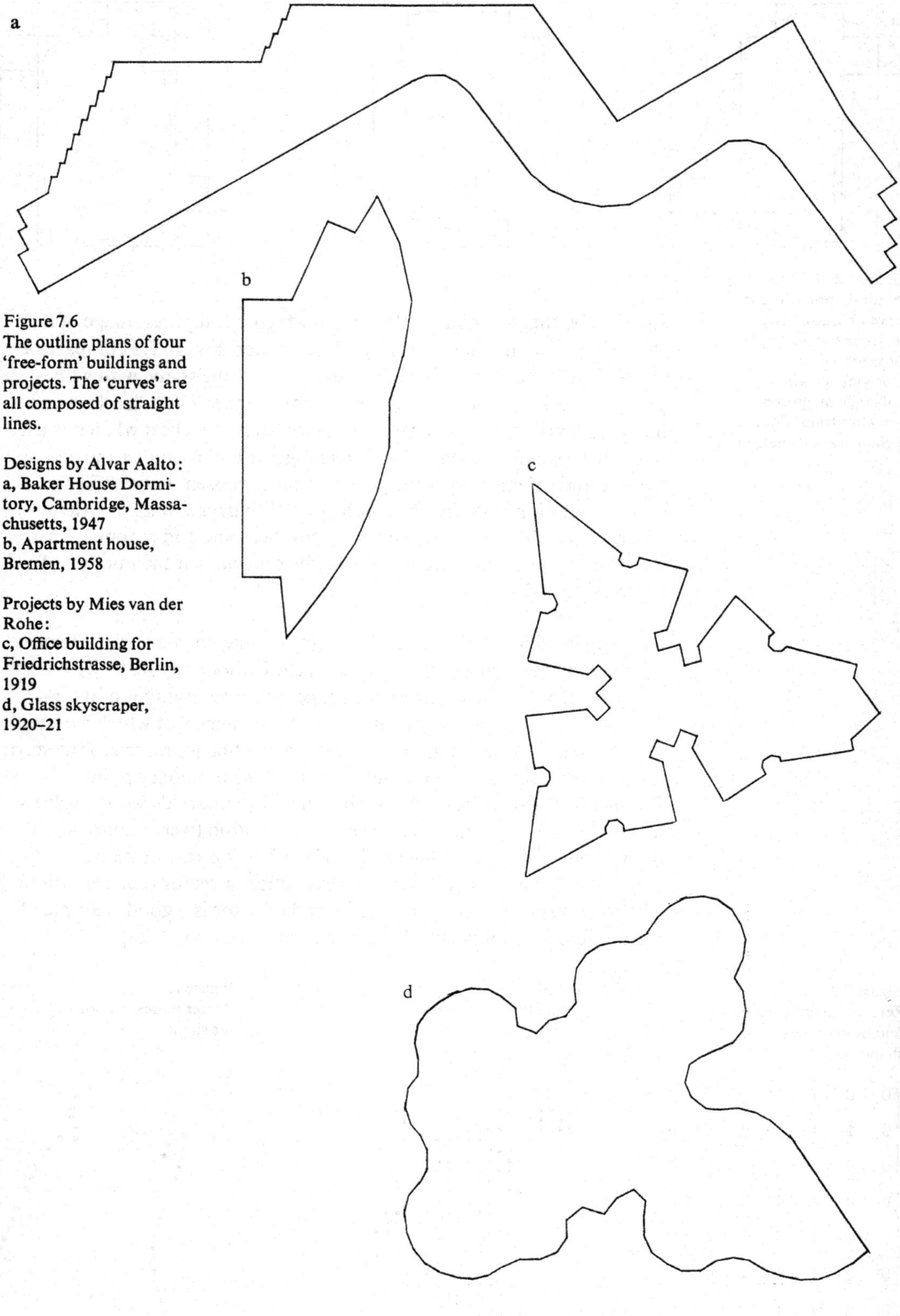

Figure 7.6
The outline plans of four 'free-form' buildings and projects. The 'curves' are all composed of straight lines.

Designs by Alvar Aalto:
a, Baker House Dormitory, Cambridge, Massachusetts, 1947
b, Apartment house, Bremen, 1958

Projects by Mies van der Rohe:
c, Office building for Friedrichstrasse, Berlin, 1919
d, Glass skyscraper, 1920–21

[1] For Aalto, see Frederick Gutheim, *Alvar Aalto*, New York, George Braziller Inc., 1960.

Thus, using a raster of length n, the element r in a vector description represents the two-dimensional point

$$(x,y) = \left(r \text{ modulo } n, \frac{r - r \text{ modulo } n}{n}\right).$$

In the example above, when the scan reaches the raster point 24 ($x = 24$) in the first line ($y = 0$) it moves onto the next point, 25, which is located in two dimensions at the beginning ($x = 25$ modulo $25 = 0$) of the next line ($y = (25 - 0)/25 = 1$). The raster point 30 in the vector is then located in this line at $(x, y) = (5, 1)$ since 30 modulo $25 = 5$ and $(30 - 5)/25 = 1$. The raster point 610 will be represented graphically at $x = 610$ modulo $25 = 10$ and $y = (610 - 10)/25 = 24$, that is at (10, 24). An alternative method of scanning is the boustrophedon, 'ox-turning', of ancient Greek writing where the first line is left to right and the next is right to left and so on.

Another way of describing planar shapes is by linear interpolation. This method takes points on the boundary of the shape and joins them together in a chain of straight lines. In fact, when architects propose irregular plans for their buildings they often compose them from rectilinear elements: the plans are then irregular polygons such as those proposed for two glass skyscrapers by Mies van der Rohe in the twenties, or the plan of a dormitory at the Massachusetts Institute of Technology where the Finnish architect Alvar Aalto[1] creates a long sinuous frontage to the Charles River out of short, but straight, runs of brick wall (Figure 7.6). Such polygons may be precisely described, while the irregular shape we illustrate is approximated more and more accurately by an n-sided polygon as n gets larger. We illustrate two approximations, one with eight sides and one with sixteen (Figure 7.7).

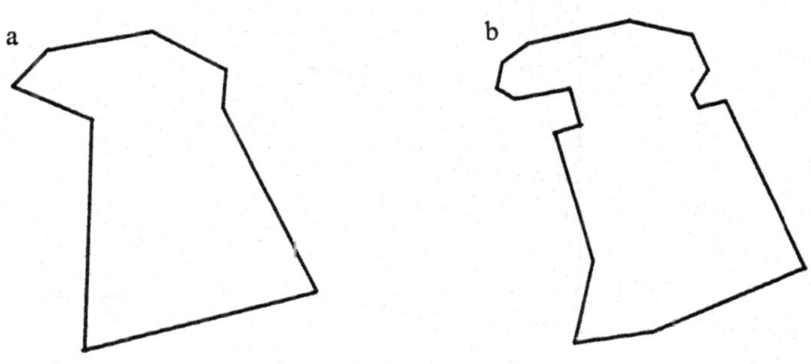

Figure 7.7
Two polygonal approximations to the shape shown in Figure 7.1a:
a, with eight edges
b, with sixteen edges

[2] '*Must* a name mean something?' Alice asked doubtfully. 'Of course it must,' Humpty Dumpty said with a short laugh: 'my name means the shape I am – and a good handsome shape it is too. With a name like yours, you might be any shape, almost.' From *The Annotated Alice*, edited by Martin Gardner, London, Penguin Books, 1965, p. 263.

We can describe an n-gon, Q, by giving it an appropriate mathematical name. One way is to list the position vectors of its vertices in cyclic order in a $2 \times n$ matrix

$$\mathbf{Q} = \begin{bmatrix} x_0 & x_1 & \dots & x_i & \dots & x_{n-2} & x_{n-1} \\ y_0 & y_1 & \dots & y_i & \dots & y_{n-2} & y_{n-1} \end{bmatrix}.$$

Where the suffix i is an integer modulo n, so that $x_n = x_0, y_n = y_0$ (Figure 7.8). Such a name, unfortunately, is certainly dependent on the position of the origin and the orientation of the unit base vectors. A change in these will give Q a new name: indeed such a transformation is known to mathematicians as *alias*[2] for this reason. The polygon has not moved, but its frame of reference has. For example, the triangle whose matrix is

$$\mathbf{T}_1 = \begin{bmatrix} 1 & 1 & 4 \\ 1 & 5 & 1 \end{bmatrix}$$

$$\text{becomes } \mathbf{T}_2 = \begin{bmatrix} 2 & 2 & 5 \\ 3 & 7 & 3 \end{bmatrix}$$

Figure 7.8

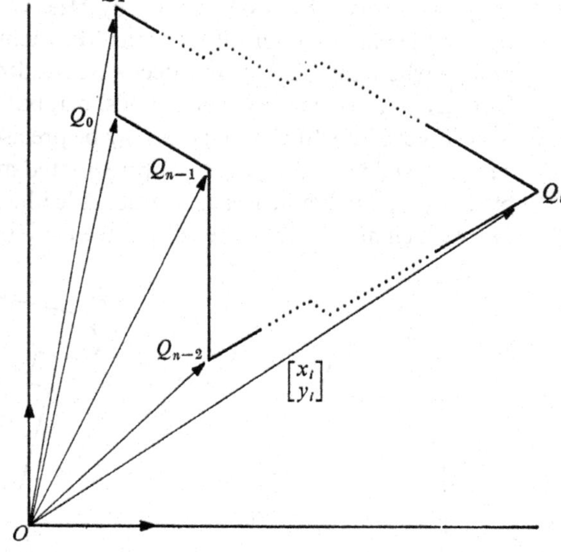

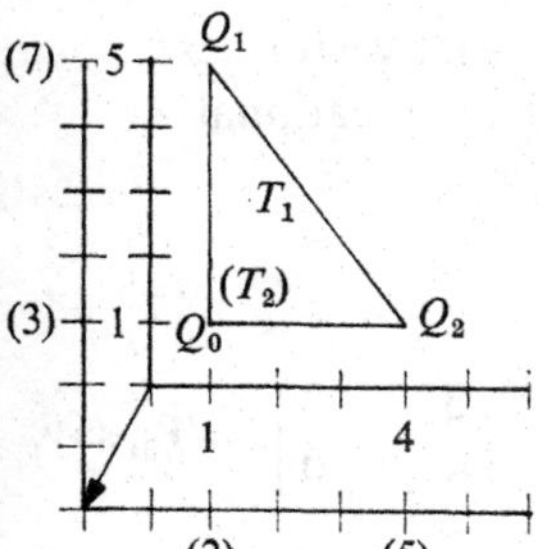

Figure 7.9

when the origin is moved to (− 1, − 2) on the old coordinate reference system (Figure 7.9). The same shape, by this $[x_i \quad y_i]^T$ notation, can be called by an infinite number of different names. Obviously it is desirable to eliminate the influence of extraneous relationships as much as possible. We can remove the arbitrariness of location by describing the shape in terms of a chain of vectors around its sides (Figure 7.10). We note that the vector

$$\begin{bmatrix} u_i \\ v_i \end{bmatrix} = \begin{bmatrix} x_{i+1} \\ y_{i+1} \end{bmatrix} - \begin{bmatrix} x_i \\ y_i \end{bmatrix} = \begin{bmatrix} x_{i+1} - x_i \\ y_{i+1} - y_i \end{bmatrix}$$

is no longer a position vector, but a vector from vertex Q_i to Q_{i+1}; that is, the vector $\overline{Q_i Q_{i+1}}$. We adopt the convention that

$$\mathbf{Q} = \begin{bmatrix} u_0 & u_1 & \dots & u_i & \dots & u_{n-2} & u_{n-1} \\ v_0 & v_1 & \dots & v_i & \dots & v_{n-2} & v_{n-1} \end{bmatrix}$$

represents Q as a chain of vectors. The components u_i and v_i are like the measurements an architect makes on a plan to specify an oblique line to aid a builder set out a plan: the difference here is that our dimensions are *directed*, being the components of a vector. If measurements upwards and to the right are positive, then downwards and to the left are nega-

Figure 7.10

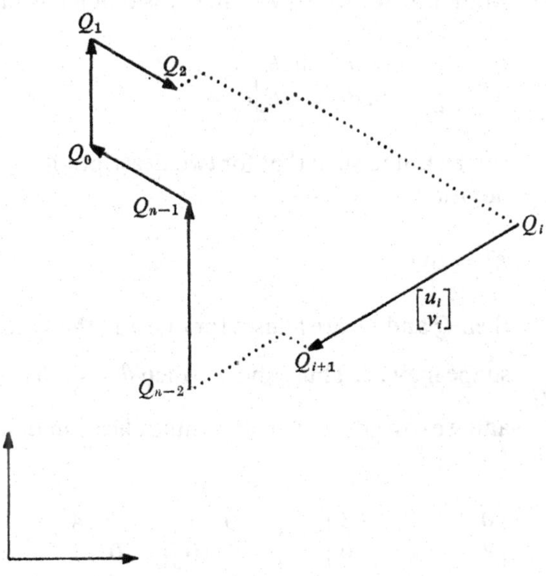

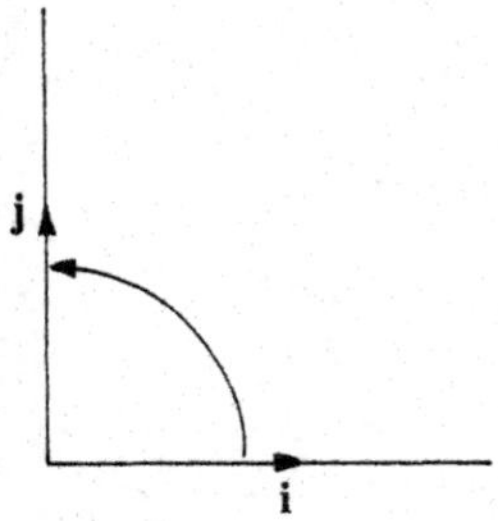

Figure 7.11

tive. In effect this implies the traditional anti-clockwise vector base (Figure 7.11). This method of description frees it from locational changes, thus for example in the case of the triangle:

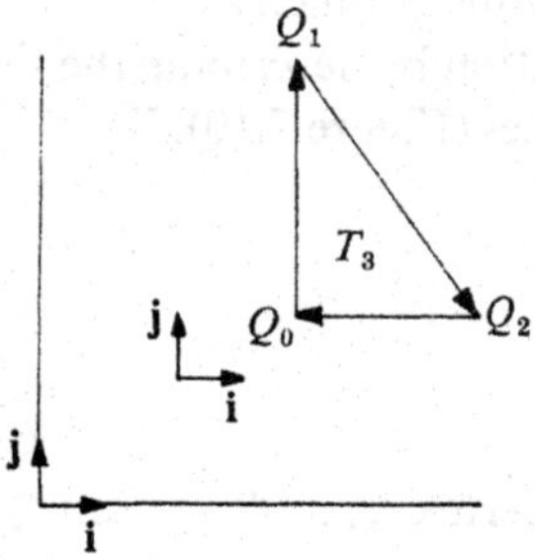

Figure 7.12

$$\left.\begin{matrix} \mathbf{T}_1 \rightarrow \begin{bmatrix} (1-1) & (4-1) & (1-4) \\ (5-1) & (1-5) & (1-1) \end{bmatrix} \\ \text{and} \\ \mathbf{T}_2 \rightarrow \begin{bmatrix} (2-2) & (5-2) & (2-5) \\ (7-3) & (3-7) & (3-3) \end{bmatrix} \end{matrix}\right\} = \begin{bmatrix} 0 & 3 & -3 \\ 4 & -4 & 0 \end{bmatrix} = \mathbf{T}_3, \text{ say.}$$

Note that $\Sigma_i u_i = \Sigma_i v_i = 0$ in this description. While the u, v notation is unaffected by *translation* (Figure 7.12), it is still altered by *rotation* (Figure 7.13) of the vector base. Thus rotation of the vector base, clockwise (Figure 7.14), through a right angle gives the alias

$$\mathbf{T}_4 = \begin{bmatrix} -4 & 4 & 0 \\ 0 & 3 & -3 \end{bmatrix}.$$

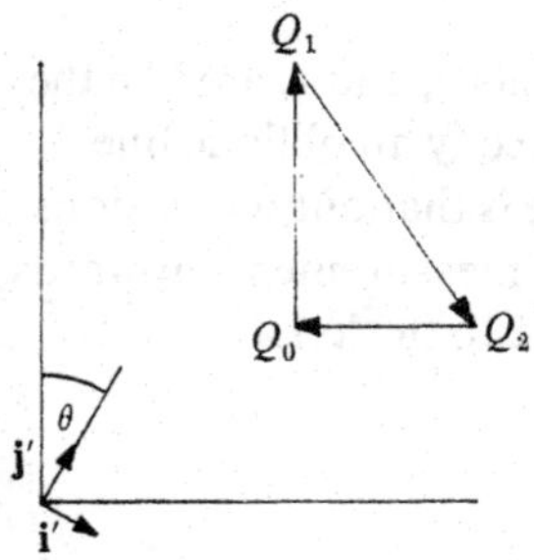

Figure 7.13

Effectively, this is the description we get if we measure positively u_i upwards and v_i right to left, whereas before we had measured u_i left to right and v_i upwards. How do we know that $\mathbf{T}_3$ and $\mathbf{T}_4$ represent the same shape when they look different? The answer, in general, is not that straightforward. All we can say is that if a rotation matrix

$$\mathbf{S}_\theta = \begin{bmatrix} \cos\theta & \sin\theta \\ -\sin\theta & \cos\theta \end{bmatrix}$$

can be found such that for two descriptions Q and Q' – using the u, v notation –

$$\mathbf{Q} = \mathbf{S}\mathbf{Q}'$$

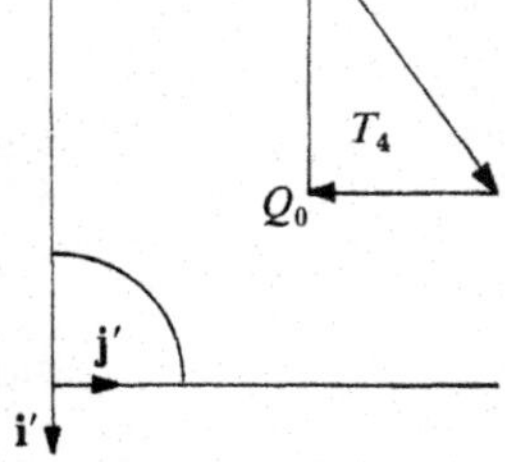

Figure 7.14

then Q and Q' are aliases for one and the same shape, whatever that shape may be. Thus, above, when $\theta = \frac{\pi}{2}$, then $\cos\theta = 0$ and $\sin\theta = 1$, and we can show by matrix multiplication that $\mathbf{T}_3 = \mathbf{S}_{\frac{\pi}{2}}\mathbf{T}_4$:

$$\begin{bmatrix} 0 & 3 & -3 \\ 4 & -4 & 0 \end{bmatrix} = \begin{bmatrix} 0 & 1 \\ -1 & 0 \end{bmatrix} \begin{bmatrix} -4 & 4 & 0 \\ 0 & 3 & -3 \end{bmatrix}.$$

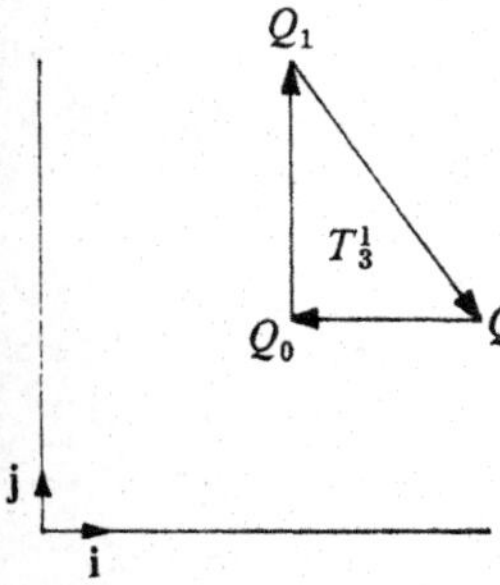

Figure 7.15a

But there are yet more disguises. It is clear that the vertex we choose to start with can be any one out of n in the general case of the n-gon. This is the same as saying that the columns of the $2 \times n$ matrix may be cyclically permuted and it will still represent the same shape (Figure 7.15). For example, using the zero-one cyclic permutation matrices which we discussed in Chapter 4 we see that our triangle is equally well defined by

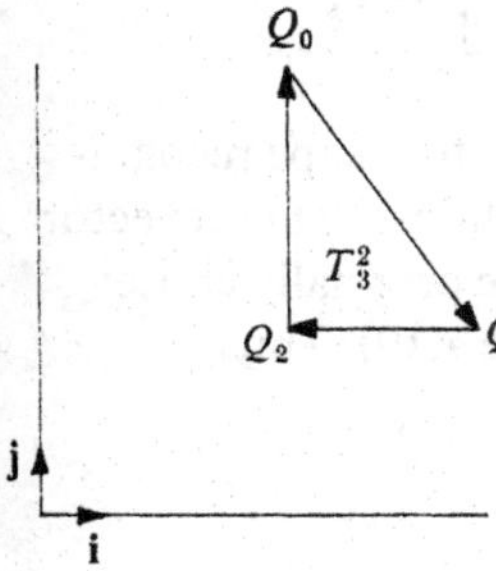

Figure 7.15b

$$\mathbf{T}_3 \mathbf{P}_{123} = \begin{bmatrix} 0 & 3 & -3 \\ 4 & -4 & 0 \end{bmatrix} = \mathbf{T}_3^1$$

$$\mathbf{T}_3 \mathbf{P}_{231} = \begin{bmatrix} 3 & -3 & 0 \\ -4 & 0 & 4 \end{bmatrix} = \mathbf{T}_3^2$$

$$\mathbf{T}_3 \mathbf{P}_{312} = \begin{bmatrix} -3 & 0 & 3 \\ 0 & 4 & -4 \end{bmatrix} = \mathbf{T}_3^3$$

$$\text{where } \mathbf{P}_{123} = \begin{bmatrix} 1 & 0 & 0 \\ 0 & 1 & 0 \\ 0 & 0 & 1 \end{bmatrix}, \mathbf{P}_{231} = \begin{bmatrix} 0 & 0 & 1 \\ 1 & 0 & 0 \\ 0 & 1 & 0 \end{bmatrix}, \mathbf{P}_{312} = \begin{bmatrix} 0 & 1 & 0 \\ 0 & 0 & 1 \\ 1 & 0 & 0 \end{bmatrix}.$$

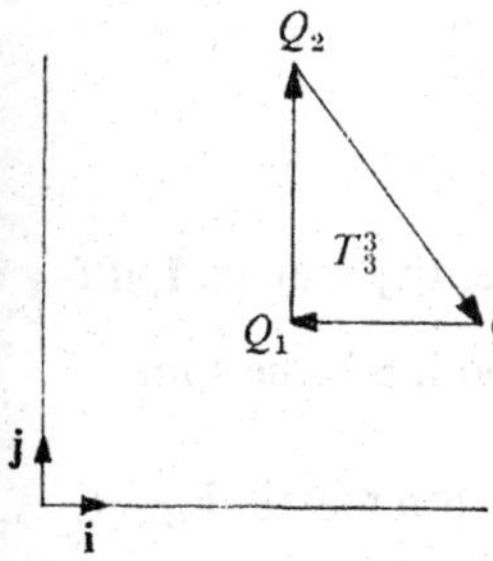

Figure 7.15c

So far we have been numbering the vectors representing the sides in a clockwise direction. If we had numbered them in the anti-clockwise direction, (Figure 7.16), our triangle would have acquired the alias

$$\mathbf{T}_5 = \begin{bmatrix} 3 & -3 & 0 \\ 0 & 4 & -4 \end{bmatrix}.$$

The zero-one permutation matrix $\mathbf{P}_{321}$ gives

$$\mathbf{T}_5 \mathbf{P}_{321} = \begin{bmatrix} 3 & -3 & 0 \\ 0 & 4 & -4 \end{bmatrix} \begin{bmatrix} 0 & 0 & 1 \\ 0 & 1 & 0 \\ 1 & 0 & 0 \end{bmatrix} = \begin{bmatrix} 0 & -3 & 3 \\ -4 & 4 & 0 \end{bmatrix} = -\mathbf{T}_3.$$

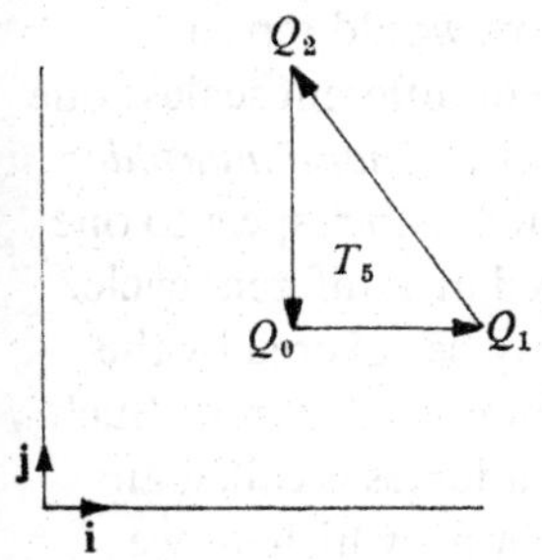

Figure 7.16

The minus sign may be considered to be the result of pre-multiplication by the 2×2 rotation matrix $\mathbf{S}_\pi$ since $\mathbf{S}_\pi = -\mathbf{I}$, thus

$$\mathbf{S}_\pi \mathbf{T}_3 = \begin{bmatrix} -1 & 0 \\ 0 & -1 \end{bmatrix} \begin{bmatrix} 0 & 3 & -3 \\ 4 & -4 & 0 \end{bmatrix} = \begin{bmatrix} 0 & -3 & 3 \\ -4 & 4 & 0 \end{bmatrix} = -\mathbf{T}_3.$$

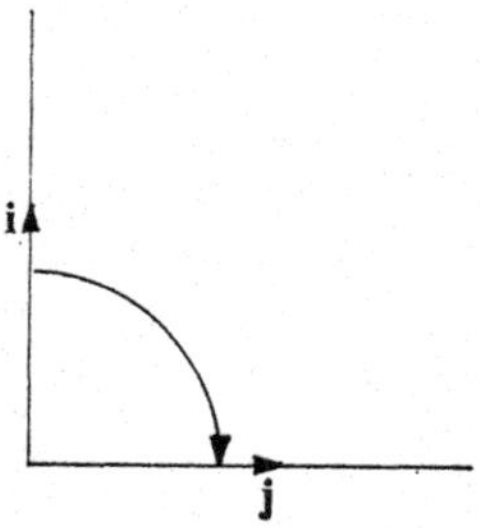

Figure 7.17

If we choose a clockwise vector base (Figure 7.17) another kind of alias arises (Figure 7.18): reflecting the base in $y = x$ changes $\mathbf{T}_3$ to

$$\mathbf{RT}_3 = \begin{bmatrix} 0 & 1 \\ 1 & 0 \end{bmatrix} \begin{bmatrix} 0 & 3 & -3 \\ 4 & -4 & 0 \end{bmatrix} = \begin{bmatrix} 4 & -4 & 0 \\ 0 & 3 & -3 \end{bmatrix} = \mathbf{T}_6.$$

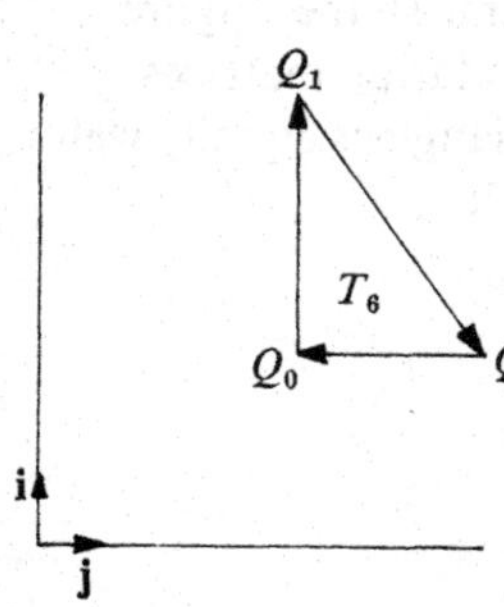

Figure 7.18

There is yet one more disguise our shape may affect to escape recognition. Its name depends on the unit of measure we choose. Thus a vector base with unit vectors half the length of the ones we originally chose will cause the terms of the matrix to double (Figure 7.19). Thus

$$\mathbf{T}_7 = \begin{bmatrix} 0 & 6 & -6 \\ 8 & -8 & 0 \end{bmatrix}$$

also represents the same shape.

Going back to the general n-gon, we see that two aliases in u, v notation $\mathbf{Q}$ and $\mathbf{Q}'$ must be related by an equation of the form

$$\mathbf{Q} = m\,\mathbf{R}^i\mathbf{SQ}'\mathbf{P}$$

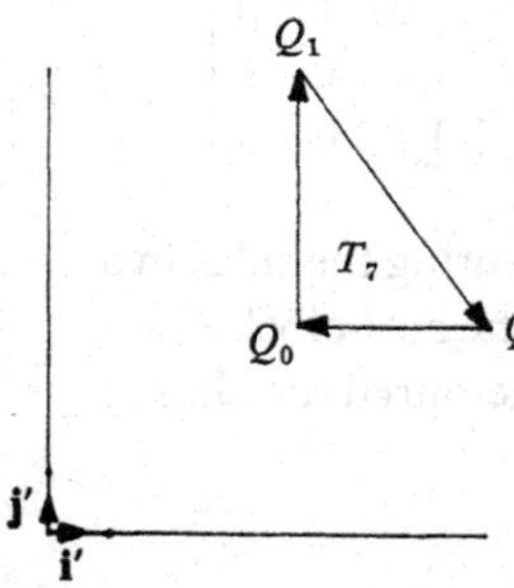

Figure 7.19

where m is a scaling factor, $\mathbf{R}^i$ is either the 2×2 identity matrix, $\mathbf{I}$, if i is even, or the reflection matrix $\mathbf{R} = \begin{bmatrix} 0 & 1 \\ 1 & 0 \end{bmatrix}$ if i is odd, $\mathbf{S}$ is the spin matrix $\begin{bmatrix} \cos\theta & \sin\theta \\ -\sin\theta & \cos\theta \end{bmatrix}$, and $\mathbf{P}$ is a $n \times n$ permutation matrix $\mathbf{P}_{ij\cdots k}$ where $(i, j, \ldots, k)$ is a cyclic permutation of $(1, 2, \ldots, n)$ or $(n, n-1, \ldots, 1)$.

This demonstrates the problems one can get into when attempting to describe shape numerically. Of course, if we were comparing two shapes to see whether they were similar the same problems would repeat themselves (Figure 7.20). The two shapes may be of different scales: one may be a *dilation* of the other. The two shapes may be *enantiomorphic*: one the reflection of the other. They may be rotated with respect to one another. And finally the vertices may be numbered in a different cycle. If the two shapes are isometric then one shape may be covered by the other by moving its position in the plane, or changing its address. Such a transformation which carries the object to a new address is called an *alibi*. Strictly speaking we must refer to the reflection – which, as we

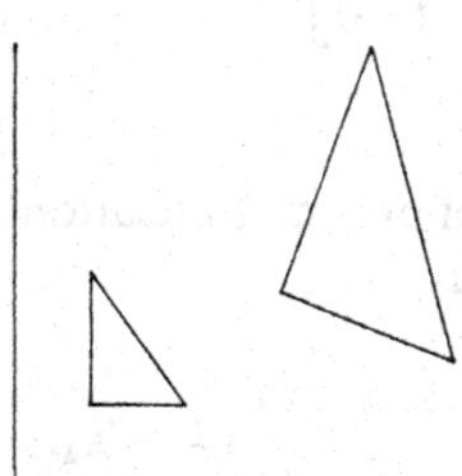

Figure 7.20

pointed out in Chapter 2, requires movement outside the plane – as an *improper alibi*. Looked at this way the equation

$$\mathbf{Q} = m\,\mathbf{R}^i\,\mathbf{S}\,\mathbf{Q}'\,\mathbf{P}$$

- translation : if $m = 1, \mathbf{R}^i = \mathbf{I}, \mathbf{S} = \mathbf{I}$
- proper isometry : if $m = 1, \mathbf{R}^i = \mathbf{I}, \mathbf{S} \neq \mathbf{I}$
- improper isometry: if $m = 1, \mathbf{R}^i \neq \mathbf{I}$
- general similarity : if $m \neq 1$

expresses the variety of transformations which are permissible between two similar shapes in a plane.

One interesting use of the u, v notation is in determining the area of any irregular polygon. The area of a polygon Q is given by $\mathbf{u}^T\mathbf{A}\mathbf{v}$ where $\mathbf{u} = [u_i]$, $\mathbf{v} = [v_i]$ and $\mathbf{A} = [a_{ij}]$ such that

$$a_{ij} = \begin{cases} 0 \text{ if } j > i \\ \frac{1}{2} \text{ if } j = i \\ 1 \text{ if } j < i. \end{cases}$$

For example, the area of the triangle $\mathbf{T}_3 = \begin{bmatrix} 0 & 3 & -3 \\ 4 & -4 & 0 \end{bmatrix}$ is given by

$$\begin{aligned} &[0 \quad 3 \quad -3]\begin{bmatrix} \frac{1}{2} & 0 & 0 \\ 1 & \frac{1}{2} & 0 \\ 1 & 1 & \frac{1}{2} \end{bmatrix}\begin{bmatrix} 4 \\ -4 \\ 0 \end{bmatrix} \\ &= [0 \quad 3 \quad -3]\begin{bmatrix} 2 \\ 2 \\ 0 \end{bmatrix} \\ &= 6. \end{aligned}$$

We may check this against the rule that the area is half the height times base. In this case the triangle has a height 3 and base 4, and the expression above is seen to provide the correct answer. In the case of shapes with holes in, building plans, for example, with light-wells and courts, the shape may be 'cut' into and providing the direction of numbering

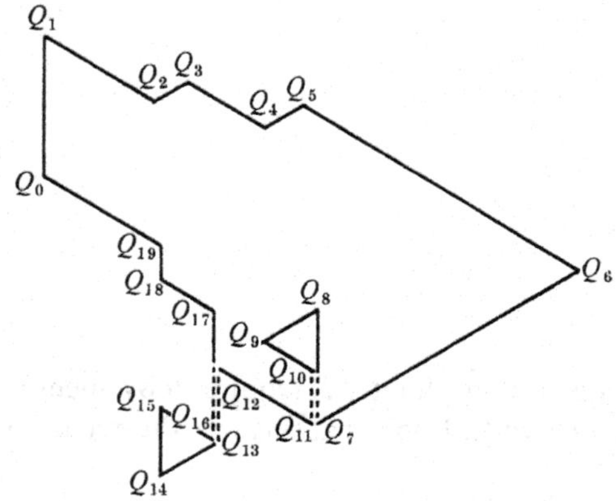

Figure 7.21

Figure 7.22

remains in the same sense, the area will be calculated for the shape less its holes (Figure 7.21). A problem arises if the polygon crosses itself. A bow-tie, for example, will have zero area calculated this way (Figure 7.22). This is because the convention gives *direction* to the areas. A shape circumscribed in a clockwise direction will have a positive area, and a shape described in an anti-clockwise cycle will have a negative area. The answer to the bow-tie problem is not to cross over lines: then a definite area will be found (Figure 7.23).

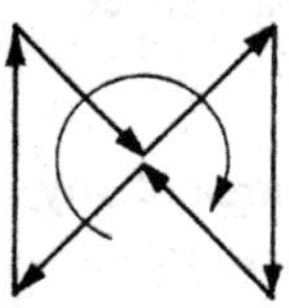

Figure 7.23

We have pursued this exercise partly to demonstrate how cumbersome it can be to recognize shape when it is expressed in algebraic terms. While the same shape may have any number of names, each alias or alibi is sufficient for us to be able to reproduce its form. Further, the descriptions allow us to calculate other measures apart from area. The perimeter, for example, of an n-gon described in u, v notation is given by the sum of the length of its sides, ${}_{+}(u_i^2 + v_i^2)^{\frac{1}{2}}$. If $\mathbf{q}_i = [u_i \quad v_i]^T$ is the vector $\overline{Q_iQ_{i+1}}$ then we write

$$\| \mathbf{q}_i \|_2 = {}_{+}(u_i^2 + v_i^2)^{\frac{1}{2}}$$

and the perimeter q, of Q, is given by

$$q = \Sigma_i \| \mathbf{q}_i \|_2$$

where $\| \mathbf{q}_i \|_2$ is known as the *euclidean norm* of the column vector $\mathbf{q}_i$. Another measure of the circumference is given by the *taxicab norm*, so named because it represents the rectangular distance travelled from point to point by taxicabs along the grid-iron roads of the United States. This norm is given by $\| \mathbf{q}_i \|_1 = (| u_i | + | v_i |)$ where $| u_i |$ and $| v_i |$ are the absolute, or positive, values of the components. The *rectangular perimeter*

$$q_r = \Sigma_i \| \mathbf{q}_i \|_1$$

is the ultimate perimeter achieved by a quadrat approximation of Q as

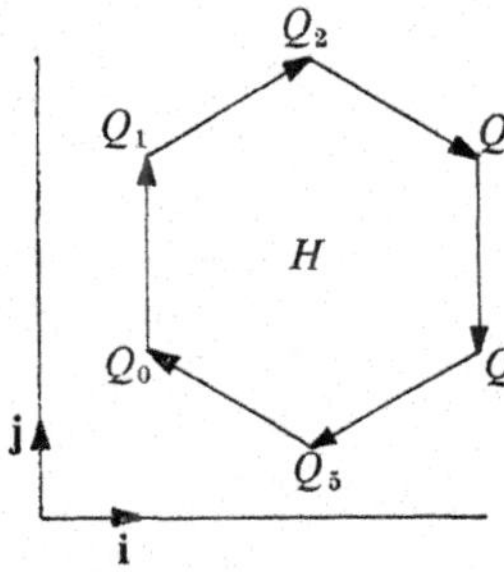

Figure 7.24

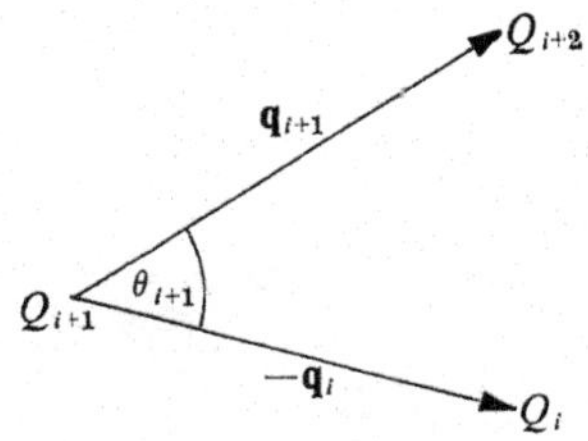

Figure 7.25

the resolution tends to zero. The euclidean norm measures airline distance between vertices, the taxicab norm provides the rectangular.

As we have already mentioned, another useful calculation is to be able to locate the centroid of an area. The *centroid*, (X, Y) of an irregular n-gon described in x, y notation is given by

$$\mathrm{X} = \frac{\Sigma_i (x_i + x_{i+1})(x_{i+1} y_i - x_i y_{i+1})}{3\Sigma_i (x_{i+1} y_i - x_i y_{i+1})}$$

$$\mathrm{Y} = \frac{\Sigma_i (y_i + y_{i+1})(x_{i+1} y_i - x_i y_{i+1})}{3\Sigma_i (x_{i+1} y_i - x_i y_{i+1})}$$

where $\frac{1}{2}\Sigma_i (x_{i+1} y_i - x_i y_{i+1})$ is the area of the n-gon. This formulation is applicable, with suitable 'cutting' or 'tying', to multiply connected regions.

One of the unfortunate characteristics of both the x, y notation and the u, v notation is the way it inevitably disguises regularity. The hexagon, in u, v notation,

$$\mathbf{H} = \begin{bmatrix} 0 & \sqrt{3} & \sqrt{3} & 0 & -\sqrt{3} & -\sqrt{3} \\ 2 & 1 & -1 & -2 & -1 & 1 \end{bmatrix}$$

exhibits some regularities (Figure 7.24), but it is only when we measure the lengths of its sides and find them equal to 2 do we begin to appreciate just how regular it is. To find out more we need to measure the angles at each vertex. These are provided by the inner products of the vectors representing adjacent sides which *radiate* from the vertices, namely the vector pairs $-\mathbf{q}_i$ and $\mathbf{q}_{i+1}$ (Figure 7.25). Thus

$$-\mathbf{q}_i \cdot \mathbf{q}_{i+1} = -(u_i u_{i+1} + v_i v_{i+1}) = -\|\mathbf{q}_i\|_2 \|\mathbf{q}_{i+1}\|_2 \cos\theta_{i+1}$$

where the angle θ_{i+1} lies between $-\mathbf{q}_i$ and $\mathbf{q}_{i+1}$, that is to say θ_{i+1} is the angle $\angle Q_i Q_{i+1} Q_{i+2}$. In the case of the hexagon we have

$$\cos\theta_1 = -\frac{(u_0u_1 + v_0v_1)}{\|\mathbf{q}_0\|_2 \|\mathbf{q}_1\|_2}$$

$$= -\frac{(0\,.\,\sqrt{3} + 2\,.\,1)}{4}$$

$$= -\tfrac{1}{2}$$

where θ_1 is seen to be $2\pi/3$ or 120°. We find in the same way that the remaining angles are also 120° so that H is a perfectly regular hexagon. This suggests an alternative description of an n-gon in terms of the lengths of its sides and its angles. We shall take angles which do not bend back on themselves so that $0 \leqslant \theta_i \leqslant \pi$. In the case of polygons with re-entrant vertices the angle measured is thus the external one

Figure 7.26

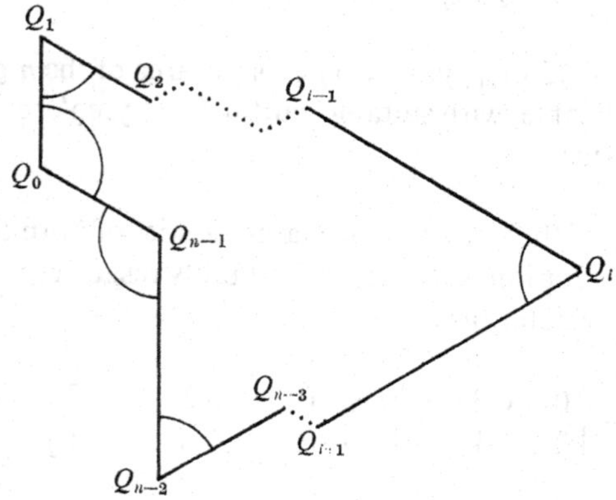

(Figure 7.26). This r, θ description takes the form, once again, of a $2 \times n$ matrix for an n-gon Q.

$$\mathbf{Q} = \begin{bmatrix} r_0 & r_1 & \dots & r_i & \dots & r_{n-2} & r_{n-1} \\ \theta_0 & \theta_1 & \dots & \theta_i & \dots & \theta_{n-2} & \theta_{n-1} \end{bmatrix}.$$

The top row sums directly to give us the perimeter, while the bottom row sums to $r\pi$ where $r \leqslant n - 2$ and the equality holds if the n-gon is convex (has no re-entrant angles). The relation between this r, θ description and the u, v description has already been given above in discussing the dot product $\mathbf{q}_i \,.\, \mathbf{q}_{i+1}$. The conversion from one to another is a trivial matter in computing terms. One additional refinement may be intro-

[3] Peter Haggett and Richard J. Chorley. *Network Analysis in Geography*, London, Edward Arnold, 1969, pp. 70–3.

[4] William Bunge. *Theoretical Geography*, Lund, C. W. K. Gleerup, 1966, pp. 72–88.

duced for the r, θ description: the angles θ_i are invariant under similarity, but the lengths r_i have real dimensions and change according to *dilation* of the figure, or choice of units for measurement. By dividing each length by the perimeter we turn the top row also into ratios, s_i, so that $\Sigma_i s_i = 1$. The s, θ notation, in this *normalized* form, provides a name for a shape which is totally self-contained and unaffected by external reference systems. Unfortunately this is only true for convex polygons, for unless the internal and external angles are clearly identified in non-convex cases, the s, θ notation is ambiguous. The reader is invited to show that different non-convex polygons can share the same s, θ name. The only aliases are due to cyclic permutation of the vertices within the figure itself. Hence two n-gons Q and Q' are similar if, in s, θ notation,

$$\mathbf{Q} = \mathbf{Q}'\mathbf{P}$$

where $\mathbf{P}$ is an $n \times n$ cyclic permutation matrix. The hexagon H in this notation is

$$\mathbf{H} = \begin{bmatrix} 1 & 1 & 1 & 1 & 1 & 1 \\ \frac{2\pi}{3} & \frac{2\pi}{3} & \frac{2\pi}{3} & \frac{2\pi}{3} & \frac{2\pi}{3} & \frac{2\pi}{3} \end{bmatrix}$$

which expresses completely the symmetry of its form.

In recent years considerable interest has been shown in describing shape by geographers. They are interested in quantifying and classifying the shapes of territories. In particular, urban studies lead to the measurement of overlapping areas such as enumeration districts for census purposes, traffic zones used for transportation surveys, and special regions defined on ecological grounds. Mostly geographers use ratios containing values for the area, perimeter, and axes of the unit compared with basic geometric forms such as the circle or ellipse. Much of this work is summarized in Haggett and Chorley's *Network Analysis in Geography*[3] where reference is made to an excellent account of the difficulties of finding a suitable *measure of shape* by William Bunge.[4] In introducing the subject Bunge reminds us that 'shape has never been measured'. 'The problem,' he says, 'is to invent a measure of shape which (1) does not include less than shape, as do the measures used by geomorphologists; (2) does not include more than shape, as do the

[5] See a special study by the Building Performance Research Unit, University of Strathclyde, 'Building Appraisal; St Michael's Academy Kilwinning', *The Architect's Journal*, 7 January, 1970, p. 26.

measures used by mathematicians; (3) is objective; and (4) does not do violence to our intuitive notion of what constitutes shape.' Bunge proposes a set of indices which compare the shape with equal-sided polygons. Haggett and Chorley refer to more complex measures involving the determination of the centroid and the length of radial lines from this centre (Figure 7.27). In our terms these radial lengths are given by

$$\| \mathbf{t}_i \|_2 = ((x_i - \mathrm{X})^2 + (y_i - \mathrm{Y})^2)^{\frac{1}{2}}.$$

Figure 7.27
A typical radial axis of an *n*-gon from its centroid to a vertex

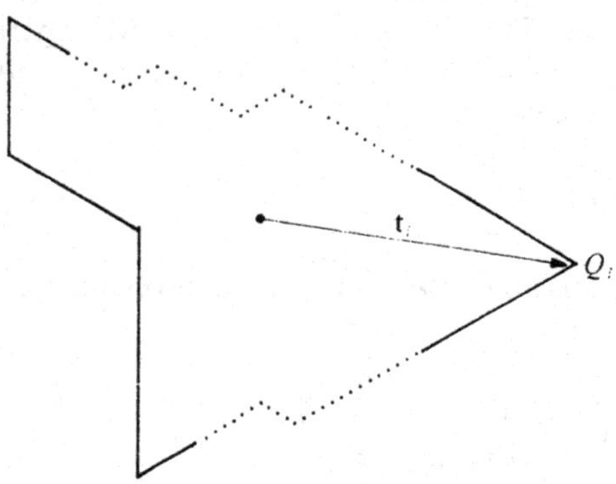

They also tabulate some of the most frequently used measures (Table 7.1), including one by Blair and Bliss designed to overcome problems of 'fragmented' and 'punctured' units.

Recently use has been made of compactness measures in the evaluation of school building. In general, it has been assumed by architects that compactness of plan relates 'somehow to convenience in circulation, lengths of service runs, amount of external walling and a number of other factors affecting cost'. The plan compactness measure devised by Markus *et al.*[5] compares the perimeter of a circle of area equal to the total floor space, with the perimeter taken floor by floor of the building. This ratio is similar to Horton's form ratio in geographical studies. But Markus has extended the notion to compactness in volume which 'probably relates to such variables as heat loss, cost of providing external skin and maintenance. [The volume ratio] can be measured by comparing the area of a curved surface of a hemisphere of volume equal to the volume of the building with the area of external skin of the building (walls and roof)'.

Table 7.1
Ratios for the comparison of the shape of closed figures

Index	Formula	Author
Form ratio:	$\frac{4A}{\pi a^2}$	Horton (1932), Haggett (1965)
Circularity ratio:	$\frac{4\pi A}{p^2}$	Miller (1953)
Elongation ratio:	$\frac{2}{a}\sqrt{\frac{A}{\pi}}$	Schumm (1956)
Radial-line ratio:	$\Sigma_i \left(t_i - \frac{1}{n}\right)$	Boyce and Clark (1964)
Ellipticity ratio:	$\frac{4A}{\pi ab}$	Stoddart (1965)
Compactness ratio:	$\frac{A}{(2\pi \int_A t^2 dA)^{\frac{1}{2}}}$	Blair and Bliss (1967)

Variables:

A Area
p Perimeter
a Diameter of minor axes
b Diameter of major axes
n Number of vertices
t_i Normalized radial axes from centroid to vertices
t Radial axes from centroid to small area dA.

This table has been adapted from Haggett and Chorley where full references are to be found. All ratios attain the value 1 for a circle.

[6] Judith Hilditch. 'An Application of Graph Theory in Problem Recognition', in *Machine Intelligence 3*, edited by Donald Michie, Edinburgh University Press, 1968, pp. 325–47.

[7] A straightforward and well illustrated account is given by L. A. Lyusternik, *Convex Figures and Polyhedra*, New York, Dover, 1963. See also Russell V. Benson, *Euclidean Geometry and Convexity*, New York, McGraw-Hill, 1966.

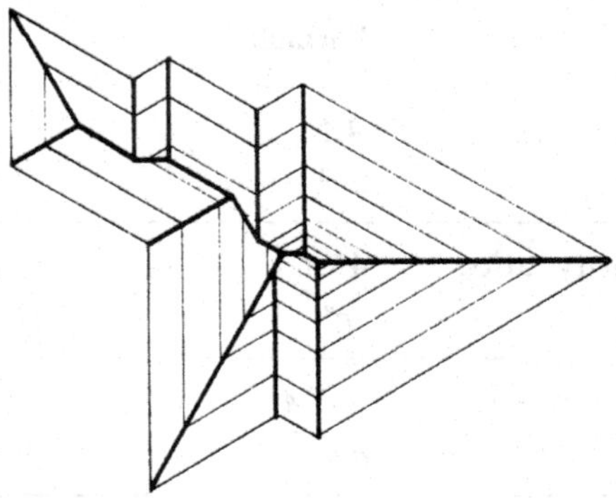

Figure 7.28
The 'skeleton' of an *n*-gon derived by the medial axis transformation

Another field where pattern recognition has become a central subject of research is machine intelligence. Judith Hilditch has described an application of graph theory in the recognition of chromosome types. We discuss some aspects of graph theory in Chapter 10, but here we may note an interesting method of reducing shape to its essential *skeleton*. This transformation was originally proposed by Blum and was named by him the 'Medial Axis Transform' (Figure 7.28).
To quote Hilditch:

> 'This produces a line drawing from a plane figure of any shape. It can probably be most simply described by the grass fire analogy – if we consider the boundary of some shape drawn out on a uniform dry grass field, and then the grass at all points on the boundary set alight at the same moment, the front of the fire will move away from the boundary at a steady rate until at certain points different parts of the front will meet. The medial axis is the locus of points at which this occurs. This set of points, plus a function giving for each point the time at which the fronts met, completely defines the original shape. This locus of points forms a type of "skeleton" for the picture.'[6]

The reader will recognize that the fire front represents the internal boundary of the *circular neighbourhood* of the figure's original perimeter.

We have had occasion to refer to *convex sets* and *convexity*. No account of the description of shape would be complete without an introduction to this important subject. Convexity now plays a fundamental role in the treatment of many geometrical studies.[7] If we denote a line segment joining A and B (excluding the points A and B themselves) by $S(A, B)$ then a *convex set* is a set of points K such that $A \in K$ and $B \in K$ implies

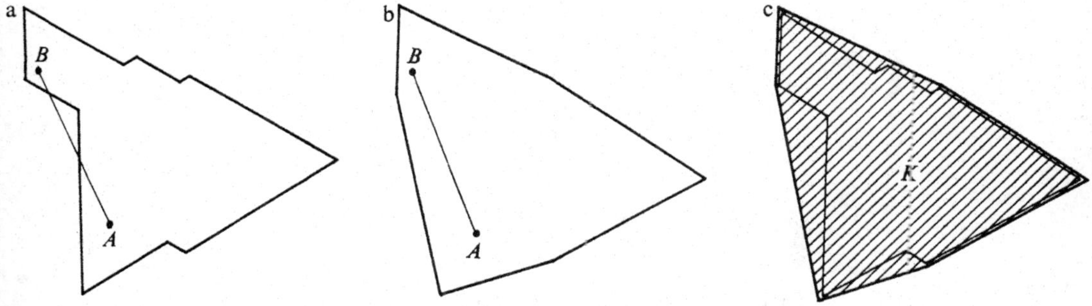

Figure 7.29

$S(A, B) \subset K$, that is to say that if A and B are in the convex set then so must be all the points that lie on the straight line between them (Figure 7.29b). As we have mentioned earlier, components, panels and blocks are examples of convex sets of points, but the irregular shape we have been using is not (Figure 7.29a).

A convex n-gon in two-dimensions may be defined as the intersection of n *halfplanes*, but now – unlike our previous usage – we allow the halfplanes to take up non-rectangular positions. In three dimensions a convex polyhedron is defined by the intersection of *halfspaces*. Halfspaces may be thought of as all the points to one side of a plane, and none on the other: such a plan divides space into two halves. We may use the vector dot product to neatly define a halfplane in R^2 and halfspace in R^3

$$H = \{\mathbf{x} \mid \mathbf{x} . \mathbf{h} \leqslant 0\}$$

where $\mathbf{x} = [x \quad y]^T$ in R^2 and $[x \quad y \quad z]^T$ in R^3. Effectively, what this definition says is that any vector $\mathbf{x}$ satisfying the condition will make an angle with the vector $\mathbf{h}$ such that its cosine will be negative.

One of the most important, and certainly one of the most elegant, theorems on convexity is due to the French mathematician Carathéodory. If we have a closed bounded set of points $\mathbf{x}_i$ then there exists a *convex cover* K (Figure 7.29*c*) given by

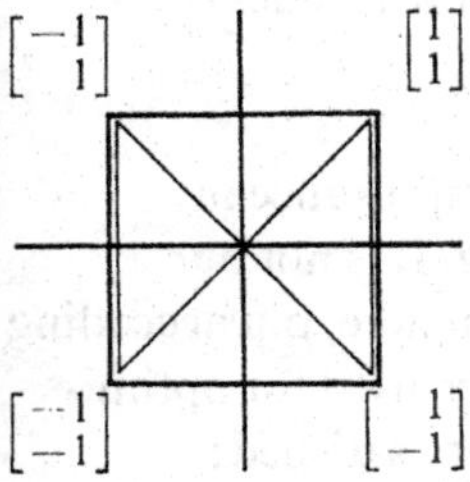

Figure 7.30

$$K = \{\mathbf{x} \mid \mathbf{x} = \Sigma_i \, a_i \mathbf{x}_i, \text{ where } \Sigma_i \, a_i = 1 \text{ and } a_i \geqslant 0\}.$$

For example, given the four vertices of the bow-tie we have used before (Figure 7.30)

$$B = \begin{bmatrix} -1 & -1 & 1 & 1 \\ -1 & 1 & -1 & 1 \end{bmatrix}$$

the *convex cover* is given by the set of points

$$\begin{bmatrix} x \\ y \end{bmatrix} = \begin{bmatrix} (-a_1 - a_2 + a_3 + a_4) \\ (-a_1 + a_2 - a_3 + a_4) \end{bmatrix}$$

where $a_1 + a_2 + a_3 + a_4 = 1$ and all of the a_i are positive or zero. Is the point $[0 \quad 1]^T$ in the convex cover? This would require

$$\begin{aligned} -a_1 - a_2 + a_3 + a_4 &= 0 \\ -a_1 + a_2 - a_3 + a_4 &= 1 \\ a_1 + a_2 + a_3 + a_4 &= 1. \end{aligned}$$

These equations are satisfied for $a_i \geqslant 0$ by $a_1 = 0$, $a_2 = \frac{1}{2}$, $a_3 = 0$, $a_4 = \frac{1}{2}$, so that the point does lie within the set. But does $[0 \quad 3]^T$ lie in the cover? This would require

$$\begin{aligned} -a_1 - a_2 + a_3 + a_4 &= 0 \\ -a_1 + a_2 - a_3 + a_4 &= 3 \\ a_1 + a_2 + a_3 + a_4 &= 1. \end{aligned}$$

Taking the last two equations and adding we find that

$$2a_2 + 2a_4 = 4$$

or that

$$a_2 + a_4 = 2$$

but this is impossible since $a_2 + a_4$ cannot equal 2 *and* satisfy the conditions that $\Sigma_i = 1$ for this requires at least a_1 or a_3 to be negative, which is not permitted. Thus $[0 \quad 3]^T$ lies outside the cover. The convex cover of the bow-tie is a square centred on the origin with sides of length 2 (Figure 7.30).

The recent interest in convexity is partly due to its importance in optimization procedures such as *linear programming*. It is not our intention to discuss linear programming in any detail here, but according to Churchman, Ackoff and Arnoff the method can be used for optimization problems in which the following conditions are satisfied:

> '1. There must exist an objective, such as profit, cost or quantities, which is to be optimised and which can be expressed as, or represented by, a *linear* function.

[8] C. West Churchman, Russell L. Ackoff, and E. Leonard Arnoff. *Introduction to Operations Research*, New York, Wiley, 1957, p. 281.

[9] However, sometimes suitable transformations can be found which will linearize the variables. Methods of non-linear programming have also been developed.

2. There must be restrictions on the amount or extent of attainment of the objective and these restrictions must be expressible as, or representable by, a system of linear equalities or inequalities.'[8]

They go on to note that the linearity assumption is inherent in the technique and simply means that if, for example, it costs ten times as much to build ten dwellings, it will cost a hundred times as much to build one hundred, and so on. If this assumption cannot be realistically made, or if the functions cannot be linearized by a suitable transformation of variables, then the techniques of linear programming will not be applicable.

The second condition quoted above implies that each constraint may be represented by a halfspace within which the restrictions will be satisfied. The *solution set*, as we saw earlier, consists then of the intersection of all the halfspaces defined for each constraint. The search in linear programming is to find the vertices of the *convex hull* bounding the solution set, for one of the fundamental theorems of the method remarks that among these extremum points the optimum solution is to be found.

In our earlier example, where we delimited the solution set of a room given certain physical constraints, the boundary of the set was not only re-entrant (so that the set was not convex) but it was also curvilinear.[9] This emphasizes a recurrent problem in spatial and building studies which may make linear programming of limited use. The dimensions of the quantities which we handle are not only linear, L^1, they are areal, L^2, and volumetric, L^3. Other measures such as flows of water in pipes are also non-linear, L^n, where $n \in R$. We cannot say that if the length of wall of a 3 m-square room is 12 m, that the length of wall of a 6 m-square room is 24 m: it certainly is not, since the perimeter, p, of the room is related to area, A, by the equation $p^2 = 4A$ which is not linear. We recommend extreme caution in giving credibility to the results of linear programming as applied to environmental problems. The real issues are too often curtailed to conform to the restrictive limitations of the analytical technique, or to those relationships which may be conveniently transformed to linear expressions.

One other important theorem of convexity states that the *intersection* of any two convex sets is a third convex set. We have already met an example of this in our previous study of panels and blocks. We are now

[10] Dean Hawkes and Richard Stibbs. *The Environmental Evaluation of Buildings 1–5*, in particular Working Papers 15, 29 and 30, University of Cambridge, Land Use and Built Form Studies, 1969–70.

in a position to describe the use of convexity in architectural problems related to sunlighting and shadowing, or *sciagraphy*, and the obstruction caused by external objects to views seen through the windows of buildings. Hawkes and Stibbs[10] in a series of papers have demonstrated the interrelatedness of environmental performance – heating, cooling, lighting, noise levels and so on – and the geometry of a building and its surroundings.

Here we shall briefly discuss the conceptual basis of their approach, which itself relies on considerable computation. In Chapter 1 we showed how sunlight arriving on earth, for all practical purposes in parallel rays, gives rise to *affine transformations* when objects cast their shadows. Consider the ray, R. Let P be an orthogonal projection of an object, O, onto a plane perpendicular to the ray. The cartesian product of R with P represents an extrusion of that object in space as we saw with Mies van der Rohe's bronze mullion. If the object is representable as a convex set then so will the extrusion be convex. Usually, in the case of a building it will be possible to *decompose* P into a number of convex polygons K_i and to treat each separately for the purposes of the exercise. $P = \cup_i K_i$. Now let us name the surface which may be cast in shadow the set S. The intersection of S with the cartesian product K_iR defines the area in shadow taken over all i, $\cup_i (S \cap K_iR)$. In practice it is only necessary to consider the set of extremum points and edges of O to obtain the shadow set.

Finally, if we are interested in the obstruction to view in a room caused by external objects and the size and shape of the window opening, we may follow a similar procedure. The cone of vision V through a plane convex opening W intersects the external objects O – described individually as convex sets – in a convex set K_i for each convex obstruction. A cone of obstruction V_{K_i} which subtends K_i and whose edges and faces contain the extremum points and edges of K_i now intersects with the convex opening W. This intersection taken over all i represents the *perspectivity* of the obstructions seen through the window:

$$P = \cup_i (V_{K_i} \cap W).$$

In both the above examples planar areas of shadowing and obstruction may be computed using the area formula for the general n-gon.

8 Modules and numbers

In the 1920s and '30s it was becoming clear to architects, to builders and to businessmen that the increasing use of the techniques of mass production in the building industry – in the manufacture of doors, windows, structural members, wall and roof panels and components of all kinds – was creating a pressing need for the control and standardization of the sizes of these various items. The advantages of repeated machine-made units in simplifying and speeding construction and in cutting costs had been shown with dramatic force in the great works of Victorian engineering: most spectacularly in the Crystal Palace, where these virtues were proved for an industrialized building system in glass and cast iron on a giant scale. What was new in the twentieth century was the growth of an industry using the principles of Henry Ford's 'production line' in the manufacture of 'off-the-peg' components, designed not for one situation only but for use in a variety of buildings and assembled with other units in a variety of different ways.

The economics of production demanded that the range of sizes manufactured should not be too great – since the 'run' for one item would become small, and the cost for tooling up for a large number of sizes expensive. On the other hand the variety of choice offered to the architect should not be so narrow as to limit his freedom of design. And especially it was important that different products made by different manufacturers should in some way be related and coordinated in their dimensions so as to fit together in various combinations. They should allow for a flexibility of arrangement, without the wasteful and troublesome problems of cutting to size on the site, which the old building methods had entailed.

One of the most forceful advocates of 'modular coordination' in building at this time was Albert Farwell Bemis, whose work has been already briefly mentioned in Chapter 5, and whose book *The Evolving House*[1] is a weighty review of the economic and social problems of mass housing, and of the methods of machine production in other industries, for comparison with the handicraft techniques of building. Bemis catalogued the American proprietary systems of prefabricated housing, which even at that time were numerous, and showed that they displayed a rich dimensional diversity. As a start to the problem of coordination he pressed for the general adoption of a smallest basic unit of size, or

[1] Albert Farwell Bemis and John Burchard. *The Evolving House.*

Figure 8.1
House structure defined within a matrix of cubelets. From Bemis, *The Evolving House*

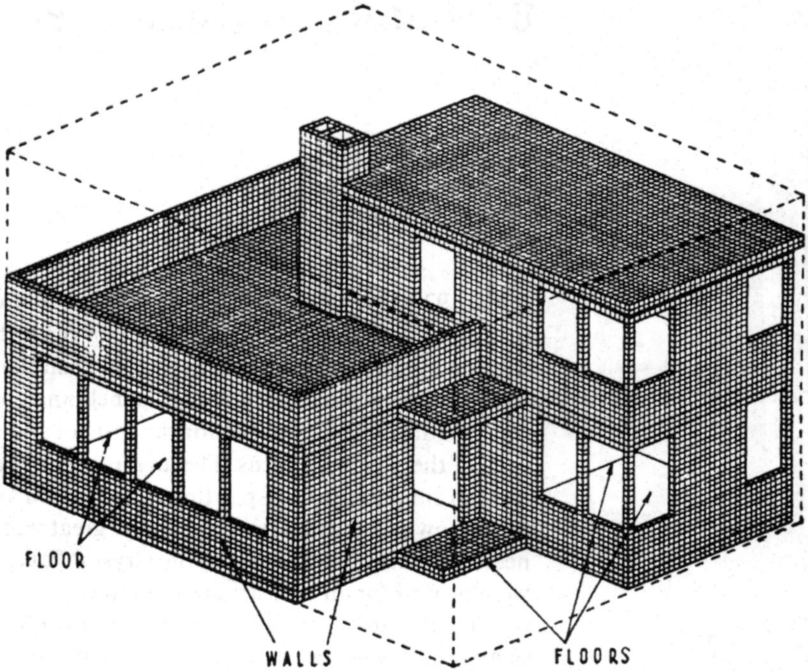

module; the dimensions of all components would ideally then be fixed as multiples of this basic unit. Dimensions in the vertical plane would conform to the module as well as those in the horizontal. One can imagine the volume occupied by the building as a stack of tiny notional 'cubes of space' – the cubelets of Chapter 5 – with every part of the building's structure exactly filling a number of these cubes within the total 'building matrix' (Figure 8.1).

Bemis proposed that the size of this basic module be 4 inches. This was partly to relate to existing practice in wood-frame construction using 4 in × 2 in studs, and to the commonly used masonry dimension of 8 in. It was sensible, he thought, to determine the minimum modular size on the basis of wall thicknesses rather than from floor or roof thicknesses, since in general the wall dimension would be the smaller. There would, of course, be some materials and small items whose dimensions would be necessarily much less than 4 in, but the intention was that these should lie within modular spaces and with their edges or faces along the lines of the modular grid. In this way all junctions between principal components would conform to the overall 4-inch spacing.

Although Bemis emphasized that the advantages for coordination, mass production and standardization would apply in principle with *any* small dimension – 3 in, $3\frac{3}{4}$ in, $4\frac{1}{2}$ in – the rightness of his original choice has been confirmed by time. The work of the British Standards Institution

[2] For a history of the development of modular coordination ideas in Britain and an account of the work of the British Standards Institution and the Building Research Station in this area, with a very full bibliography of the subject, see Bruce Martin, *The Co-ordination of Dimensions for Buildings*, London, Royal Institute of British Architects, 1965.

[3] J. W. Harding. 'Co-ordination by Design Modules' in *The Builder*, 22 September 1961, pp. 544–8.

and of the Building Research Station in Britain[2] has resulted in the official adoption of the 4-in module, and the same unit has been recommended in other countries on the imperial standard; while in Scandinavia and Europe the metric module of one decimetre (100 mm) is as close an equivalent as is possible in round figures. This neat correspondence of the metric and imperial sizes is in fact one of the reasons behind the choice of the 4 in/100 mm unit. Also some round number multiples of the basic module size give convenient dimensions for architectural planning. 10 modules gives 3 ft 4 in and 1 metre respectively, the width of a door opening for example, while 30 modules (10 ft or 3 metres) could be a floor-to-ceiling height, or a useful small room dimension in plan.

But the agreement of a 4-in or a 100-mm unit as a basis for dimensional coordination is only a beginning; for the number of possible sizes allowed for bigger components by taking all multiples of the basic module, is still very great. J. W. Harding[3] has coined the term 'Bemis set' to describe the set B of dimensions ia, where $a = 4$ in and i is some positive integer. We express this in the notation of set theory as

$$B = \{x \mid x = ia, i \in Z_+\}$$

where Z_+ is the set of all positive integers; so that the mathematical sentence says that the Bemis set comprises all those dimensions x such that x is some positive integral multiple of the basic modular dimension $a = 4$ in.

Note that the lower case letter a will be used here throughout to signify a basic modular *dimension*, although capital M is conventional in the modular coordination literature. In our case we wish to use M to signify a particular *set* of modular dimensions, that set which comprises all multiples of the internationally accepted 100-mm standard, which we shall call the 'Modular set'; in the same way that capital B is used for the 4-in Bemis set.

To limit the range of possibilities which the Bemis type of set offers, it is desirable to select from the set certain *preferred sizes* for components. In particular, there may be established a hierarchy of modules at different scales, (or 'multi-modules'), all related. A major module governs the spacing of the structural system, for example, within this a

Figure 8.2
The 'trade mark' of Le Corbusier's proportional system, the *Modulor*. 'A man-with-arm-upraised provides, at the determining points of his occupation of space – foot, solar plexus, head, tips of fingers of the upraised arm – three intervals which give rise to a series of golden sections, called the Fibonacci series.' (See pages 234–7 and Figure 9.13.)

'planning module' fixes the sizes of such elements in the middle range as doors, windows and partition units; and the detailed construction of the parts is controlled by the basic 4-in or 100-mm module, at the smallest scale.

There are very many criteria on which the larger dimensions can be chosen. Konrad Wachsmann[4] has listed twelve types of module, or rather, in effect, factors affecting the choice of modular dimension. These factors overlap, and in the end the final choice of sizes would depend on a synthesis of (or compromise between) their various demands. They are the outcome of raw material sizes, of the various structural characteristics of materials, and of production engineering requirements; they are governed by 'transportation, storage and erection procedures' – the capacity of loading equipment, techniques of handling in warehouses, and the limits of weight and size imposed on those elements moved by hand; or they may depend on the methods of jointing and assembly of units. These are technical considerations. More important perhaps is the relation of the planning module to those features of the design of buildings which take their size directly from the measurements of the human body. The height and width of a door, the height of such fixtures as worktops, cupboards, seats or wash basins, the length of a bath or a bed, the width of a laboratory bench or an office desk; all these depend on the proportions of the limbs, or the space swept out by the body in movement.

It is the standard *tatami* or floor mat, designed for sleeping and sitting, which acts as the module in Japanese domestic architecture. And the key dimension of Le Corbusier's Modulor set of preferred sizes is chosen as equal to the height of 'those good-looking men, such as the policemen, in English detective novels, who are always six feet tall'[5] – though this is more to relate the proportional system to the measure of man in a symbolic way, than for any strict ergonomic reason – (Figure 8.2). Figure 8.3 is an illustration from Pierre Bussat's *Die Modul-Ordnung im Hochbau* which tabulates in graphical form some modular

[4] Konrad Wachsmann. *The Turning Point of Building; Structure and Design*, New York, Reinhold, 1961.

[5] Le Corbusier. *The Modulor*, London, Faber & Faber, 1954.

Figure 8.3
'Functional modular sizes'. From Bussat, *Die Modul-Ordnung im Hochbau*

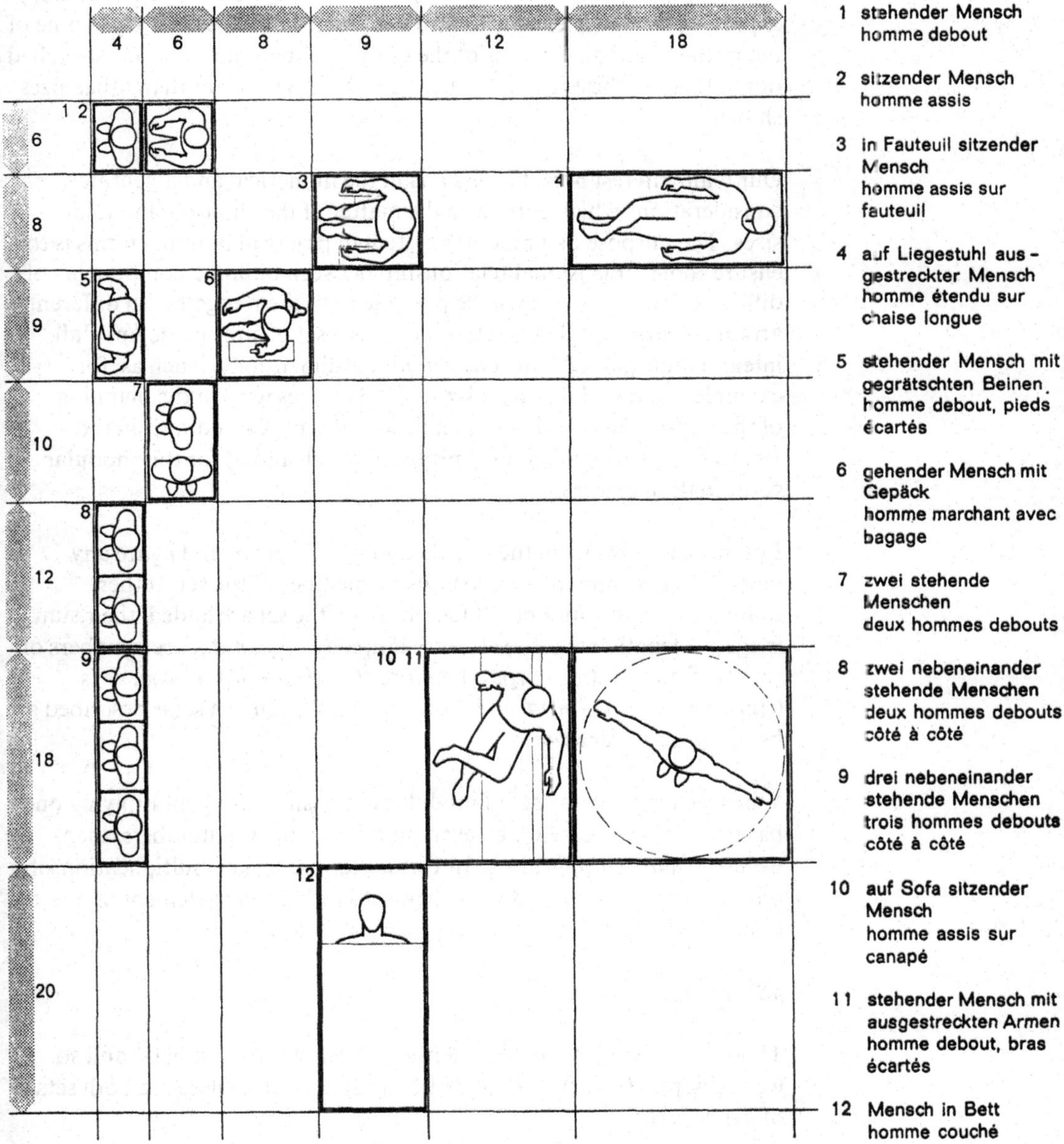

[6] Described in J. D. Kay, 'Modular Co-ordination in Herts School Design', *The Architects' Journal*, 8 December 1955, pp. 783–8.

sizes relating to furniture and to the human body. The dimensions are expressed as multiples of a 100-mm module. It will be clear that some of these (the 'plan' dimensions of the body, the span of both arms stretched out to their furthest reach, the bed) are more significant than other sizes shown.

Our main interest here, however, is in the numerical and algebraic considerations which enter into the matter of the choice of modular sizes. The purpose of a modular system in practical building terms is to ensure some dimensional relationship between building components of different sizes, so that it will be possible to fit them together in different arrangements. The Bemis set, or any other set of sizes made up of all integral multiples of some basic modular dimension a (such as, for example, the $a = 3$ ft 4 in and $a = 8$ ft 3 in sizes used in the planning of the famous Hertfordshire prefabricated school buildings[6] in the 1950s) has a particular simple property which allows for this modular coordination of sizes.

For any two sizes from the set, n_1a and n_2a (where n_1 and n_2 are any integers), their sum $(n_1 + n_2)a$ is also a member of the set. It then follows that if *any* number of lengths from the set are added, their sum is again a length from the set; since if $n_1a, n_2a, n_3a, n_4a...$ are members of the set, then so is $(n_1 + n_2)a$, therefore so is $((n_1 + n_2) + n_3)a$, so is $(((n_1 + n_2) + n_3) + n_4)a$, and so on. Recall that such a set is described as being *closed under addition*.

When we produce any set of modular dimensions D by multiplying one basic modular dimension by every number in a set of numbers, then we use *scalar multiplication*. In formal terms, scalar multiplication of a set S by a number n, produces a new set in which each element of S is multiplied by n, that is

$$nS = \{ns_i \mid s_i \in S\}.$$

Thus we distinguish sets of dimensions from sets of numbers; and so we might put $D(4 \text{ in}) = B$, or $D(100 \text{ mm}) = M$, for these are both sets of *dimensions*.

The sizes comprising the Bemis set B, the modular set M, and all similar series, are generated by multiplying the basic module a in each case by the set Z_+ of all positive integers {0, 1, 2, 3...}; and Z_+ is the most obvious example of a set which is closed under addition. Because of this property, of closure under addition, it tends to be possible in such a system to equal some dimension determined by a large component with a suitable combination of smaller components and in general to coordinate the different sizes together.

The system based on multiples of a single modular size a – be it 100 mm, 3 ft 4 in, 8 ft 3 in, or whatever – gives a notional grid underlying a building's design, made up of square units (if we look at a plan or an elevation, or cubic units – as in the Bemis drawing – if we take the whole volume of the building) with the edges of components lying along grid lines, and their principal dimensions filling exactly some number of grid intervals.

We can give a formal description of this grid by making use of the concept of the cartesian product, which was introduced in Chapter 6. The set of real numbers R can be represented by the number line, extending indefinitely from zero in the two, positive and negative, directions. We are only interested here in the positive half of the set, R_+, since there is no meaning to negative numbers or dimensions in this context. The cartesian product R_+R_+, or R_+^2, is the set of all points (x, y) in two-dimensional space, such that x and y are positive real number coordinates of those points. R_+^2 is thus a shorthand expression signifying that part of the continuous two-dimensional plane which has positive coordinates (Figure 8.4a). In the same way we use the treble

Figure 8.4

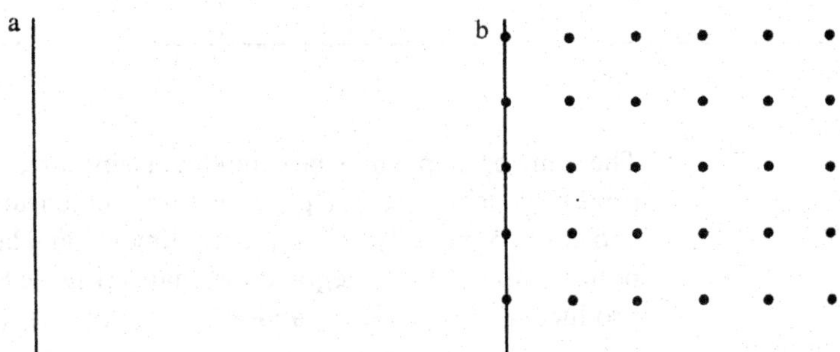

[7] Bruce Martin, *The Co-ordination of Dimensions for Building*, p. 25. Compare Martin's diagrams with the set-theoretic representation of grids here using cartesian products, and in our Figures 8.8 and 8.11.

product R_+^3 to signify positive three-dimensional space, by the set of points (x, y, z):

$$\{(x, y, z) \mid x \in R_+, y \in R_+, z \in R_+\}.$$

If we now represent the positive integers Z_+ set out along the positive half of the number line in a similar way, then by taking the cartesian product Z_+^2 we derive the *lattice of modular points* (Figure 8.4b). Since Z_+ contains only positive integers this lattice comprises only those points (x, y) whose coordinates x and y take positive integral values.

The cartesian product R_+Z_+ contains all those points (x, y) where the x coordinate may take any positive real value, but the y coordinate is some integer. This gives the pattern of *modular lines* at unit spacing parallel with the x-axis (Figure 8.5a); and the corresponding product Z_+R_+ gives the equivalent pattern at right angles, parallel to the y-axis (Figure 8.5b). It follows that the union of these two product sets $R_+Z_+ \cup Z_+R_+$ form a regular square grid of lines, or what Martin[7] calls in the language of modular coordination theory, the *basic module grid* (Figure 8.5c).

Figure 8.5

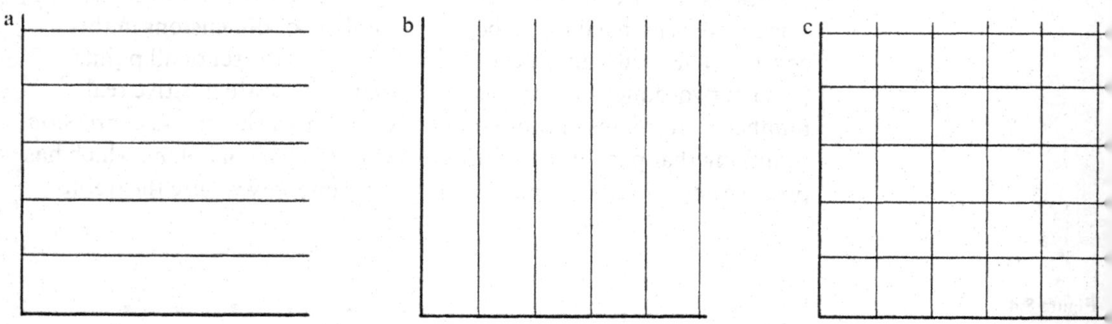

The same principle may be extended, using cartesian products still, into three dimensions. R_+^3 is 'positive' three-dimensional space (Figure 8.6a), and correspondingly Z_+^3 is a three-dimensional lattice of modular grid points with positive integral coordinates (Figure 8.6b). The three products $Z_+R_+^2$, $R_+Z_+R_+$ and $R_+^2Z_+$ are the sets of parallel modular planes in the three orthogonal orientations (Figure 8.6c–e); while their union $Z_+R_+^2 \cup R_+Z_+R_+ \cup R_+^2Z_+$ form a cellular structure of hollow cubes of space, similar to the Bemis 'grand matrix' (Figure 8.6f).

Figure 8.6

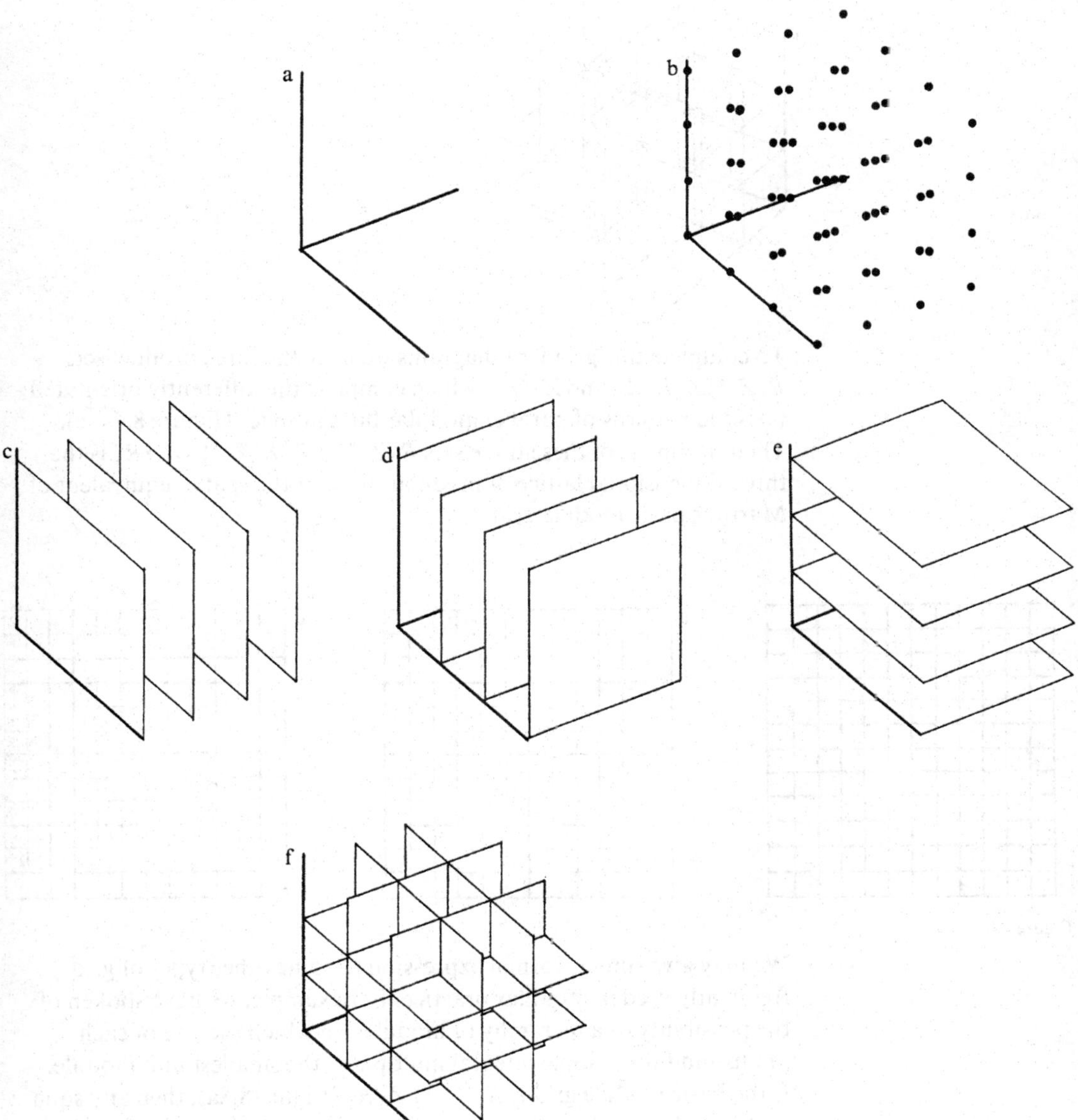

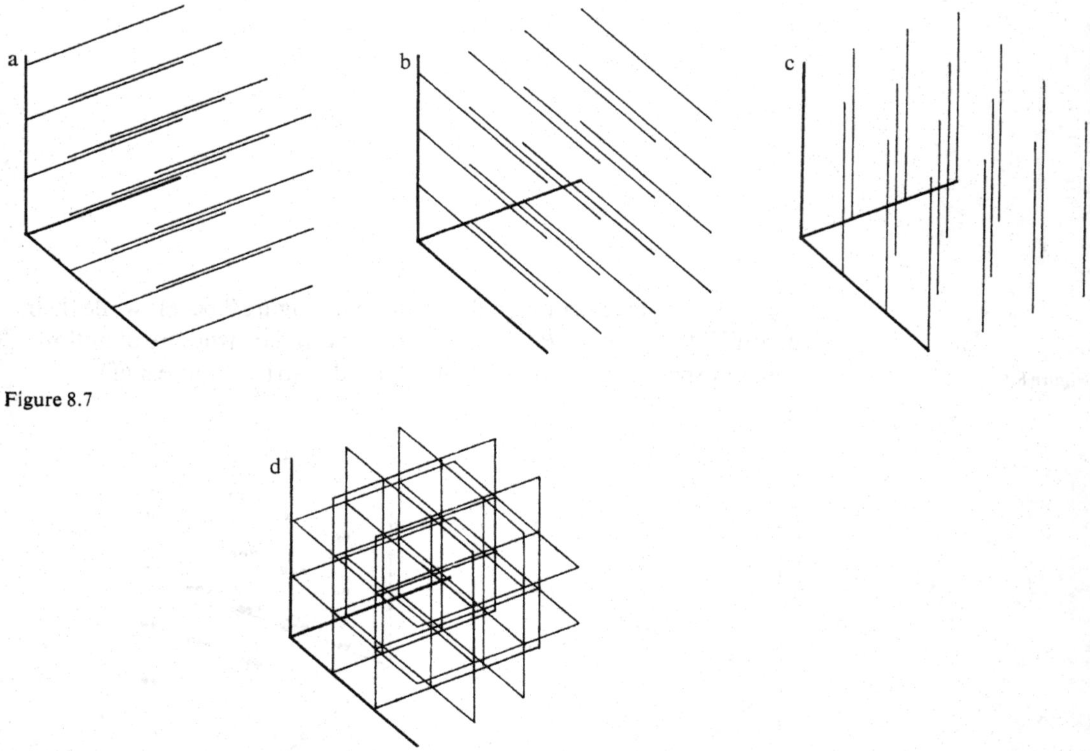

Figure 8.7

To complete this group of diagrams we have the three product sets $R_+Z_+^2$, $Z_+R_+Z_+$ and $Z_+^2R_+$, which comprise the differently orientated possible patterns of parallel modular lines in space (Figure 8.7a–c). Then the union of these three sets $R_+Z_+^2 \cup Z_+R_+Z_+ \cup Z_+^2R_+$ is the three-dimensional lattice of modular lines, or the spatial equivalent of Martin's basic module grid.

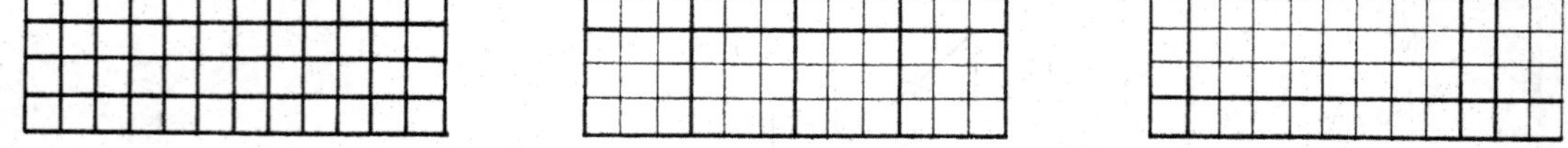

Figure 8.8

We may give similar formal expression to some other types of grid frequently used in modular practice. For example, we have spoken of the possibility of a hierarchy of modules, in which the size of each 'multi-module' is some integral multiple of the smallest unit module. If the basic module grid is $R_+Z_+ \cup Z_+R_+$ (Figure 8.8a), then any square multi-module grid will be given by $R_+(nZ_+) \cup (nZ_+)R_+$, where nZ_+ is the scalar product $nZ_+ = \{0, n, 2n, 3n...\}$. In the figure we illustrate the two grids superimposed, for the case where $n = 3$ (Figure 8.8b). It is clear that if we were now to give dimensions to the units, this figure could represent the basic module grid for a 4-in module, with a 1-ft planning grid overlaid.

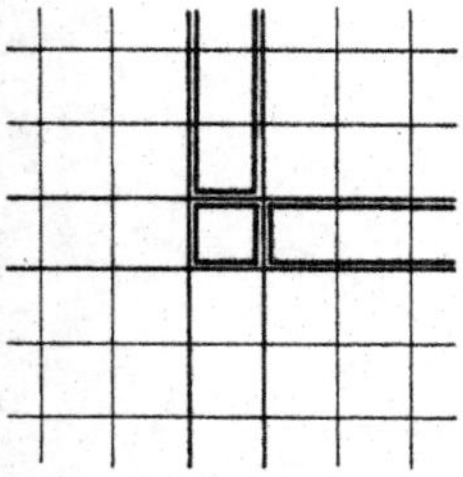

Figure 8.9

One special type of grid which has a particular usefulness is the *tartan grid*, shown in Figure 8.8c. It is produced in effect by taking an alternating pair of modular dimensions in each direction. Its main purpose is to determine the regular spacing of a building's structural frame, where the beams or walls would lie within the narrow modular bands, and the columns at their intersections (Figure 8.9). This meets the so-called 'thickness problem' which arises when components whose length is modular, but not their thickness, come together at a corner. Otherwise, without the tartan the two must overlap to give continuity of structure, and the thickness of one must be added to the length of the other, so making that length non-modular (Figure 8.10). A square tartan grid in which the smaller dimension is the basic unit size, (as in the figure) is expressed by

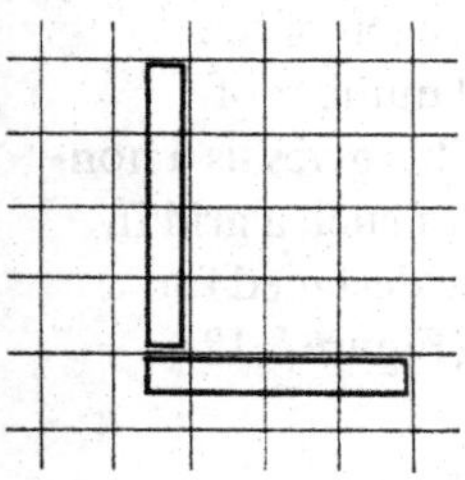

Figure 8.10

$$R_+(nZ_+) \cup (nZ_+)R_+ \cup R_+(nZ_{+1}) \cup (nZ_{+1})R_+$$

where Z_{+1} is the set $\{1, 2, 3, 4...\}$.

This corresponds, therefore, to the union of a multi-module grid with that same grid 'stepped on' one unit in both positive directions. For a general tartan where the smaller of the pair of dimensions is a, then the equivalent form is

$$R_+(nZ_+) \cup (nZ_+)R_+ \cup R(nZ_{+a}) \cup (nZ_{+a})R_+$$

where $Z_{+a} = \{0 + a, 1 + a, 2 + a...\}$; and so the multi-module grid is stepped on by a units.

Finally, there is no necessity for a multi-module grid to be square, and it may be convenient in certain design problems to take different multi-modular intervals in the two perpendicular directions, to give a rectangular grid $R_+(n_1Z_+) \cup (n_2Z_+)R_+$, where n_1 and n_2 are some integers (Figure 8.11).

Figure 8.11

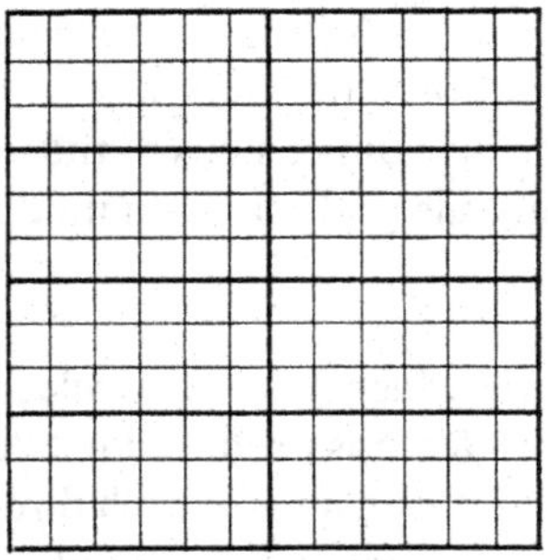

By extension, there is no theoretical objection to the idea of taking two different *basic* modular dimensions, to be used one in each of the perpendicular directions – other than that the overall number of different modular sizes is thus very much increased. This gives us a non-square basic module grid; and an example of the use of such a grid in practice is to be found in the famous 'glass box' house designed for Dr Edith Farnsworth in 1950 by Mies van der Rohe (Figure 8.12).

Figure 8.12
Plan of Farnsworth House by Mies van der Rohe

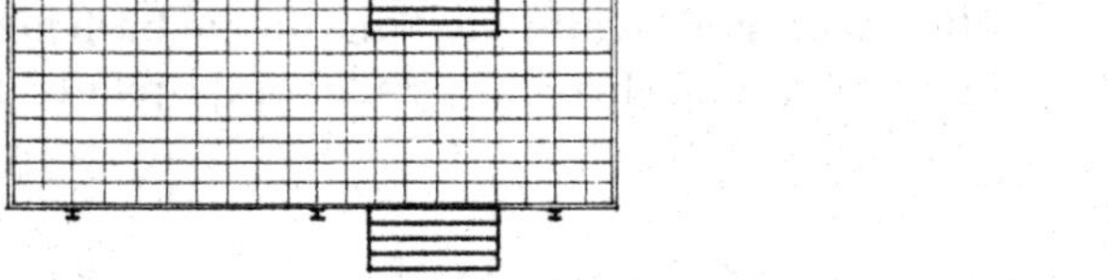

The dimensions of the grid here are given by the shape and size of a single component, a paving slab or tile, which measures 24 in by 33 in. Since the slab is used in one orientation only, it thus establishes two distinct sets of modular dimensions $D(a)$ and $D(b)$ where $a = 24$ in and $b = 33$ in, given by integral multiples of these two basic sizes; and the basic module grid is represented by $R_+D(a) \cup D(b)R_+$. The two sets may not be mixed or combined. The sheets of glass which form the walls, for example, must strictly be of two different widths, and a sheet from one of the long walls could not be used in a short wall, if the points where the sheets meet are to fall on the grid. Nor can there be any modular dimension in the plan which is some combination of the two basic units ($n_1a + n_2b$).

Figure 8.13

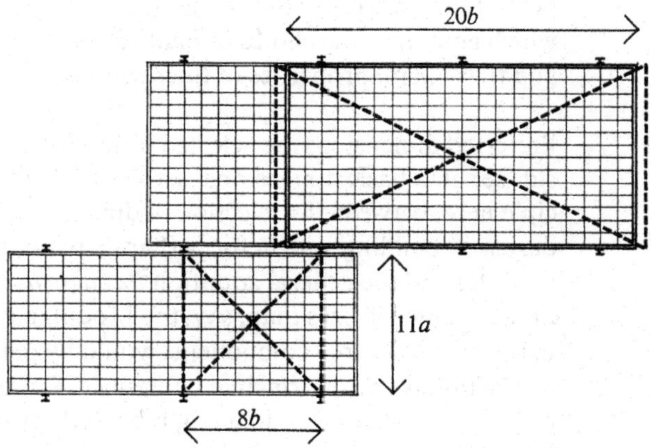

Except in a few special cases. If we look at the smaller terrace platform we see that Mies produces a square spacing of supporting stanchions, with a bay that is $11a$ wide and $8b$ long. Since the ratio of a to b (24 in : 33 in) is 8 : 11, it follows that $11a = 8b$ (= 22 ft, which is the *lowest common multiple* of a and b). Clearly, we could superimpose a large square multi-module grid whose dimensions would be higher multiples of $a = 22$ ft, $R_+D(a) \cup D(a)R_+$ onto the smaller grid of the repeated paving tiles. Note that the enclosed area of the house itself, which might at first sight appear to be of 1 : 2 proportion, does not, however, fall on any large square grid. It in fact measures $14a$ by $20b$, and the actual proportion thus works out at 112 : 220, just slightly off the double square. The dimensions involved are not common multiples of the two basic modular sizes (Figure 8.14).

Figure 8.14

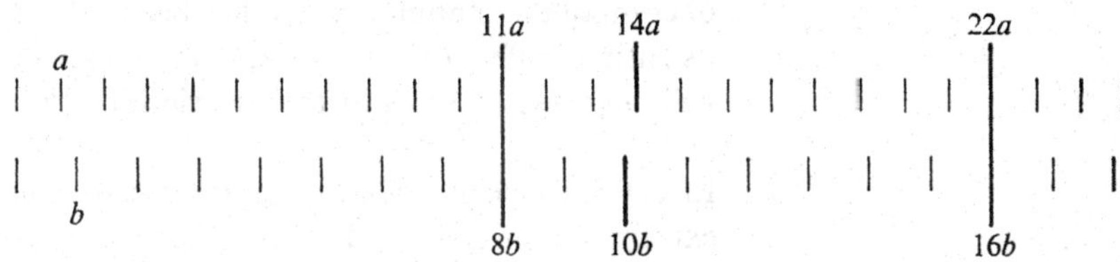

There is nothing especially significant in the proportions of the other rectangles which Mies's grid produces; only the square spacing of stanchions, which depends on the existence of common multiples of the two basic dimensions. We shall talk more about the question of proportions later. Meanwhile let us go back to those general considerations which enter into the choice of multi-modular or preferred dimensions, and the general properties of their common factors.

Preferred dimensions are selected in some way from, and thus form a sub-set of, a basic dimensional set $D(a)$. There is a distinction in principle here, between the selection of dimensions to be used in a particular design or building system for perhaps a planning or structural multi-module; and the general adoption *throughout the industry* of some more limited range of sizes than, say, the Modular set M offers, and to which many manufactured components would by preference conform. The distinction is one of principle only, since, for example, the planning module for a given building might be fixed by the size of some commercially available component, or by the preferred sizes made use of in a proprietary building system.

A combination of interrelated planning and structural module grids serves to organize the layout and dimensions of a building. Such a hierarchy of modules has been compared to the common systems of coinage or weight, where particular units of value, of different orders of magnitude, can be combined to make up any desired quantity; and where often large quantities are rounded to whole numbers of the larger units of measurement. But while, for the aims of the modular coordination movement to be achieved, all buildings must conform to the universal 100 mm module, the dimensions of larger types of grid might, unlike units of coinage, quite properly vary in size, for good practical and economic reasons, from one building or building type to another. It is not feasible or desirable to fix standard planning or structural modules. And yet on the other hand, in the broader problem of component manufacture, it has been argued that the Modular set in its entirety offers too many sizes. Significant economies in production will only result from a further restriction of the range of sizes allowed.

Efforts have been made to find agreement, somewhere between the two extremes, on a preferred range which would comprise enough suitable sizes to cope with most of the typical manufactured items. The actual

[8] Ezra Ehrenkrantz. *The Modular Number Pattern; Flexibility through Standardisation*, London, Tiranti, 1956.

choice must be governed by a whole variety of pragmatic considerations – what are the commonest forms of construction? what are the most frequently used dimensions in current practice? – including all the technical and anthropometric factors we mentioned earlier. But it is reasonable to ask that, overall, the sizes be grouped closer together at the smaller end of the series; and that there be some underlying rationale to the range, governing the ways in which sizes may be added and combined together.

In the middle 1950s the European Productivity Agency launched an international project to study problems of modular coordination and standardization of dimensions for building; and most of the suggestions made to the project in the various national reports were for ranges of sizes constructed essentially upon different *geometric series*. A geometric series has the general form:

$$ar^0\ (= a),\ ar^1\ (= ar),\ ar^2,\ ar^3,\ ar^4 \ldots$$

that is, it consists of successive powers of some number r, multiplied by a constant a. We express the set of numbers $aG(r)$ in geometric series, then, by

$$aG(r) = \{x \mid x = ar^i,\ i \in Z_+\}.$$

The proposal of the group from the British Building Research Station, for example, consisted of a set of numbers which they called the 'modular number pattern' (described in Ezra Ehrenkrantz's book *The Modular Number Pattern: Flexibility through Standardisation*)[8] which embodied as its principal features amongst other properties a simple doubling series:

1, 2, 4, 8, 16, 32...   (successive powers of 2)

giving the set of numbers $G(2) = \{x \mid x = 2^i,\ i \in Z_+\}$; and a simple tripling series:

1, 3, 9, 27, 81, 243...   (successive powers of 3)

giving the set $G(3) = \{x \mid x = 3^i,\ i \in Z_+\}$.

[9] European Productivity Agency Project 174. *Modular Co-ordination in Building*, Paris, OEEC Publications, 1956.

The geometric series has the desired property, of giving smaller sizes more closely spaced, with successively larger intervals occurring between successively larger sizes. It appears that any one geometric series on its own does not provide enough sizes to meet practical needs, and so the problem is to select several which are simply interrelated, but do not duplicate the same sizes too frequently. In the first publication[9] of the European Productivity Agency project the various recommendations of the national reports were summarized in a single diagram in which doubling and tripling series were set out in triangular form, then a further pair of doubling and tripling series formed from multiples of 5 laid out inside the first two:

				1				
			2	5	3			
		4	10		15	9		
	8	20				45	27	
16	40						135	81

and finally the whole triangular space filled in with series of the same kinds running diagonally, the doubling series down and to the left, the tripling series down and to the right:

				1				
			2	5	3			
		4	10	6	15	9		
	8	20	12	30	18	45	27	
16	40	24	60	36	90	54	135	81

Figure 8.15

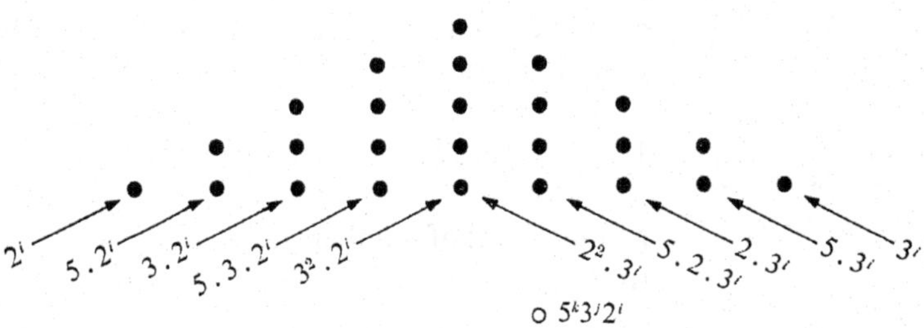

Figure 8.15 represents the terms in the series by dots, and gives the general term for each; an expression for the general term of the whole pattern is given by $2^i 3^j 5^k$, where $j = 0$ for the doubling series, $i = 0$ for

the trebling series, and k may take only the two values 0 or 1, so as to multiply each whole series by 5, or not, as the figure illustrates. The set of numbers contained by the whole pattern, which we might perhaps call the European Productivity Agency or 'European' set E is thus expressed by

$$E = \{x \mid x = 2^i 3^j 5^k, \quad i,j \in Z_+, \quad k \in \{0,1\}\}.$$

The numbers were intended to be multiplied by the basic 100-mm unit to give actual recommended sizes for components.

There is nothing magical or mysterious here. All this means is, again, that the smaller sizes are more closely spaced; that in the nature of the doubling and tripling series based on 5, we have fragments of a decimal system in the range; that because $2 + 3 = 5$, the multiples of 5 can always be made up of combinations of equal numbers of multiples of 2 and 3; and that in general larger sizes can always be made up from multiples of smaller ones, and specifically, that a large dimension can be matched by either two or three units of a smaller size from the range. Note, however, that a geometric series is not closed under addition; and in the diagram above there are many pairs or groups of sizes we can take whose sum is not a dimension in the range.

Curiously, although the aesthetic interest of the architect may be in reducing the numbers of dimensions used in a building, so as to give his design some simple rhythm or some set of proportions which can be readily appreciated, it will be in the commercial interest of the manufacturer of building components to work almost in the opposite direction. The manufacturer wants certainly to limit the number of sizes of different items which he produces. But he would like to choose their sizes such that by different combinations of component units the greatest number of different larger dimensions can be filled; so that his products are useful in the greatest variety of situations, and he can so please the most customers.

To take an example. Suppose the manufacturer's interest concerns fitted cupboard units. He will probably produce them to dimensions from the now generally adopted Modular set M of multiples of 100 mm. But beyond this he would like it to be possible for a great number of larger (modular) dimensions to be filled exactly by some combination

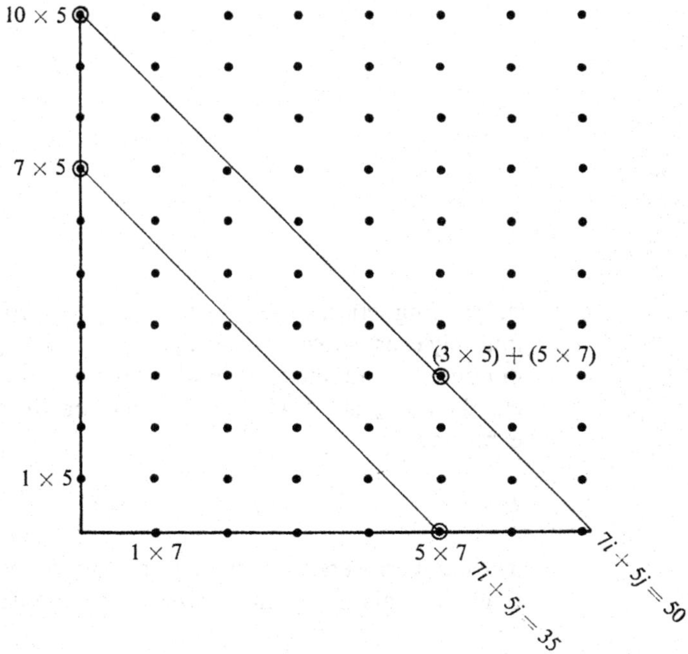

Figure 8.16
Graph to illustrate combinations of units of length 5 modules and 7 modules, and in particular how these units may be combined to fill spaces of 35 and 50 modules.

or other of units. Thus, in the greatest possible number of situations, by a judicious choice of widths for the cupboards, units can be placed along the complete length of a wall, say, without gaps being left.

The question is, what widths should he choose? For the sake of example, let us suppose that the manufacturer makes only two widths of cupboard, 5 modules and 7 modules (500 mm and 700 mm). We draw a graph in which the two axes represent different numbers of the different width units; the 5-module unit on the vertical axis, and the 7-module unit on the horizontal (Figure 8.16). The *lattice points* indicate all the combinations of units – one 5 and one 7, two 5s and one 7, two 7s and one 5, and so on.

Figure 8.17

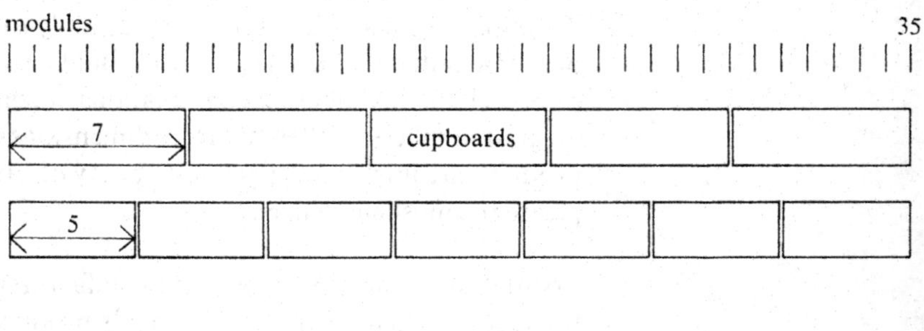

Now let us take some specific dimension to be filled. It is clear that a space 35 modules wide could be filled either with five 7-module cupboards or seven 5-module cupboards (Figure 8.17).

We draw a straight line joining 5×7 on the one axis with 7×5 on the other axis. This is the graph of the equation $7i + 5j = 35$, where i and j are integers. It intersects no lattice points, and this means that there are no intermediate combinations of some number of 5s with some number of 7s which will make up 35. If we look at the graph of $7i + 5j = 50$ on the other hand, the solutions which the diagram gives for this equation are $i = 0$, $j = 10$ (ten 5-module cupboards), and $i = 5$, $j = 3$ (five 7s and three 5s). (These are of course the positive solutions of the equation only, although algebraically the equation would have solutions involving negative quantities, which have no meaning for our practical application.)

Problems of this kind are called *linear indeterminate* problems, and equations of the form $ia + jb = n$ are *linear Diophantine equations*, after the Greek mathematician Diophantos whose known works, in particular the *Arithmetics*, are devoted to the study of rational and integral numbers. For the Greeks 'arithmetic' had the special meaning of the systematic investigation of the properties of numbers, today termed *number theory*. Diophantos was interested in finding rational (i.e., including simple fractional) as opposed to just integral solutions to various algebraic problems, and so from his point of view our linear equations would have been rather trivial.

Problems to which integral solutions are required, on the other hand, crop up in the ancient folklore of mathematical puzzles in anecdotal form, in Arabic, Chinese or medieval European sources, often virtually the same problem appearing in widely separated places dressed up in different guises. Typically, one of these puzzles might ask 'A man spends £100 buying horses and cows. A horse costs £11, a cow £7. How many of each did he buy?' ($11i + 7j = 100$). Or the problem might be to do with the way a bill at an inn is divided between people who pay different amounts, or with the mixing of wine.

In the architectural context we can imagine a variety of situations in which linear Diophantine problems might arise. What sizes should prefabricated wall panels be produced in, so as to allow the greatest flexibility in design? In an office block of rectangular slab form, with rows of rooms down either side of a corridor in a 'band' plan, what combinations of widths of offices can be fitted along the length of the building? (Considerations of this kind could perhaps give some measure

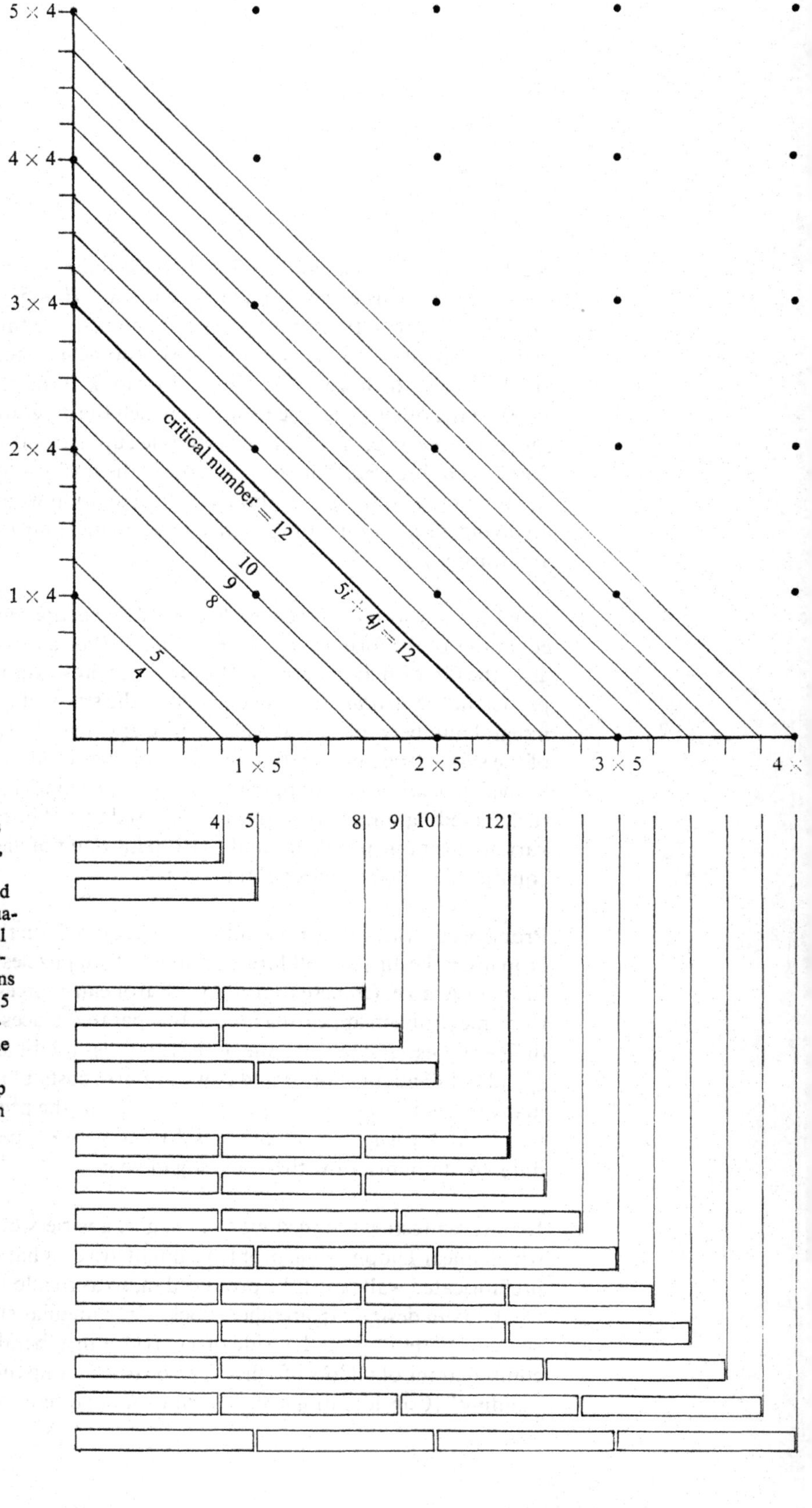

Figure 8.18
Graph of combinations of dimensions ($5i + 4j$), showing integral solutions to $5i + 4j = n$, and illustrating how the equation has solutions for all $n \geqslant 12$, the *critical number*. Actual combinations of units of length 4 and 5 modules are shown diagrammatically below the graph, filling progressively larger intervals up to the critical dimension and beyond.

of the flexibility of the plan in accepting different rearrangements of rooms over the building's life.) In a housing layout comprising rectangular street 'blocks' of fixed size, what are the ways in which plots of varying frontage can be put together to fill the block length?

Let us examine some more diagrams of the same kind as Figure 8.16, which show up certain features of these Diophantine problems. We take first two small modular sizes, 4 modules and 5 modules, and illustrate what combinations they offer to fill increasingly larger dimensions (Figure 8.18).

We may represent the series of equations $5i + 4j = n$, where n takes various integral values, by a series of parallel lines at 45° to the axes, as before, each line corresponding to the equation for a particular value of n. It hardly needs saying that 4, and then 5, are the lowest values of n for which the equations have solutions. Above that, neither $n = 6$ nor $n = 7$ give integral solutions for i and j; $n = 8$, $n = 9$ and $n = 10$ give solutions; and $n = 11$ again does not. For $n = 12$ and above, as Figure 8.18 suggests, and as may in fact be shown theoretically, every higher dimension can be filled with some combination of 4-module and 5-module units. This dimension, 12, above which, for these two sizes of unit, every greater length can be filled, is called the *critical dimension* (or, put another way, 12 is the *critical number*, for the number pair 4 and 5). Below the graphs of the equations the actual combinations of units are illustrated in diagrammatic form, arranged in rows and in order of increasing overall length.

In the next example, we take the two sizes 3 and 6 (Figure 8.19). The dimensions which this pair of sizes will fill are 3, 6, 9, 12, 15.... Any dimension which can be filled by combinations of the pair can be filled by 3-module units alone, since every 6 modules may be replaced by two 3s. And as the pattern of the diagram shows, it is *only* dimensions which are some multiple of 3 which can ever be filled. We are looking for solutions of the equation $2ai + aj = n$ (where $a = 3$), i.e., $a(2i + j) = n$. Since $2i$ and j must be whole numbers (i and j are the numbers of units of either kind), it follows that the equation has solutions only when n is divisible by a. In our example, therefore, there can be no critical dimension for the sizes 6 and 3; and in general we can say that for a pair of numbers to give a critical number they must be *relatively prime*, they must share no common factors (beyond the 'trivial' factor 1).

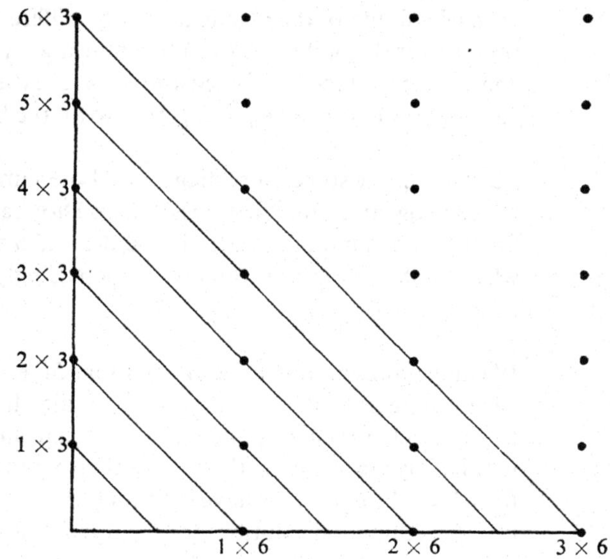

Figure 8.19
Graph of combinations of dimensions (6*i* + 3*j*)

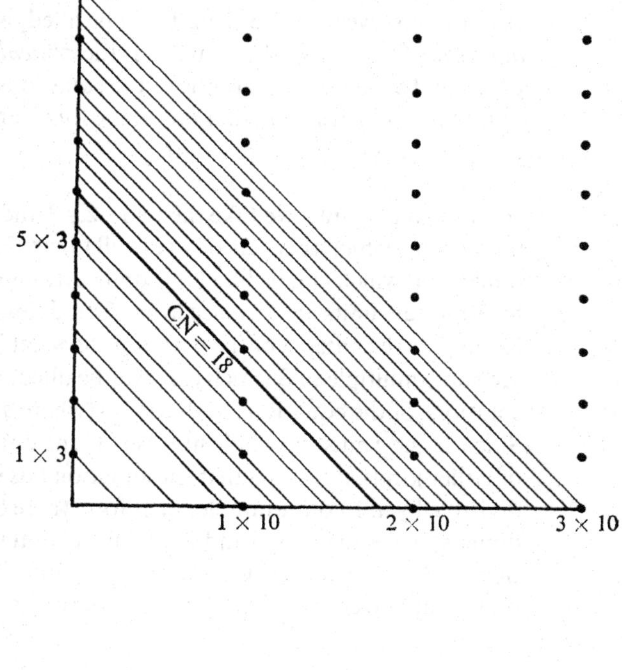

Figure 8.20
Graph of combinations of dimensions (10*i* + 3*j*)

[10] P. H. Dunstone. *Combinations of Numbers in Building*, London, Estates Gazette, 1965.

For a final illustration, we take the two sizes 3 and 10 (Figure 8.20). This pair has no common factor, and the critical dimension at which the two sizes 'spark', is 18.

The 'critical number' for any pair of sizes a and b which are relatively prime, is given by the expression $(a - 1)(b - 1)$; and below their critical number CN, two component sizes can (where CN is even) fill half the number of dimensions less one, or $CN/2 - 1$. In the example of Figure 8.20, the critical number for the sizes 3 and 10 is $(3 - 1)(10 - 1)$, which is 18. And below 18 there are $18/2 - 1$, or 8 smaller dimensions which can be filled, as the graph verifies. Where CN is odd, the number of dimensions filled is $(CN - 3)/2$.

The most exhaustive study of these problems is P. H. Dunstone's book *Combinations of Numbers in Building*,[10] in which he not only describes the theory of combinations but also tabulates the modular dimensions which can be filled by a great variety of pairs of sizes, and of groups of three sizes as well.

9 Proportions and series

The principal concern of modular coordination is with the ratios of the *lengths* of components in either one of the two (or three) directions of the grid: the fact that a dimension between columns may be filled exactly with some number of wall panels of given width, or that the height of a window unit be equal to the height of some exact number of brick courses, for example. Of course, with the square or cubic grids the same component may be placed equally well in any of the perpendicular directions. But there is no special concern with the *proportions* of components – of, say, a rectangular door or window which might be some dimension n_1a in width, and n_2a in length – or with the proportions of rooms, spaces or elevational arrangements which the grid determines.

This concern with lengths and ratios of lengths, with linear relationships, distinguishes modular coordination from many of the historical number systems employed in architectural design, where the greater interest tends to be in the proportions of the shapes which the system produces. Some of the proportional systems have incidental properties useful in modular coordination, and vice versa. It is P. H. Scholfield's view, expressed in his book *The Theory of Proportion in Architecture*,[1] that the common denominator of most systems used in fixing architectural proportion, is an attempt at 'the creation of order apparent to the eye by the repetition of similar figures, and that this is accompanied by the generation of patterns of relationships of mathematical proportion between the linear dimensions of the design'. But it is best that these two aspects are kept clearly distinct.

In this chapter we illustrate some of the various relationships which exist between modules, grids, ratios of lengths, and proportions of rectangles.

The basic module grid is illustrated again in Figure 9.1. To the left is the basic square areal unit whose side length is a. The set of dimensions $D(a)$ allowed by taking all integral multiples of a is depicted below the grid arranged out along the number line. In the lower half of the figure is a representation, in increasing order, of their two perpendicular dimensions, of the shapes of rectangular components (or areas, or rooms) which the series of modular lengths allows. The proportions of these

[1] P. H. Scholfield. *The Theory of Proportion in Architecture*, Cambridge University Press, 1958.

Figure 9.1

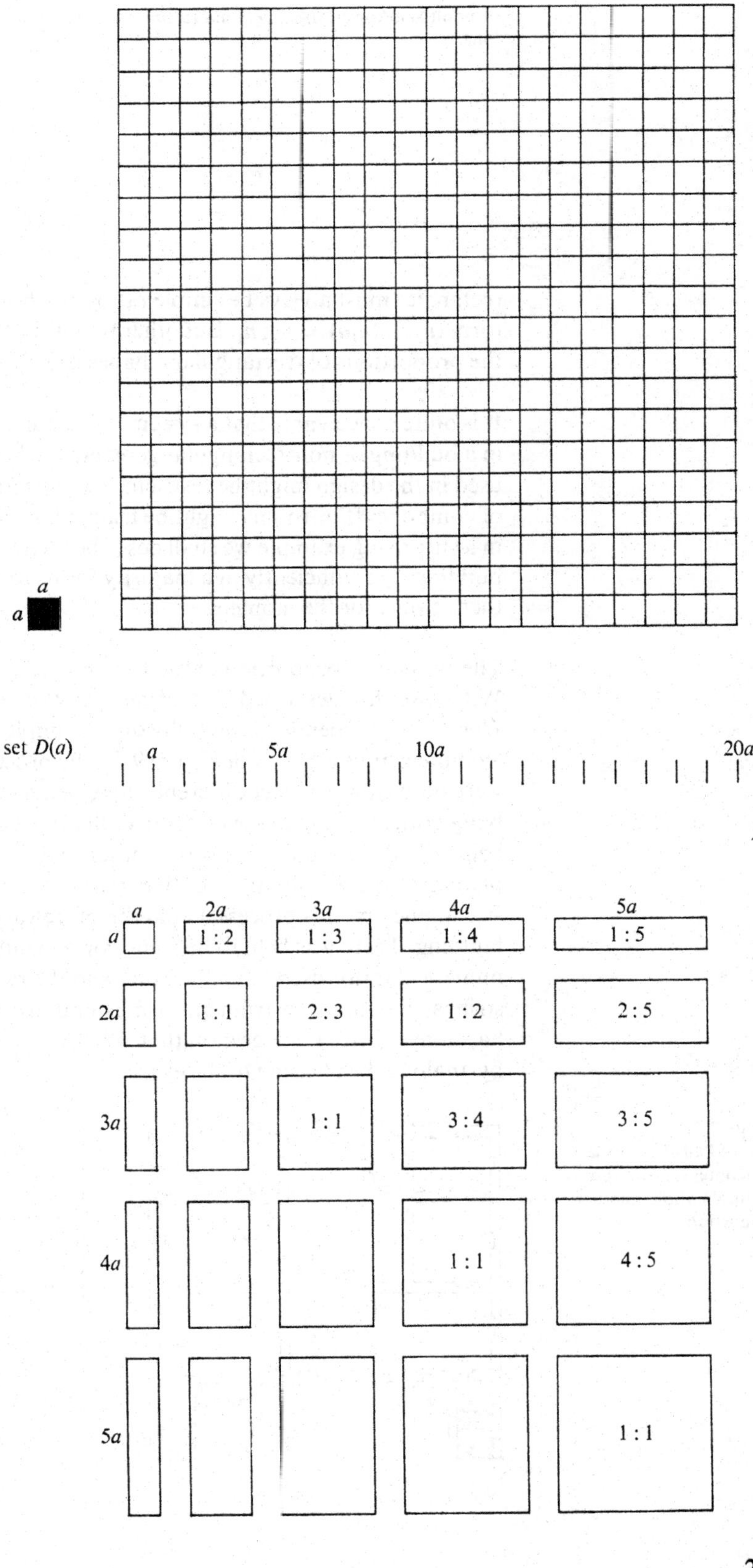

a
a
set D(a)
a
5a
10a
20a
a
2a
3a
4a
5a
a
2a
3a
4a
5a
1 : 2
1 : 3
1 : 4
1 : 5
1 : 1
2 : 3
1 : 2
2 : 5
1 : 1
3 : 4
3 : 5
1 : 1
4 : 5
1 : 1

[2] Rudolf Wittkower. *Architectural Principles in the Age of Humanism*, London, Tiranti, 1962.

rectangles must always be simple ratios of whole numbers, of the general form $(n_2a):(n_1a) = n_2:n_1$. But other than this, there is no restriction on the proportions of (rectangular) shapes possible.

It is quite conceivable that a system of modular coordination be applied in a building of non-rectangular geometry: where the linear dimensions used in the design might be restricted to some limited set, but the shapes of components or rooms might be triangular, say, hexagonal, or even as in a surprising example we shall describe later, circular. But rectangular buildings are sufficiently in a majority for us to restrict discussion to them alone, for the moment.

The systems of proportion of the Renaissance, as Professor Rudolf Wittkower has described in *Architectural Principles in the Age of Humanism*,[2] depended essentially on the simple whole number or 'commensurable' ratios of Figure 9.1. The process by which the ratios were derived was rather different, there being no suggestion of an underlying grid, but instead the larger overall dimensions of the building being divided in suitable ratios in order to determine the breakdown of a plan or elevation into its subsidiary parts. The effective basic modular size implied by this process would therefore vary from building to building. There was held to be some correspondence between the simple numerical ratios used in architecture and the ratios (of the lengths of strings, in stringed instruments, for example) underlying musical harmony; with the suggestion that what was pleasing to the ear would, by analogy, be pleasing to the eye.

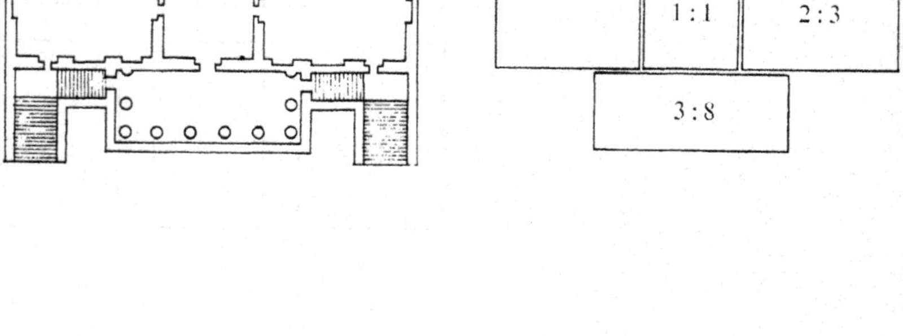

Figure 9.2
Plan of Palladio's Villa Malcontenta, with rooms of simple whole-number proportion

[3] Leone Battista Alberti. *Ten Books on Architecture*, p. 142.

Later, especially in the work of Palladio, this method of dividing larger dimensions into smaller parts according to the 'harmonic' ratios of whole numbers which the musical analogy suggested, was turned the other way about. Instead, Palladio would, in his villa plans (Figure 9.2), first determine a series of proportions for individual rooms – typically, for example, in such ratios as 1 : 1 (square), 1 : 2 (double square), 2 : 3, 3 : 4 or 3 : 5 — and then assemble these rooms in an essentially *additive* procedure, so that the resulting overall dimensions would be somewhat fortuitous. The illustration of the series of possible rectangular shapes in Figure 9.1 shows just those additive properties which Palladio's method depends on: for just as the lengths from the *D*(*a*) series can be added to make larger members from the series, so it follows that any two (or more) rectangles from any row or any column of the diagram can be put together to form another rectangle from that row or column (Figure 9.3). This is analogous to the condition for the stacking of two panels that we prescribed in Chapter 6: that they exist within the same band, where in this case the rows and columns of the figure are, in effect, bands.

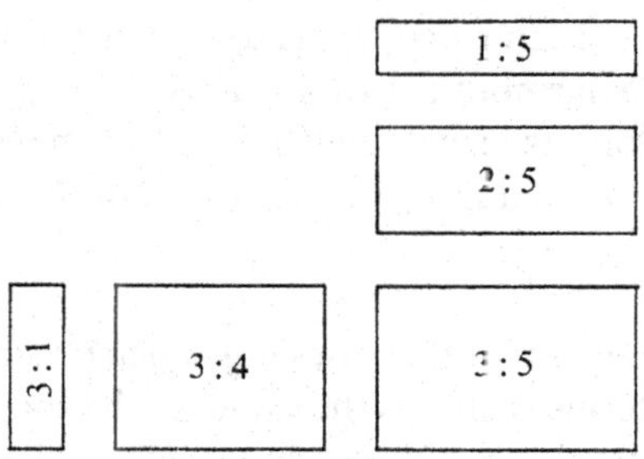

Figure 9.3

The way in which, in the proportional systems of the Renaissance, smaller parts of the design were determined in size by successive *subdivisions* of the whole or of larger parts, tended to give rise to geometric progressions like the ones forming the European Productivity Agency diagram – the European set – illustrated earlier. Alberti[3] described the base of the Doric column: 'Thus the height of the whole base was three times that of the die, and the breadth of the die was three times the height of the base.' (1 : 3 : 9).

In measuring the proportions of the human figure, which had such significance as a model for the ideal relationship of the parts of a design to the whole in Renaissance art, geometric progressions were also

[4] J. P. Richter and I. A. Richter. *The Literary Works of Leonardo da Vinci*, London, Oxford University Press, 1939. vol. 1, p. 245.

[5] Leone Battista Alberti. *Ten Books on Architecture*, p. 199.

[6] Le Corbusier. *The Modulor*, p. 26.

[7] Heinrich Wölfflin. 'Zur Lehre von den Proportionen' 1889, reprinted in *Kleine Schriften*, Basle, 1946.

sought. Leonardo finds doubling and tripling series in the proportions of the human face.[4] He also subdivided the whole body into three *braccia* or arm-lengths, the *braccio* into three faces, and the face into three smaller units still: part of the descending geometric series, in effect

$$\frac{1}{3^0}(=1),\ \frac{1}{3^1}\left(=\frac{1}{3}\right),\ \frac{1}{3^2}\left(=\frac{1}{9}\right),\ \frac{1}{3^3}\left(=\frac{1}{27}\right)\ldots.$$

In the series of ratios proposed in theoretical treatises, and in those found employed in actual building practice, different geometric series occur again and again, sometimes fragments only of one series, sometimes a doubling and a tripling series or more complex combinations, used together.

There is no doubt that Alberti, for one, recognized that a simple process of halving or division by three would give rise to relationships in the design which would be most readily appreciated visually. He makes a plea that the ratios chosen should not be used 'confusedly and indistinctly, but in such a manner as to be constantly and in every way agreeable to harmony; as for instance, in the elevation of a room which is twice as long as broad, they [architects] make use, not of those numbers which compose the triple, but of those only which form the duple...'.[5]

Figure 9.4

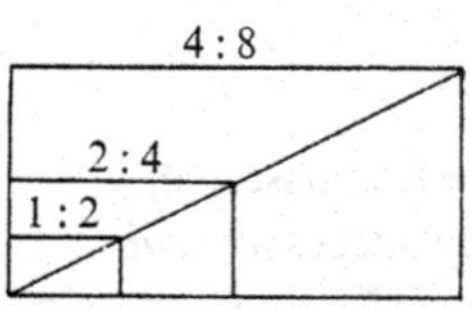

It is in this sense that the series is *geometric*: since the ratio of two successive terms is always the same, so the shapes of rectangles with these dimensions are always similar. Herein lies the significance of those familiar proportional analyses of paintings, or the façades of buildings, where a net of diagonal lines is placed over the design, joining key points in the composition – what Le Corbusier[6] calls (following Auguste Choisy) the *tracé regulateur* (regulating lines). The purpose is generally to demonstrate the repetition of similar rectangular figures in the design's underlying structure. When two adjacent rectangles of the same proportion are found in two orthogonal orientations, then their two diagonals come together at a right angle; and Le Corbusier describes the composition as being 'commanded' from these points, *les lieux de l'angle droit*. The art historian Heinrich Wölfflin,[7] among others, has made such analyses of classical and Renaissance buildings, where it is well established that geometric systems of proportion were applied consciously in their original design. Others have studied the shape of

[8] As for example Jay Hambidge, *Dynamic Symmetry; the Greek Vase*, Yale University Press, 1920, or Matila Ghyka, *The Geometry of Art and Life*, New York, Sheed and Ward, 1946.

[9] P. H. Scholfield. *The Theory of Proportion in Architecture*, p. 51.

vases in the same way, or the proportions of the human face and figure;[8] but here it is more questionable what the supposed findings of analysis might mean.

Curiously, the number $\sqrt{2}$ occurs occasionally in Renaissance proportion, despite the musical analogy and the general emphasis on whole number ratios. As Scholfield says, 'The occurrence of the $\sqrt{2}$ rectangle is embarrassing if we are trying to attribute to the Renaissance a consistent theory based wholly on commensurable proportions.'[9] It seems, however, that the lingering influence of some remarks of Vitruvius may be responsible for the inconsistency, and that the simple *geometrical* construction by which the rectangle may be produced, by making the long side equal to the diagonal of the square on the short side, may account for the interest in the $\sqrt{2}$ proportion.

Figure 9.5

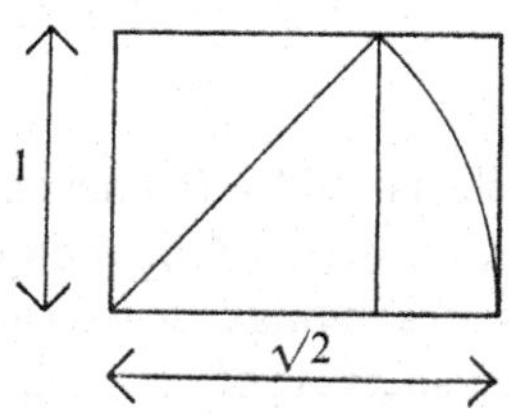

$\sqrt{2}$ is an *irrational* number, that is, it cannot be represented exactly by any *ratio* of two integers. The proportion of the rectangle whose sides are in the ratio $1:\sqrt{2}$ is nevertheless interesting. We take a $1:\sqrt{2}$ rectangle and double its shortest side; and so produce a new rectangle whose proportion is $\sqrt{2}:2 = 1:\sqrt{2}$. This is therefore a rectangle *of the same proportion*, but twice the area. Looked at another way, $\sqrt{2}$ is the solution to the equation $1/x = x/2$.

Figure 9.6

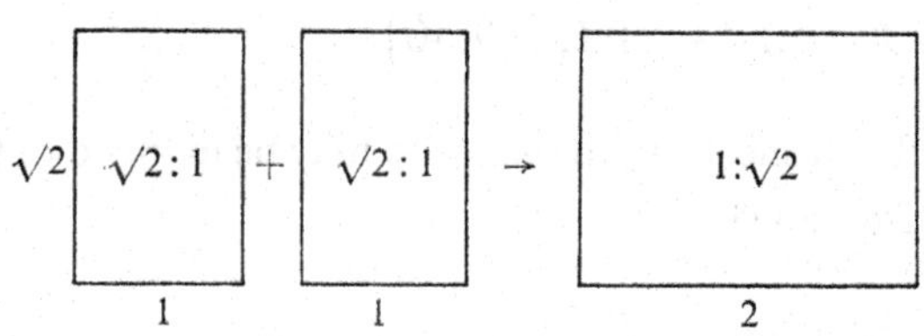

In the same way we can double and double again indefinitely and derive a whole series of similar-shaped rectangles whose *areas* this time form a geometric series, 1 unit, 2 units, 4, 8, 16....

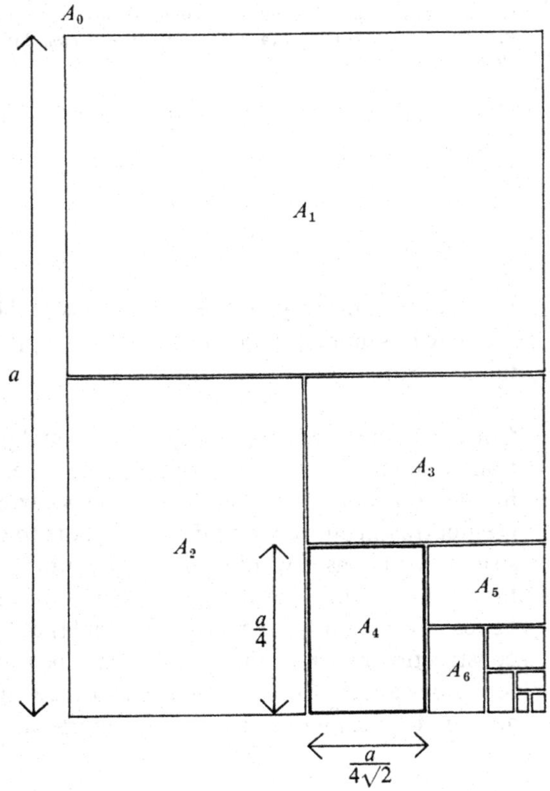

Figure 9.7
International standard A series paper sizes, of 1 : $\sqrt{2}$ proportion

It is this proportion which is used for the new internationally adopted standards for paper sizes, known as *A* sizes, in which all sheets are of 1 : $\sqrt{2}$ proportion. The series in this case is in actual fact constructed *downwards*, from a largest sheet A_0 of one square metre in area. The A_0 sheet is halved and halved again to give smaller sizes A_1, A_2, A_3... down to A_{10} which is about the size of a postage stamp. A_4 is a sheet slightly smaller than foolscap (Figure 9.7).

If we take the dimension of the longer edge of the biggest, A_0 sheet as a, then its shorter side measures $a/\sqrt{2}$, and thus the area $a\,(a/\sqrt{2}) = a^2/\sqrt{2} = 1\text{ m}^2$. The whole set of *A*-size sheets can be expressed in the usual notation as:

$$A = \{A_i \mid i = 0,1,..., 9, 10\}.$$

Each sheet, A_i, in the series may be mapped onto its dimensions x_i and y_i, that is,

$$A_i \rightarrow (x_i, y_i) \in R_+{}^2$$

where $x_i = a/(\sqrt{2})^i$ and $y_i = a/(\sqrt{2})^{i+1}$.

For example, A_4 has dimensions $(a/(\sqrt{2})^4, a/(\sqrt{2})^5) = (a/4, a/4\sqrt{2})$.

There is no question here of several sheets fitting together in any way, like building components. The practical significance of the restriction of dimensions of paper sizes is rather that printers can produce the smaller sizes from standard large sheets without waste, by simply cutting or folding them in half and half again. And one practical purpose of the constant proportion of the shape of sheet, besides the possible aesthetic significance, is that in any kind of enlargement or reduction of photographs, maps or plans, for example, what fits onto an *A*-size sheet at one scale will fit onto either the sheet two sizes larger or the sheet two sizes smaller at exactly double or half that scale, respectively.

Incidentally, there is nothing unique about the number $\sqrt{2}$ in giving rise to patterns of nested rectangles like the *A*-sizes. A very similar kind of system with a basic unit of proportion $1:\sqrt{3}$ would mean that when the smaller side was tripled a larger rectangle of the same proportion $1:\sqrt{3}$, but three times the area, would result $(1/x = x/3)$. And similarly for systems based on $1:\sqrt{4}\ (= 1:2)$, $1:\sqrt{5}$, and so on.

There is one number, however, with rather special properties, which occurs repeatedly in the nineteenth- and twentieth-century literature of architectural proportion, around which a great mystique has arisen, and which also produces a set of nesting rectangles like the $1:\sqrt{2}$ series. This is the celebrated 'golden number' ϕ, the basis of the 'golden section' and the 'golden rectangle'. It is another irrational number, it is the solution to the equation $x^2 - x - 1 = 0$, and its approximate value is $1\cdot618$.

The golden rectangle, with sides whose lengths are in the ratio $1:\phi$ has this property: that this ratio of the length of the smaller side to the greater is equal to the ratio of the length of the greater side to the sum of the lengths of the two sides, i.e.,

$$1/\phi = \phi/(1+\phi)$$

so that $1 + \phi = \phi^2$

whence $\phi^2 - \phi = 1$.

Figure 9.8

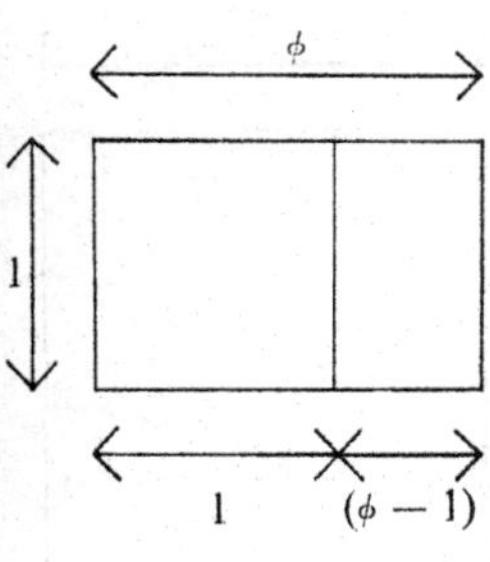

If we divide the golden rectangle into two parts by a line perpendicular to its longer side, such that one of the two smaller resulting rectangles is a square, then it follows that the proportion of the second rectangle is $(\phi - 1):1$ (Figure 9.8).

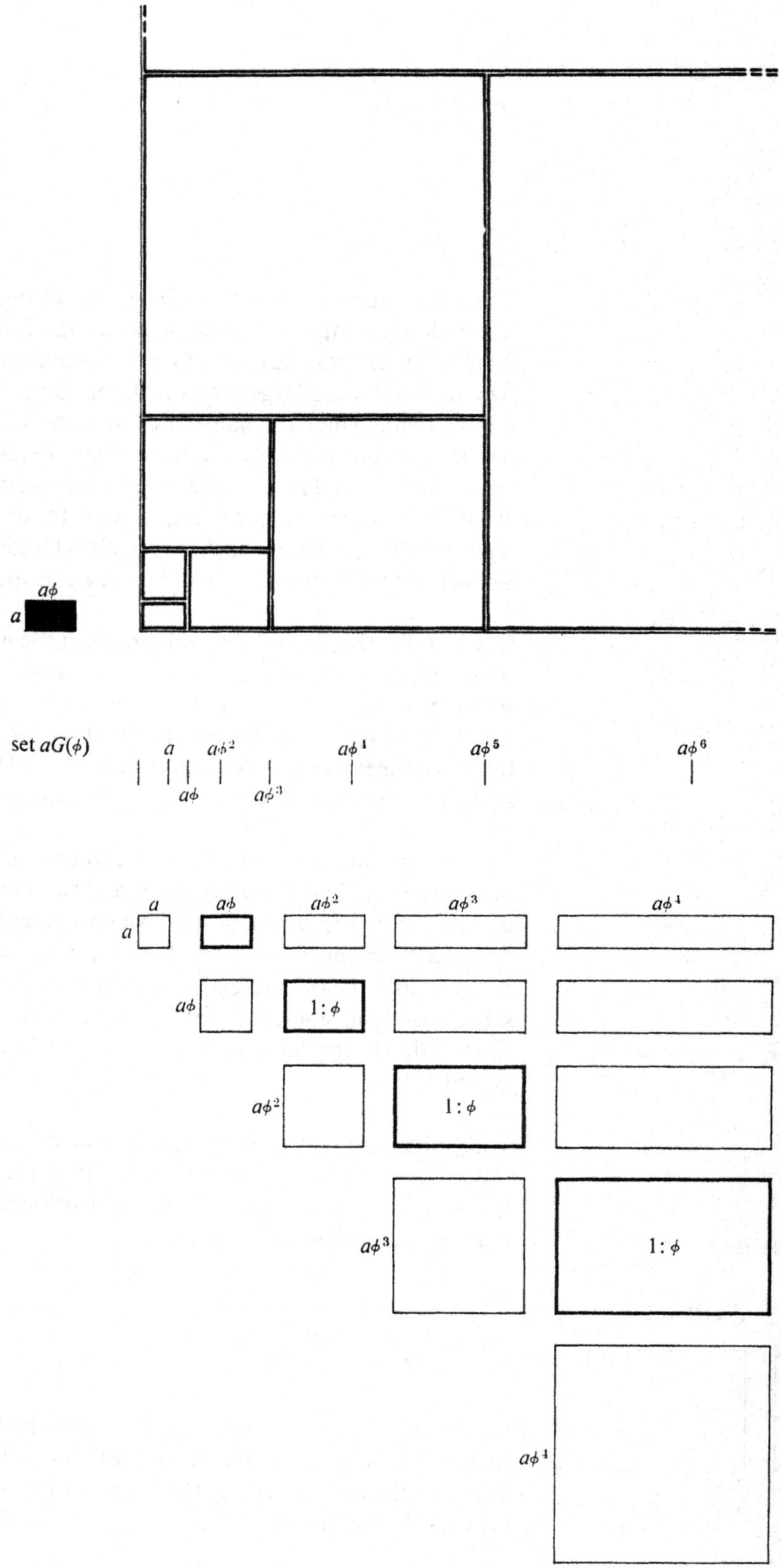

Figure 9.9
'Whirling squares' pattern produced by nesting golden rectangles, or stacking squares

[10] Jay Hambidge. *Dynamic Symmetry; the Greek Vase*, Yale, 1920, p. 18. See also *The Elements of Dynamic Symmetry*, Yale University Press, 1926 (new edition 1948).

[11] P. H. Scholfield. *The Theory of Proportion in Architecture*, p. 138.

But by multiplying both top and bottom of the expression $(\phi - 1)/1$ by ϕ, we get $(\phi^2 - \phi)/\phi$; and from the equation above we know that $\phi^2 - \phi = 1$. So $(\phi - 1)/1 = 1/\phi$, i.e., this smaller rectangle is itself a $1:\phi$ rectangle.

The golden rectangle may thus be divided into a square and another, smaller, golden rectangle. On this depends the nesting property which we refer to. Starting with a 'unit' golden rectangle, we add a square to its longer edge to give a larger golden rectangle. The process can be repeated indefinitely, and a pattern of the kind illustrated in Figure 9.9 results.

Some writers seem to have become mesmerized by this pattern of 'whirling squares' (as Jay Hambidge[10] calls it). It is worth noticing, however, that we can achieve a similar effect with any proportion of rectangle we care to choose. We take a rectangle of proportion $1:x$. We add to it a second rectangle of proportion $x:y$. To satisfy our condition for nesting or stacking we require that the new larger rectangle formed from the two is itself of proportion $1:x$ again, i.e., that $x/(y+1) = 1/x$ or $x^2 - y - 1 = 0$ (Figure 9.10).

Figure 9.10

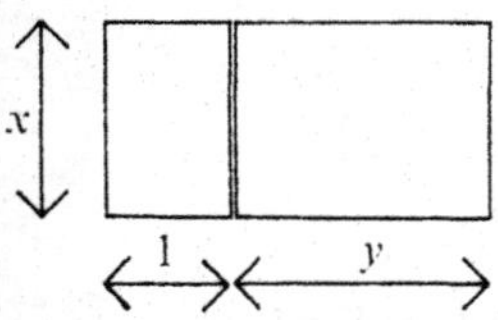

Any value of x substituted in this equation will give some resulting value for y; and the golden rectangle is only the special case where $x = y$, and therefore where the pattern is made up by adding squares. We take a $1:2$ rectangle just for example, i.e., $x = 2$. This gives $y = 3$, and we can then create a stacking pattern by the successive addition of $2:3$ proportion rectangles (Figure 9.11).

Figure 9.11

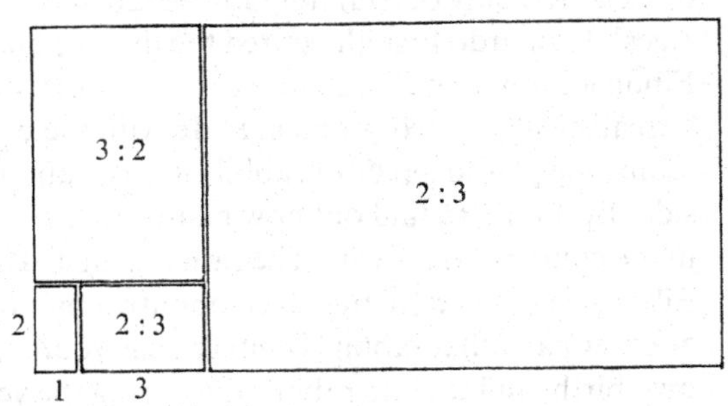

Scholfield[11] has given the name θ to an irrational number related to ϕ which has cropped up on occasion in proportional systems, and which corresponds to a pattern produced by the addition of *double squares*,

[12] Leonardo's great work is *Liber Abacci*, a compendium of arithmetical and algebraic knowledge, which survives in a manuscript from 1228. The rabbits problem is quoted from the English translation of N. N. Vorob'ev, *Fibonacci Numbers*, Oxford, Pergamon Press, 1961, p. 2.

i.e., where $y = 2x$ in our equation. The resulting value is approximately 2·414. This is the beginning of another series of numbers, where y is equal progressively to x (giving the $1:\phi$ proportion), $2x$ (giving $1:\theta$), $3x$,... nx corresponding to the stacking of squares, double squares, triple squares and so on.

To go back to Figure 9.9: there is no suggestion of a grid of repeated dimensions underlying the nesting of golden rectangles. Nor are the lengths of sides of the rectangles formed from integral multiples of either a or $a\phi$, the sides of the 'unit' rectangle, in the way which we saw in Figure 9.1. There are, however, interesting *additive* properties in the series which these lengths constitute. If we arrange the dimensions in order of increasing size, irrespective of their orientation, we have:

1, 1·618, 2·618, 4·236, 6·854, 11·090...

(to three places of decimals only).

If we add the first two terms together (1 + 1·618) we get the third, 2·618. In the same way the second and third terms added (1·618 + 2·618) give the fourth, 4·236; and so on. This follows from the way the golden rectangles nest in the pattern. Each successive term is the sum of the preceding two, and in general if u_n is the nth term, then $u_n + u_{n+1} = u_{n+2}$.

This is a *Fibonacci series*, named after Fibonacci, otherwise Leonardo of Pisa, 'the only outstanding European mathematician of the Middle Ages'. Leonardo first illustrated the development of the series of Fibonacci numbers in relation to an ostensibly practical, if somewhat unrealistically stated problem to do with the breeding of rabbits.[12] 'Someone placed a pair of rabbits in a certain place, enclosed on all sides by a wall, to find out how many pairs of rabbits will be born there in the course of one year.' The rather artificial assumptions which Fibonacci made were that every month a pair of rabbits produces another pair; that rabbits begin to bear young two months after their own birth, and that no rabbits die during the year.

At the start of the experiment there is *one* pair only. In the first month they produce young, so that there are now *two* pairs. In the second month the younger pair have not begun to breed, but the original pair produce another pair to bring the total to *three* pairs. In the third month the pair born in the first month reach breeding age. The original pair are still producing too, so the total now reaches *five* pairs. Of these five pairs three produce offspring in the fourth month, bringing the total to *eight* pairs. Of the eight pairs, five have offspring in the next month, giving *thirteen* pairs, and so it goes on. By the end of the year the total has reached 377 pairs.

If we put these numbers down in order

1, 2, 3, 5, 8, 13...

we have the Fibonacci series based on the first two positive integers for its first two terms. But equally we could produce series of the Fibonacci type with some quite arbitrarily chosen pair of numbers, for example 7 and 10, the rule being that each successive term is formed from the sum of the previous two:

7, 10, 17, 27, 44, 71...

If the first two terms of any Fibonacci series are u_0 and u_1, it is clear then that successive terms will be of the form:

$$u_0,\ u_1,\ (u_0 + u_1),\ (u_0 + 2u_1),\ (2u_0 + 3u_1),\ (3u_0 + 5u_1)\ldots$$

so that the whole set $F(u_0, u_1)$ is determined by the initial choice of u_0 and u_1, and is given by the expression

$$F(u_0,u_1) = \{x \in R_+ \mid x = u_0, u_1 \text{ or } u_i + u_{i+1} = u_{i+2},\ i \in Z_+\}.$$

All the terms are of the general form $(n_1u_0 + n_2u_1)$ where n_1 and n_2 are positive integers. But it does not follow that the Fibonacci series is a set which is closed under addition, since there are many 'linear combinations' of u_0 and u_1 which are not found in the series. It does not contain, for example, the term $2u_0$ or any higher integral multiples of u_0; and similarly for u_1. Nor does it contain such combinations as $(2u_0 + 2u_1)$, $(3u_0 + 2u_1)$, $(3u_0 + 3u_1)$.... Indeed n_1 and n_2 are themselves successive terms of a Fibonacci series.

[13] The inception, development, principles and application of Le Corbusier's system are described, with a mass of anecdote and digression by the way, in Le Corbusier, *The Modulor*.

To return to the Fibonacci series generated from the pair of numbers 1 and ϕ: we wrote these out before as numerical approximations in decimal form. But we know from our earlier demonstration that the third term in the series, $(1 + \phi)$ must equal ϕ^2. The fourth term $(1 + 2\phi)$ can be rewritten $\phi + \phi^2$, which is $\phi(1 + \phi)$ or ϕ^3. The fifth term is similarly ϕ^4 and so on: and the whole series can be represented more economically in geometric series as successive powers of ϕ:

$1, \phi, \phi^2, \phi^3, \phi^4, \phi^5 \ldots$

Le Corbusier's proportional system, the Modulor, consists of two interrelated series of preferred dimensions.[13] Again there is no suggestion of a unique repeated grid unit and any of the dimensions in either of the series may be used together with any other. Le Corbusier calls the two series the Red and the Blue; they are both Fibonacci series. The Red contains as one member the key dimension corresponding to the height of the 'six foot detective', and which Le Corbusier converts into the metric equivalent of 1830 mm (in round figures). A second dimension is introduced, 2260 mm, which Le Corbusier shows (in his famous sketch of the human figure (Figure 8.2) which is the Modulor's 'trade-mark') as being the height a man can reach with his arm stretched upwards. This second basic size belongs to the Blue series. It is halved, to give 1130 mm; and 1130 constitutes the next lowest term in the Red series. We can begin to see how the two series interlock.

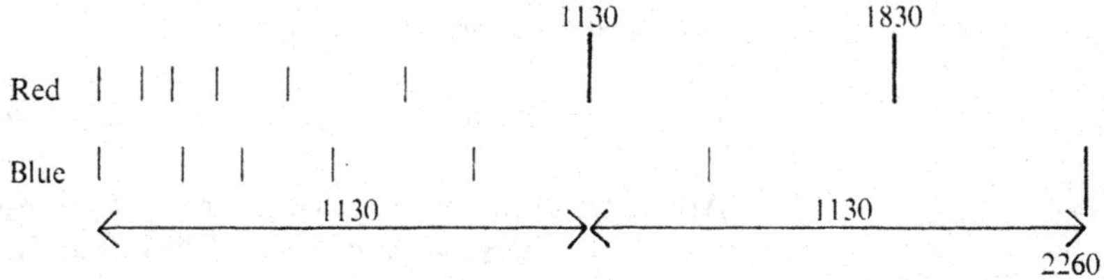

Figure 9.12

With two consecutive members of the Red series given, 1130 and 1830, we can construct the whole series both upwards and downwards. (1830 — 1130) gives the next lowest member, 700; and by a similar process of subtraction the series is given in descending order as 1830, 1130, 700, 430, 270.... Working upwards from 1830 by adding terms, we get 2960, 4790, 7750 and so on. The relationship of the Blue series to the Red series, as was indicated in halving the basic Blue dimension to give a Red one, is that each of the Blue sizes is twice some corresponding Red dimension. Besides 2260 mm, the Blue series thus contains 1400, 860, 540, 320... in descending order; and can be constructed upwards beyond 2260 by addition of successive terms as before, or by doubling appropriate Red sizes.

To put this in formal notation, if b_i is the general term in the Blue series, and r_i is the general term in the Red series, then

$$b_i + b_{i+1} = b_{i+2}$$
$$r_i + r_{i+1} = r_{i+2}$$
$$\text{and } 2r_i = b_i.$$

But the Modulor is more cunningly constructed yet. We have given the dimensions in the two series rounded to the nearest whole number of centimetres. But by a very careful choice of the two key sizes, as measured to a greater accuracy, Le Corbusier has in point of fact ensured that each of the two Fibonacci series is a ϕ series. If we call the key size in the Red series d (d = '6 foot detective'), then the next largest size is ϕd, and the series working upwards is given by

$$d,\ \phi d,\ (1+\phi)d,\ (1+2\phi)d,\ (2+3\phi)d...$$

or rewritten, as before:

$$d,\ \phi d,\ \phi^2 d,\ \phi^3 d,\ \phi^4 d... \qquad \text{We put } C_{\text{Red}} = dG(\phi).$$

The Blue series then becomes simply:

$$2d,\ 2\phi d,\ 2\phi^2 d,\ 2\phi^3 d,\ 2\phi^4 d... \qquad C_{\text{Blue}} = 2dG(\phi).$$

The dimensions in the series below d and $2d$ respectively are given by d/ϕ, d/ϕ^2, d/ϕ^3... and $2d/\phi$, $2d/\phi^2$, $2d/\phi^3$... and the complete 'Corbusier set' C of all dimensions from the Modulor is written

$$C = d(G(\phi) \cup 2G(\phi)).$$

In Figure 9.13 we show the series of different shaped rectangles which the Modulor produces, distinguished (by shading in the figure) into three separate groups. There are rectangles whose side lengths are drawn only

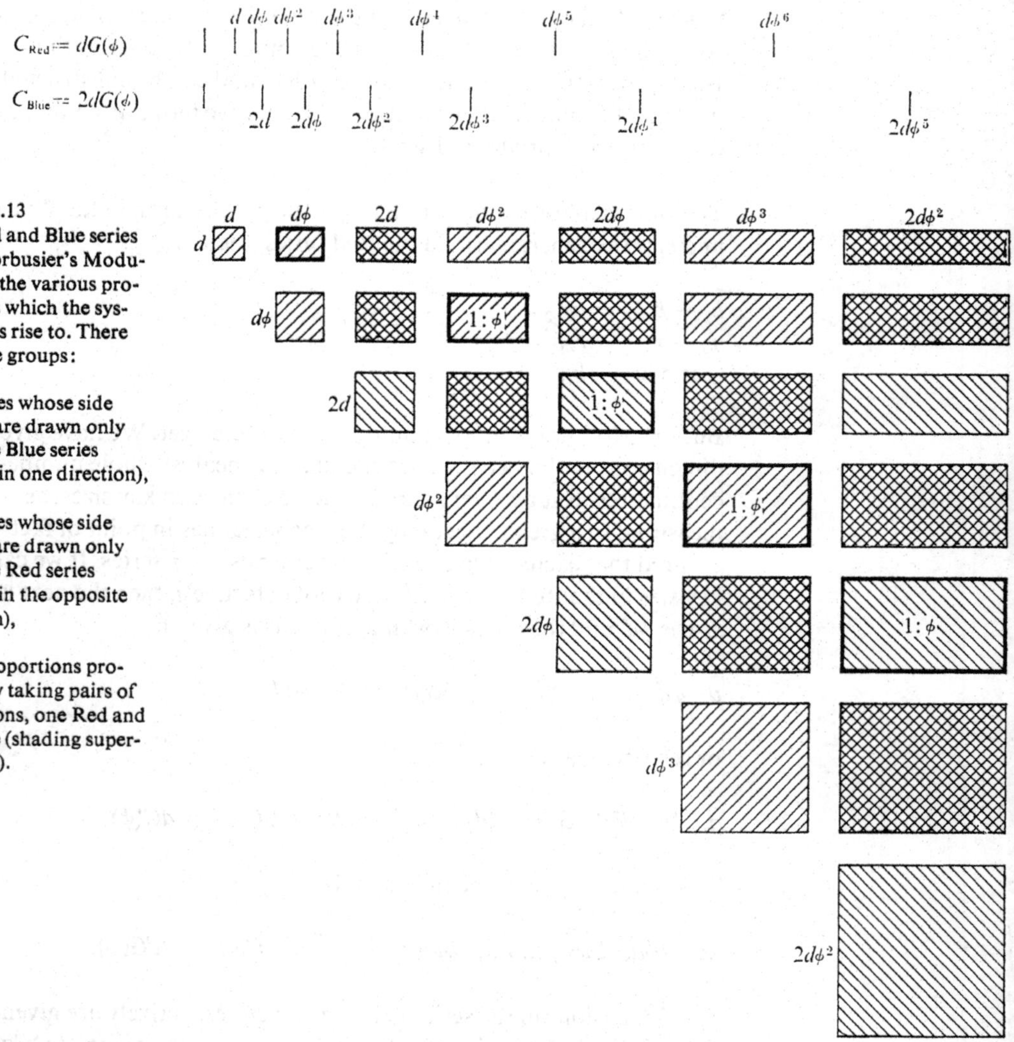

Figure 9.13
The Red and Blue series of Le Corbusier's Modulor, and the various proportions which the system gives rise to. There are three groups:

rectangles whose side lengths are drawn only from the Blue series (shaded in one direction),

rectangles whose side lengths are drawn only from the Red series (shaded in the opposite direction),

those proportions produced by taking pairs of dimensions, one Red and one Blue (shading superimposed).

[14] Christopher Alexander. 'Perception and Modular Coordination', *RIBA Journal*, October 1959, pp. 425–9.

[15] Christopher Alexander quotes the experiments of G. T. Fechner reported in *Vorschule der Aesthetik*, Leipzig, 1876, pp. 190–202; and of T. R. Austin and R. E. Sleight, *Journal of Applied Psychology*, vol. 35, 1951, pp. 430–1.

from the Blue series (shaded in one direction), and the same for the Red series (shaded in the opposite sense); those shapes produced by taking pairs of dimensions, one Red and one Blue, to form the two sides of each rectangle, are shown with both types of shading superimposed.

Viewed as a *proportional system*, the Modulor gives repeated golden rectangle proportions, as we see, with the 'Blue only' and 'Red only' combinations, and double squares where the two series are mixed – besides many other proportions of a more complex nature. The additive properties of the Modulor which would be useful in *modular coordination* are also shown up in the figure. For each separate group of rectangles, since their side-lengths in increasing order of size along any column or row are in Fibonacci series, it follows that any two adjacent rectangles in a column or row can be 'stacked' together to form the next in that row. (It is *not*, however, possible to combine a rectangle from one group with any rectangle from one of the other two groups, and produce another rectangle contained within the whole repertoire.) Looked at the other way round, it is a consequence of taking the Fibonacci series based on ϕ that any large dimension in one of the series may be broken down, first into the golden section $1:\phi$ ratio, and then each of these two subdivisions filled with some whole number of small modular dimensions. Finally, the Modulor – because of its original derivation from sizes which are, somewhat loosely, determined from measurements of the human body – does acknowledge the ergonomic factors which must enter into a practical choice of modular dimensions.

Christopher Alexander, in a most lucid and penetrating article, 'Perception and Modular Coordination', has suggested that the aesthetic significance of systems of architectural proportion like the Modulor and others depends on rather general properties which they share in common, rather than any special properties of the numbers found in each system.[14] In particular, he debunks much of the mythology and superstition which has surrounded the golden number ϕ. As he says, it is 'an unaccountable empirical fact' that the particular shape of the golden rectangle pleases the eye (as psychological experiments have certainly indicated),[15] but that nevertheless people are unable to distinguish in practice between a golden rectangle and a rectangle which differs in proportion from it by 3% or 4%; so that one can make no appeal to the special numerical properties of ϕ as an explanation of people's preference for the shape. It has been suggested in the past that

[16] Edgar Kaufmann, ed. *An American Architecture.* New York, Horizon Press, and London, Architectural Press, 1955.

the fact that ϕ is an irrational number means that the incommensurable ratios which it gives rise to result in 'dynamic' proportions as opposed to the 'static' appearance produced by integral lengths. But this of course is sheer play with words. An object in the real world can have a length which approximates to ϕ but must, in the sense that it is measured, be a rational quantity. Irrational numbers are purely abstract concepts. There is no sense in which one can *see* or *measure* the 'irrationality' of a proportion or ratio based on irrational numbers.

The only real effects of the use of a proportional system which may be appreciated visually, in Alexander's view, result from the general ways in which the limitation of the number of dimensions used in a building's design to some preferred set can produce an appearance which – because of the repetition of similar shapes, lengths, simple whole number ratios – we must, in a broad sense, call 'ordered'. 'There is a lack of confusion. A certain simplicity. Relations between the parts.'

Lest we should leave the subject of proportional systems and modular planning and give the impression that their use is confined uniquely to rectangular geometry, we shall end this chapter with some startling evidence to the contrary. This is the promised example of a system based on the circle; in a design which on the face of it would appear to represent the very extreme of romantic arbitrariness.

The design in question is Frank Lloyd Wright's project for a house for Ralph Jester in Palos Verdes, dating from 1938 (Figure 9.14). The plan is presented in the anthology *An American Architecture*[16] over the motto 'Exuberance is Beauty'. The house consists of a series of separate cylindrical pavilions for the rooms, grouped around a patio which is open to the outside, with a protecting rectangular roof linking the composition together. Two principal materials are used: thin curved plywood sheets for the circular rooms and concrete in the substructure of the curved terraces and swimming pool. The very massive cylindrical columns, 4 ft in diameter, would have been built from Wright's 'poured masonry' – large local stones set roughly in concrete – as used in many of his Western houses, including his own Taliesin West.

The basic unit module for the plan is given by the dimension of the standard 4-ft wide plywood sheet – the size in which it comes from the manufacturer. Because plywood is a thin supple material the same sheet can be readily bent into curves of different radius; and indeed this gives the sheet structural strength, where used flat it would require bracing.

The plywood, however, has a double function. It is used as a structural material in its own right for the rooms. And it is also used to make the cylindrical *formwork* into which the concrete walls of the swimming pool and other parts are cast; including the formwork for the 'poured masonry' columns. The one unit gives the measure, therefore, for the

Figure 9.14
Frank Lloyd Wright, project for Ralph Jester House (1938), plan

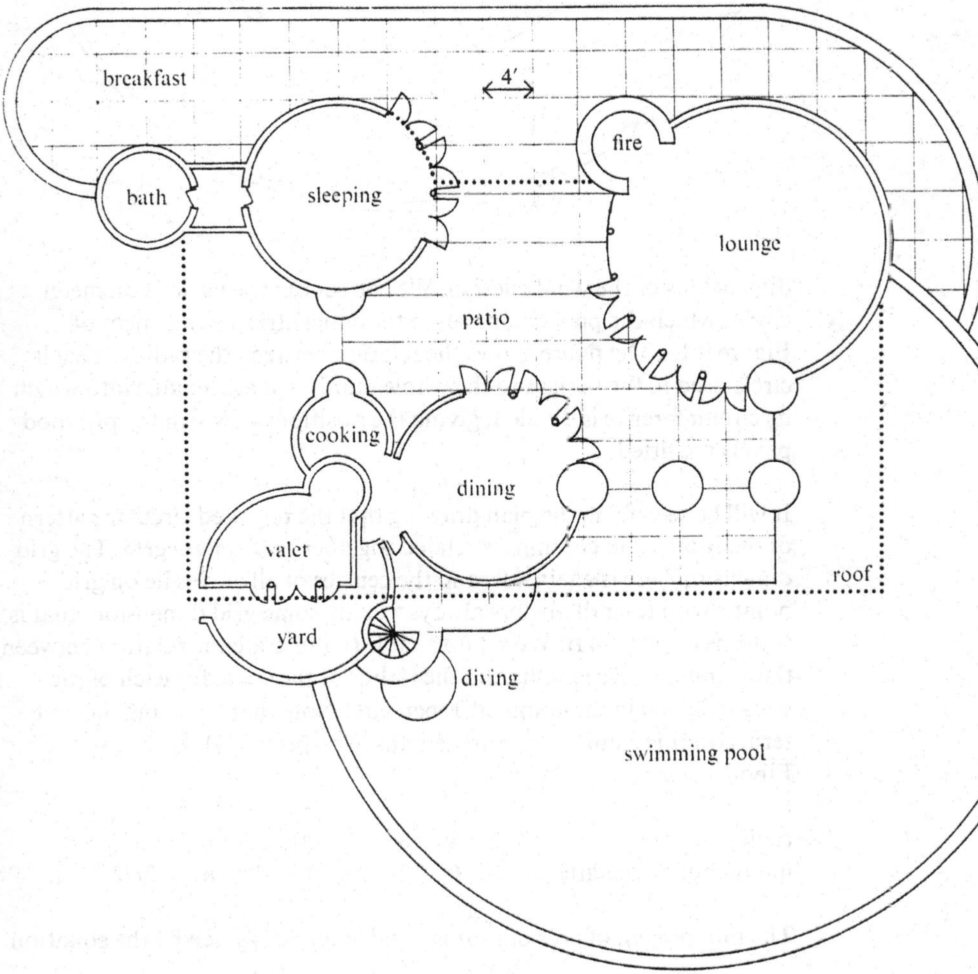

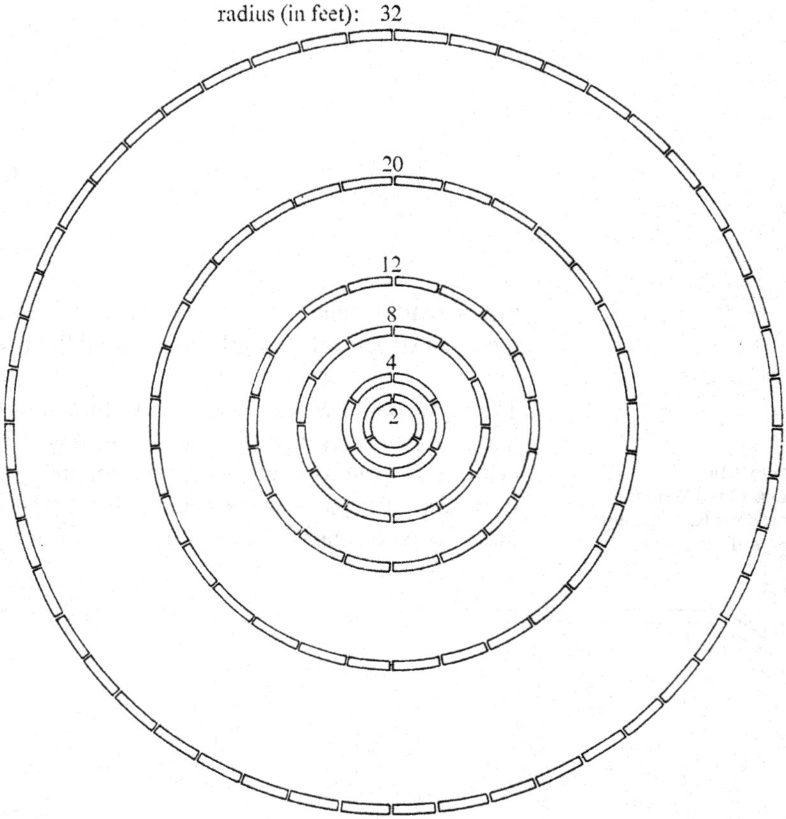

Figure 9.15
The set of circles employed in the plan of Wright's Jester House, indicating the whole number of arc lengths into which the circumference of each is divided, to correspond to the standard dimension of the bent plywood panels

dimensions of the whole design. We can set out the various diameter circles which the plan comprises in the concentric arrangement of Figure 9.15. The figure shows the relation between the radius of each circle used in the design, to the whole number of arc lengths into which its circumference is divided, giving the positions in which the plywood panels are fitted.

It will be seen from the plan drawing that the repeated circular pattern of the rooms and columns is related together on a *square* grid. The grid dimension here is again 4 ft, and the centres of all circles lie on grid points, with their diameters always exactly some grid dimension, that is, some multiple of 4 ft. We set up a table to illustrate the relation between the number of arc lengths and the radius of the circle for each of the sizes included in the complete repertoire. Note that, ignoring the first terms (radius 2, number of arc lengths 3, respectively) these are Fibonacci series.

radii	2	4	8	12	20	32	r
number of arc lengths	3	6	12	18	30	48	$n_r = 3r/2$

The number, n_r, of arcs of radius r and length a is given by the equation

$$2\pi\, r = n_r a.$$

When $n_r = 3r/2$, as we have it in the Jester House, then

$$2\pi\, r = 3ra/2$$

so that the arc length is given, in feet, by

$$\begin{aligned} a &= 4\pi/3 \\ &\doteqdot 4\cdot 18. \end{aligned}$$

Wright uses the standard 4-ft wide plywood panel, formed to the radii we have listed above, to fill each arc intended. This leaves a tolerance of about 2 in for joints, door frames and mullions.

Let S be the set of all 4-ft wide plywood sheets and J be the set of plywood sheets, s_r, of radius r (in units of 1 ft) employed for the walls and formwork of the Jester House. Then we may express the Jester set concisely as

$$J = \{s_r \in S \mid r \in 2 \text{ or } F(0,1)\}.$$

This shows that the system neatly marries the 4-ft square planning grid, governing the location of all straight walls (s_∞) and the centres of radii, with the Fibonacci series $F(0,1)$. For detailed planning Wright also permits the use of a half-module radius of sheet and a simple halving division, $G(2)$, of the planning grid. The ingenuity of the Jester set, J, is such that we are inclined to credit Wright with having 'solved' the impossible mathematical problem of squaring the circle in this exceptional project.

10 Planar graphs and relations

The Swiss mathematician Euler is acknowledged as the father of the *theory of graphs*, and some of the popular puzzles to which the early theory was applied, for instance the problem of the Königsberg bridges, with which Euler introduced his first paper,[1] are perhaps not too remote from matters of architecture, or at least town planning. Euler's problem depended on the particular layout of the city of Königsberg (now Kaliningrad) in Prussia, which stands on the River Nagel. The different parts of the city which lie on either bank, between a fork in the river, and on the island of Kneiphof, were joined at that time by seven bridges. Was it possible to take a walk around the city starting from any point and, crossing all the bridges, each only once, to arrive back home at the same place?

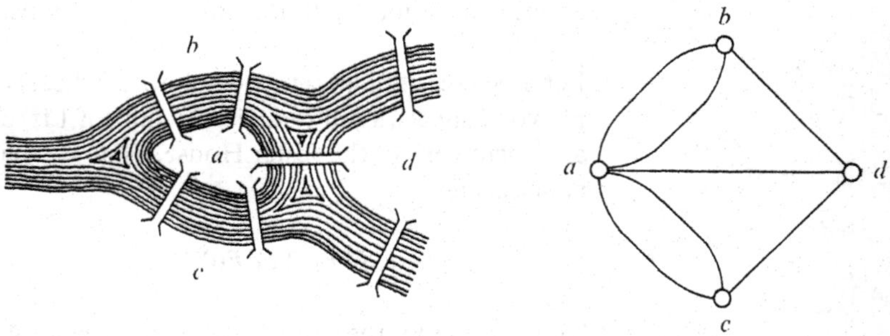

Figure 10.1
The Königsberg bridge problem, with Euler's representation of the city and its bridges as a *graph*

Since the bridges are, for the sake of the problem, the most important features of the plan, Euler devised a simplified diagram in which the four parts of the city were represented by points and the bridges by appropriate lines joining these points.

This type of simple geometrical figure of points and lines is the 'graph' of graph theory. (The term has two distinct meanings in mathematics. More usually, of course, it describes a diagram showing the variation of one quantity with another of which it is a function, in relation to some system of coordinates – a graph of speed against time for example, to show acceleration.) The points in a graph are termed *vertices*, and the lines, or more strictly line segments, joining the vertices, are *edges*.

[1] In the 1736 volume of the publications of the Academy of Science in St Petersburg.

In Euler's diagram the graph is in effect a simplified map, and in general it is possible to represent any kind of route map as a graph, where the edges are roads (or railways, or footpaths) and the vertices junctions. But there are some important points here. The actual way in which the graph is drawn is of no significance to the theory. The vertices may be placed anywhere in the plane, and the edges need not necessarily be straight, or of any fixed length. In the case of a road map, the edge in the graph indicates simply the existence of a road between two points, and not the exact direction which it takes or how long that road is. Euler's graph shows simply which bridges join which parts of Königsberg, and nothing more. In general, the edge represents some relationship, of whatever kind we may be interested in, between the two objects signified by the vertices. We could use a graph to represent a chess tournament – the players by vertices, and the games played between different opponents by edges. Graphs have been used in the theory of electrical networks, to denote the structure of human organizations or social groups, and even in the scientific study of decision-making – where vertices signify the decisions to be made and the edges determine the necessary relationship which one decision bears to another.

The value of the graph lies in the capacity it has for showing up the essential structure of a set of relationships. In the Königsberg problem we know that we must arrive in each part of the city (*a*, *b*, *c* or *d*) the same number of times as we leave it – otherwise we would never get back to the starting point. Since we may only cross each bridge, or traverse each edge of the graph, once, this means that for there to be a possible route the number of edges joined to or *incident* with each vertex must be even. This is true for none of the four vertices *a*, *b*, *c* or *d*, and so the problem has no solution, as Euler demonstrated.

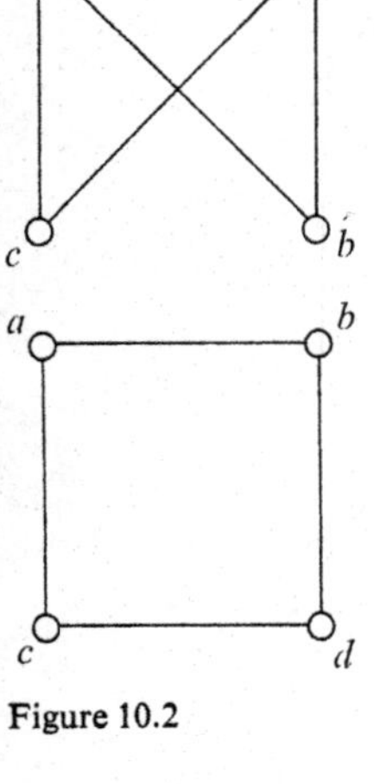

Figure 10.2

It is quite permissible to draw a graph with some edges crossing (without their intersections being vertices) – although it may perhaps be confusing. Graphs that it is possible to draw without the edges intersecting are called *planar* and it is often convenient to redraw such graphs in this form, for the sake of clarity. (To take an elementary example, the graph of four vertices *a*, *b*, *c*, *d* of Figure 10.2 is shown in such a way that the edges (*a*, *b*) and (*c*, *d*) cross. Despite this, the graph is planar and may be redrawn in the form beneath.) Since the same graph may be drawn in a variety of ways, it follows that two graphs (G_1 and G_2) which are apparently at first sight dissimilar, may prove to be identical when analysed. In this case they are termed *isomorphic*: that is, if they have the same number of vertices, and wherever two vertices in G_1 (say a_1 and b_1) are connected by an edge, then there are two corresponding vertices a_2 and b_2 in G_2 also connected.

A classic graph theory puzzle, and another with almost architectural aspects, is the so-called 'utilities' problem. In a typical form the puzzle concerns three houses and three wells. The wells (x, y and z) are apt to dry up, so the occupants of the houses (a, b and c) who do not, incidentally, enjoy each others' company, wish to be able to use whichever well is full; but without meeting their neighbours on the way. The problem consists therefore in planning nine footpaths, one from every house to every well, in such a way that no footpaths cross.

Figure 10.3
Graph of the 'utilities' problem, or the problem of houses and wells

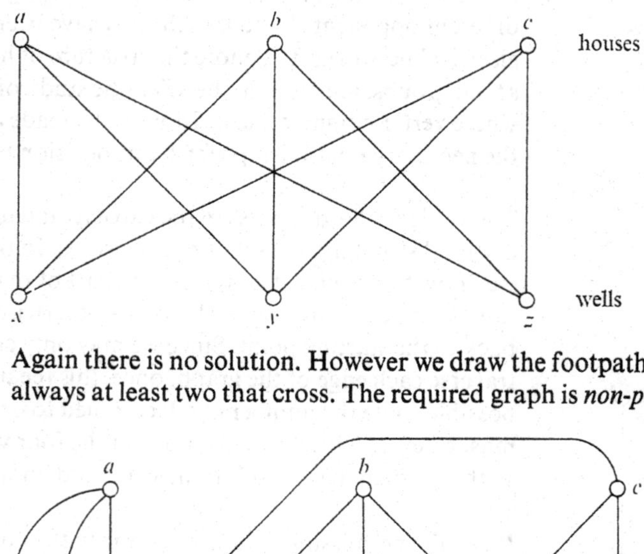

Again there is no solution. However we draw the footpaths, there are always at least two that cross. The required graph is *non-planar*.

Figure 10.4

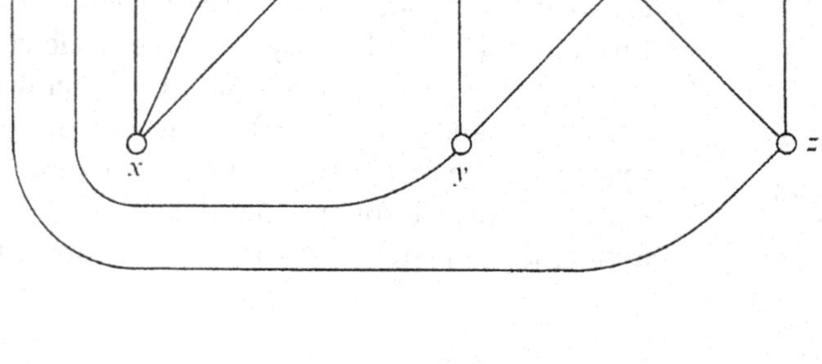

We can show the reasons for this by invoking the *Jordan curve theorem* which in itself might appear trivial. Suppose *K* is a continuous closed curve in the plane. Then *K* divides the plane into an outer and an inner part so that whenever any point *p* in the inner part is connected to a point *q* in the outer part by a continuous curve *L*, then *L* intersects *K*. This is perfectly self-evident, and has the equally clear implication that if two points *p* and *q* on a closed curve *K* are connected by a curve *L* which does not otherwise intersect *K*, then *L* lies either completely inside or completely outside *K*.

Figures 10.5 and 10.6
The Jordan curve

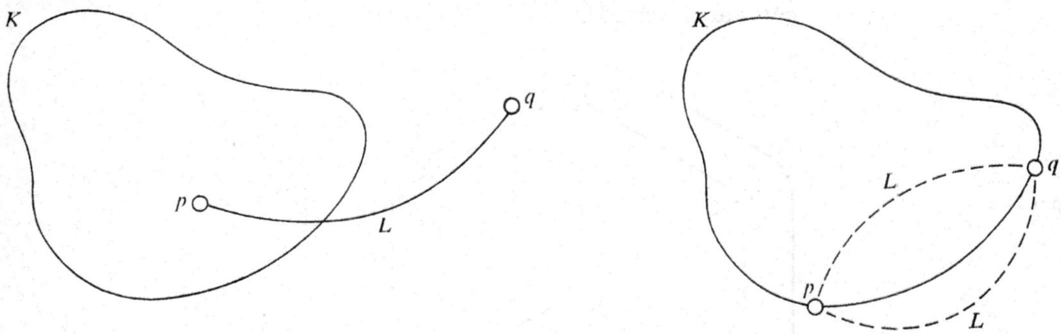

Now let us redraw the houses and wells in such a way as to make use of these facts. The footpaths (*a*, *z*), (*z*, *c*), (*c*, *x*), (*x*, *b*), (*b*, *y*), (*y*, *a*) form the closed Jordan curve. This *cycle* of edges is expressed more compactly by the notation (*a*, *z*, *c*, *x*, *b*, *y*, *a*).

Figure 10.7

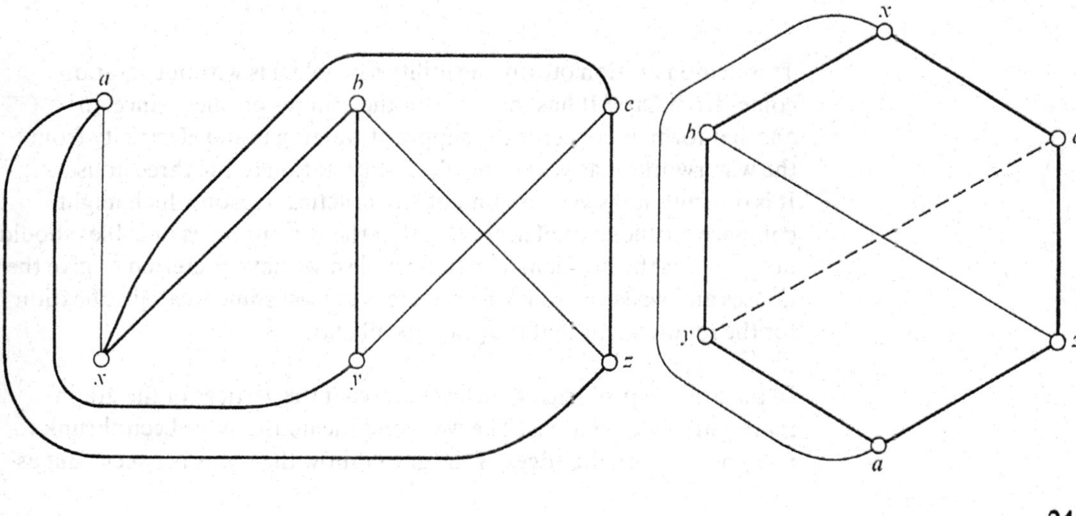

[2] Oystein Ore. *Graphs and their Uses*, New York, Random House, 1963, pp. 14–17.

It remains to join a to x, b to z and c to y. (a, x) can lie either inside or outside the curve, as can (b, z). In order that they shall not intersect, one must lie inside and the other out. (This too is capable of proof by the theorem.) Suppose (a, x) is on the outside, as in Figure 10.7; the edges (a, x), (x, b), (b, z), (z, a) now form a second closed curve (a, x, b, z, a), with y on the inside and c on the outside of the curve. By Jordan's theorem (c, y) must therefore intersect one of the four edges making up this second curve, as Figure 10.8 makes clear. Exactly similar arguments apply if the positions of (a, x) and (b, z) inside and outside the first curve are reversed.

Figure 10.8

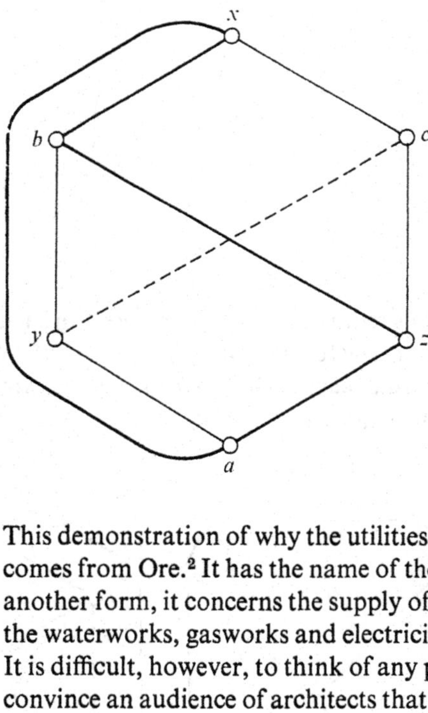

This demonstration of why the utilities problem is without solution comes from Ore.[2] It has the name of the utilities problem since, in another form, it concerns the supply of water, gas and electricity from the waterworks, gasworks and electricity station to the three houses. It is difficult, however, to think of any practical reason which might convince an audience of architects that the various pipes or cables should not cross, as the problem demands; and so we have preferred to give the houses and wells version, where there is at least some weak justification for the requirement that the graph be planar.

In the utilities problem as we have drawn it the vertices of the graph represent places or areas. The wells and the houses have been shrunk to mere points, and the edges of the graph show the paths between. But as we have said, a graph may be used to denote other kinds of relationship, and we can draw graphs corresponding to maps and plans which show properties other than the network of routes.

Let us take a more realistically architectural problem, in the design of one floor plan of a small terrace house. On the ground floor of this house there will be three rooms: a kitchen *k*, a dining-room *d* and a living-room *l*. There will also be some circulation space *c* – a hall or corridor giving access to the rooms, and to the stairs. We shall specify a set of requirements which the plan of the house is to fulfil. These requirements are all stated in terms of *adjacency*, that one room be next to another. The living-room is to be next to the dining-room, and the dining-room next to the kitchen (but the plan will still be quite acceptable if the kitchen is next to the living-room, for instance, although this is not one of the stated requirements).

The circulation space *c* we specify must be adjacent to *k* and *l*, but not *d* – it will be permissible to give access to the dining-room via the living-room, or via the kitchen, rather than direct. *c* must also be adjacent to the area outside the house on the street side. By 'adjacent' here we mean that it must border on, or have some wall in common with the area in question. The purpose of this requirement in real terms is of course to allow for a front door into the house from the street. We shall label the areas outside the house on the four sides (the street, the garden and the two neighbouring houses) *n*, *s*, *e* and *w* for the points of the compass, so as to identify them as in Figure 10.9.

Figure 10.9

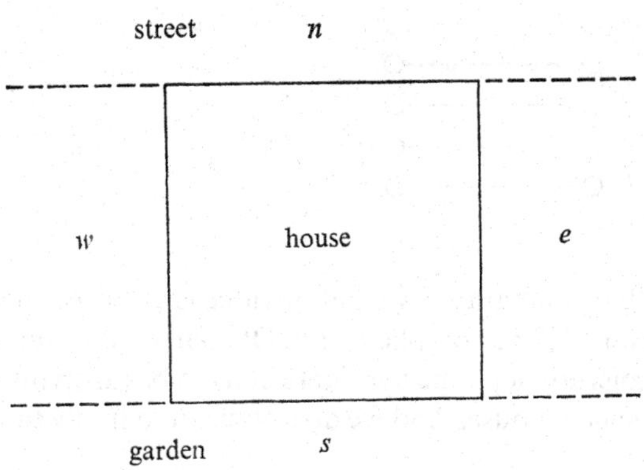

So far the implication behind the 'adjacency requirements' is that when satisfied they allow for direct access (by means of a door) from one room to another. Note that we are placing no restriction on the size or shape of rooms; we simply demand that they share some length of wall or boundary in common (although in practice, to provide space for a door, this length of common wall would naturally have to be about a metre or more). We can imagine other adjacencies which would ensure the satisfaction of other practical needs. Rooms may be required to be adjacent to outside walls so as to allow for windows and natural ventilation. We shall specify that in our example the living-room *l* and the kitchen *k* both face onto the garden front *s*, so as to catch the sun. (The dining-room may either be internal, or else face *n* onto the street.) One can imagine other reasons why one might wish to specify that two rooms be adjacent. One can even imagine a contrary requirement, that two rooms should *not* be next to each other – for reasons of noise, or privacy, for example – but we shall not introduce any such requirements into our present example.

We can summarize the requirements made so far in a simple graph theory notation (Figure 10.10), where the different rooms or areas are vertices, and the pairs required to be adjacent are joined by edges. Note that the edge here, unlike in the graphs of the Königsberg bridges, and the houses and wells, signifies not necessarily a path or route, but a relationship between areas in the plan. Two areas, *e* and *w*, do not feature in any of the required adjency pairs.

Figure 10.10

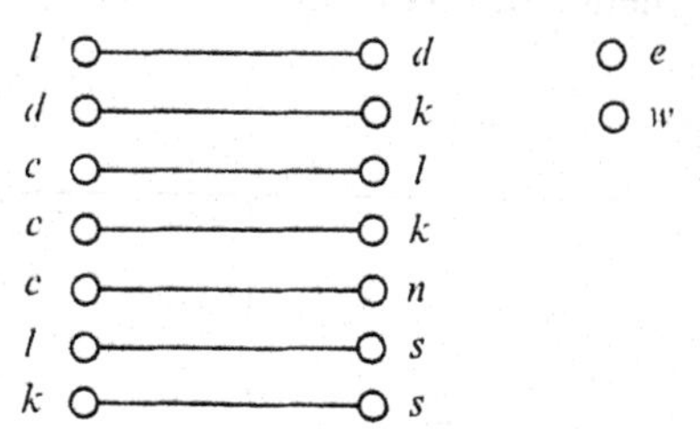

l living-room

d dining-room

k kitchen

c circulation space

n, *e*, *s*, *w* areas around house

No comparative weight or value is attached to these requirements – they must all be complied with. Of course these hardly represent a complete picture of all the considerations to be taken into account when planning such a house. And we do not mean to imply that the specifications we

Figure 10.11

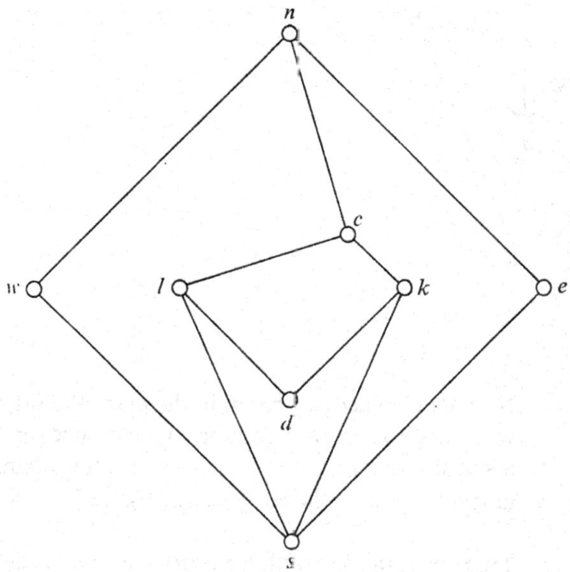

have laid down for our example have any universal value. In terms of some real house design, one could well argue over the pros and cons of this or that relationship in the plan. Our stated problem is not, however, unrealistic in the choice of some set of requirements that *might* be made; and these kinds of requirements do cover some of the broader questions of the disposition of rooms in relation to each other and to the plan as a whole which tend to be the determining factors in a real layout. In a less artificial example one can imagine that requirements might be placed in some order of importance, and some plan relationships be considered essential, others just desirable.

Fig. 10.11 is a planar graph showing all the requirements satisfied. *n* has been joined to *e*, *e* to *s*, *s* to *w* and *w* to *n*, since clearly these areas are adjacent. Note the important distinction between these relations of required room and space *adjacency*, and the relations of room *linkage* (i.e., which room is accessible from another directly, via a door) of the kind illustrated for the three Wright houses in Figures 1.13 and 1.14.

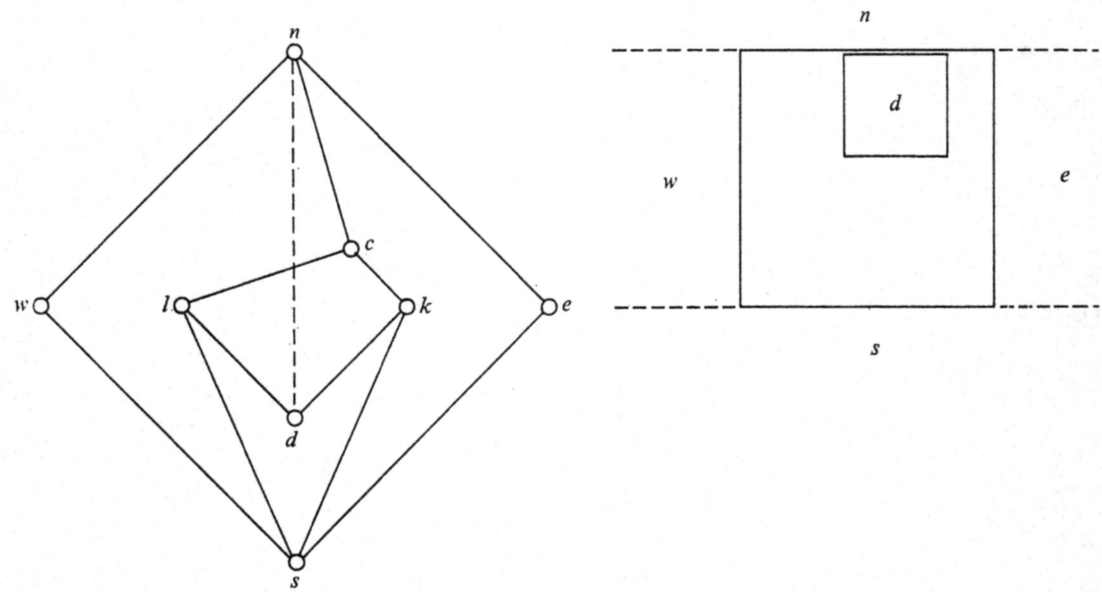

Figure 10.12

Now let us place one room in the plan. We take the dining-room and, without choosing any precise shape or size for the room, we decide that it should look onto the street side *n*. This means we must amend the graph to include the edge (*d*, *n*) (Figure 10.12).

Two edges in the graph now cross, and once again, however we redraw the figure, this is always the case. This graph too is non-planar.

Figure 10.13

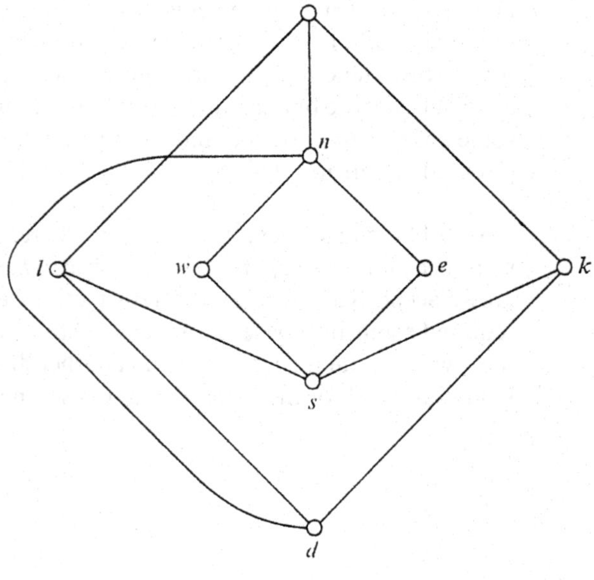

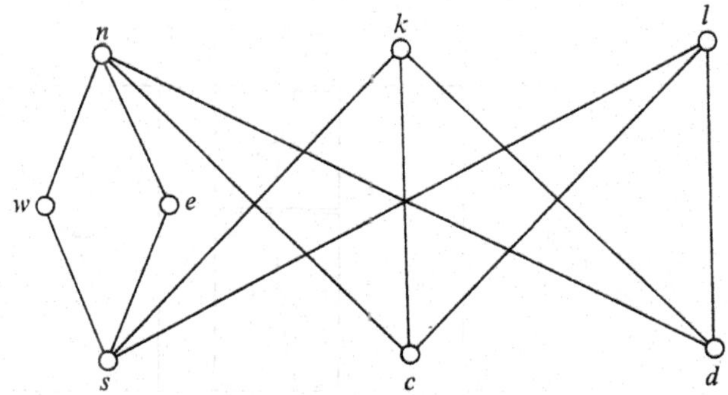

Figure 10.14

But if we organize the vertices in one particular arrangement (Figure 10.14), then the features of the problem appear suddenly strangely familiar.

With the exception of the intermediate vertices *w* and *e* lying between *s* and *n*, the graph is similar to that of the houses and wells. Note that there are only two edges incident with *e* (or *w*). We can define the process of *contraction* of a graph, as that of replacing such a vertex and its two adjoining edges with a single edge. Thus we may contract (*n*, *e*), (*e*, *s*), for example, to a single edge (*n*, *s*). It is clear that the contracted graph now 'contains' exactly the graph of the utilities problem.

Figure 10.15

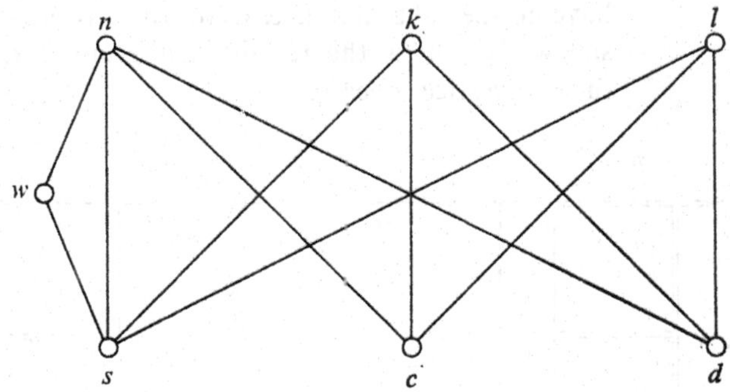

If one graph *H* is contained within another *G*, that is to say if *H* comprises some of the vertices of *G* and comprises, for those vertices, all their associated edges in *G*, then it is called a *sub-graph* of *G*. The original graph of our house plan (Figure 10.12) when contracted thus contains a sub-graph which is isomorphic to the utilities graph. It is for this reason that it is not planar.

The implication for the plan of the house is that once having placed the dining-room *d* in the position against the north wall that we have chosen, we will be unable to comply with all the stated requirements by any subsequent arrangement of the other rooms whatever. If we go back to the plan we can appreciate intuitively why this is so, although of course any number of examples will not provide conclusive proof of the

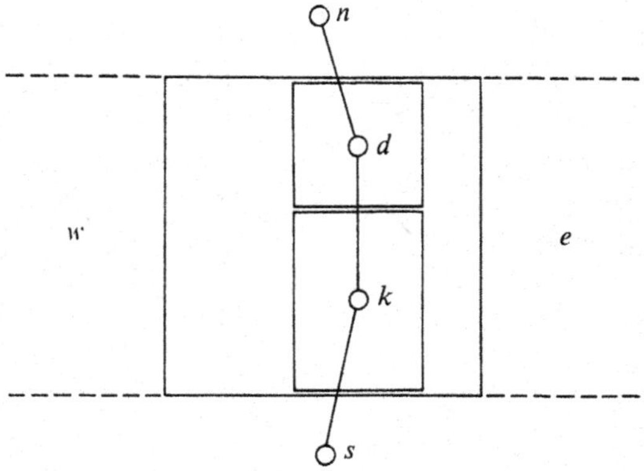

Figure 10.16

matter. With *d* in its present position, and since we know that *k* must both adjoin *d* and look out onto *s*, the garden front, then (*n*, *d*, *k*, *s*) must be joined in a *chain*: the dining-room and kitchen form a 'barrier' across the plan, isolating the two remaining unfilled areas from each other (Figure 10.16).

Put the living-room on one side, and the circulation space on the other. The kitchen/dining-room 'barrier' prevents the circulation giving access to the living-room: the link (*c*, *l*) cannot be made. The topological properties of the left-handed and right-handed versions of a plan – symmetrical by reflection as here – will always be the same. (Since we have no special preferences as to which rooms are adjacent to the two side walls *e* and *w*, the 'requirement graph' is not affected by swapping the two vertices over.)

Figure 10.17

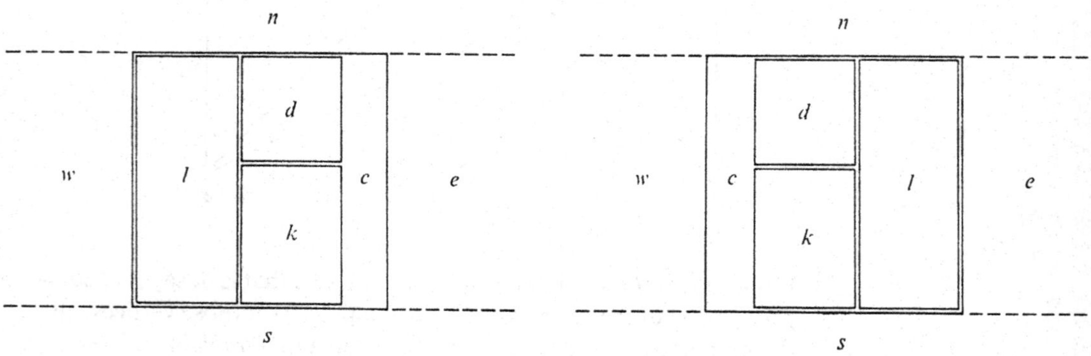

The answer might seem to be to move *d* and *k* to the side of the plan, and to place *l* and *c* together in the remaining area, *c* in the form of an entrance hall between living-room and dining-room. This hall must extend far enough to give access to *k*; but now the chain (*n*, *c*, *k*, *s*) blocks the plan and prevents the making of the link (*l*, *d*), to give direct access from living-room to dining-room.

Figure 10.18

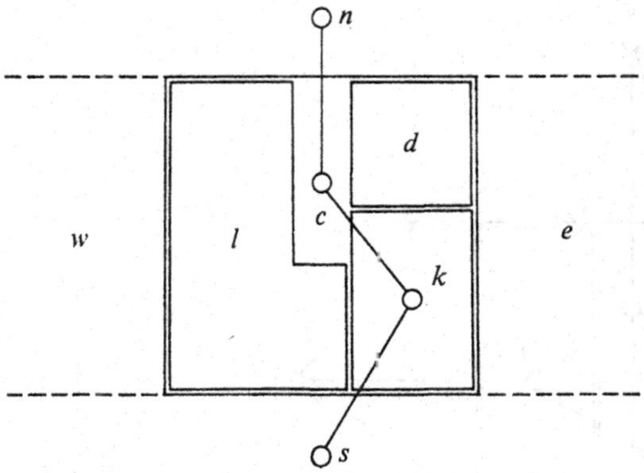

We can resort to more extreme measures. We put *l* next to *d*, and run a passage (*c*) down the side of the house and along the garden front to meet *k*. Now the living-room has no wall onto the garden. The requirement (*l*, *s*) cannot be met. And so on.

Figure 10.19

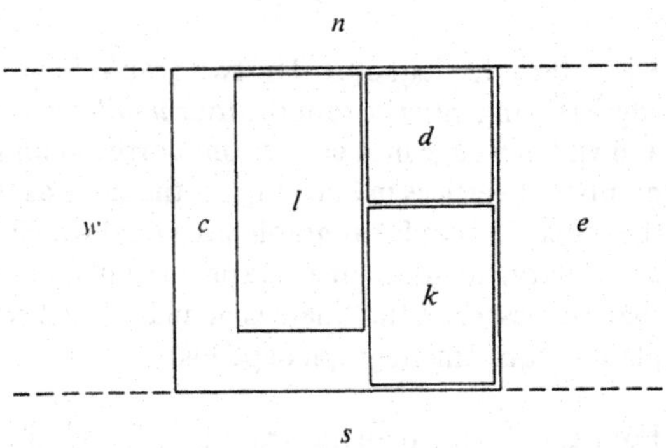

The fact is that *d* must not adjoin the north wall. Put it on the garden side and we can find a plan which meets all the specifications. Figure 10.20 shows the requirement graph superimposed on a permissible solution, with those extra edges added which correspond to the other (non-specified) adjacencies which occur in this particular arrangement ((*l*, *n*), (*k*, *e*), (*l*, *w*), (*k*, *e*) and (*d*, *s*)). Let us call this augmented graph the 'adjacency graph'; it indicates those rooms which actually are adjacent in a plan, as opposed to just those which are required to be.

The plan itself can be regarded as a graph, where the walls or boundaries of rooms are the edges, and the corners where the walls meet are vertices. The adjacency graph *A* is said to be the *dual* of this plan graph *P*. We refer to each separate area bounded by the edges of a graph as a *face*. (In the plan graph *P* each room is a face.) To each face f_P of the plan graph *P* there corresponds one vertex v_A in *A*, and the number of edges at a vertex v_A in *A* is the same as the number of boundary edges of the corresponding face f_P in *P*. (This is in effect how we defined the original requirement graph, which is now 'contained' in *A*.)

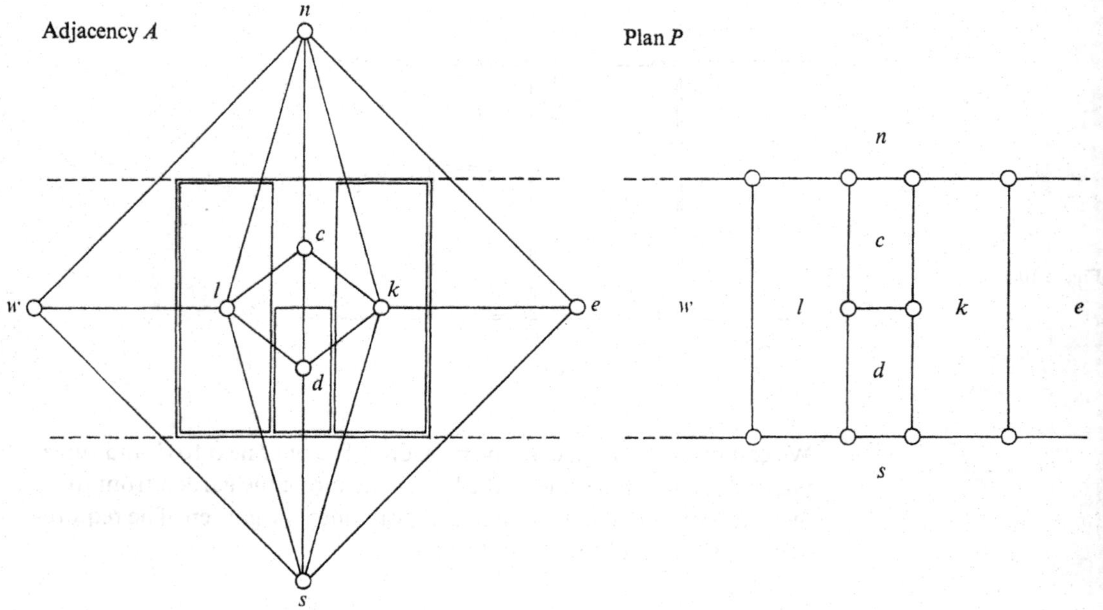

Figure 10.20
Adjacency graph *A*, and plan graph *P*

P is in turn the dual of *A*. The two graphs have the same number of edges (taking into account the four *infinite edges* in *P* separating *n*, *e*, *w* and *s*). Each edge in *A* crosses one corresponding edge in *P*. And the number of vertices in one graph is the same as the number of faces in the other. Every planar graph has a dual which is also planar; and, as a corollary, no non-planar graph has a planar dual. It is on this theorem that our assertion that there is no plan which corresponds to a non-planar requirement graph depends.

For a final house plan illustration of this kind we will slightly revise the terms of the problem, by assuming now that access to the house is from the garden side *s* rather than from the street. We shall demand in addition that the living-room be adjacent to the kitchen, and that the circulation space *c* shall give direct access to all rooms. The adjacency requirements for the new problem are then these:

Figure 10.21

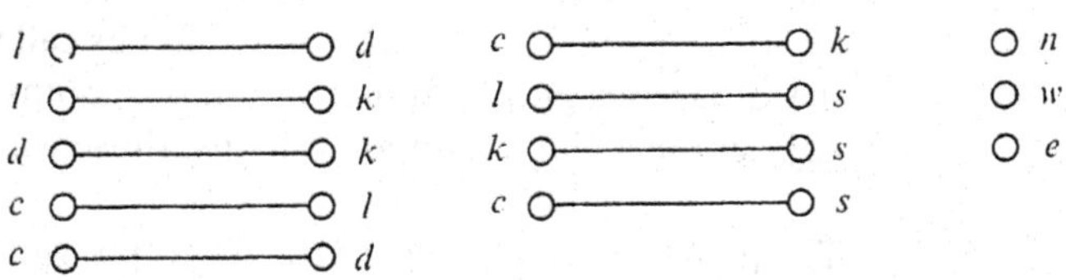

And the requirement graph is as follows (Figure 10.22):

Figure 10.22

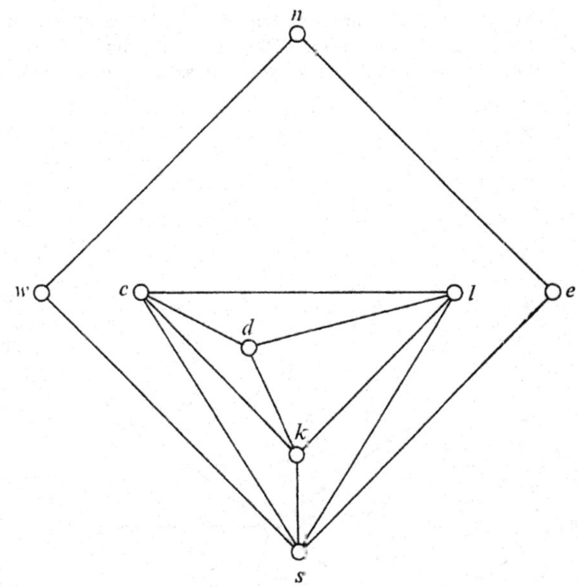

We place the dining-room first, as before, but this time on the garden side, adjacent to *s*. With the addition of the edge (*d*, *s*) the graph is no longer planar.

Figure 10.23

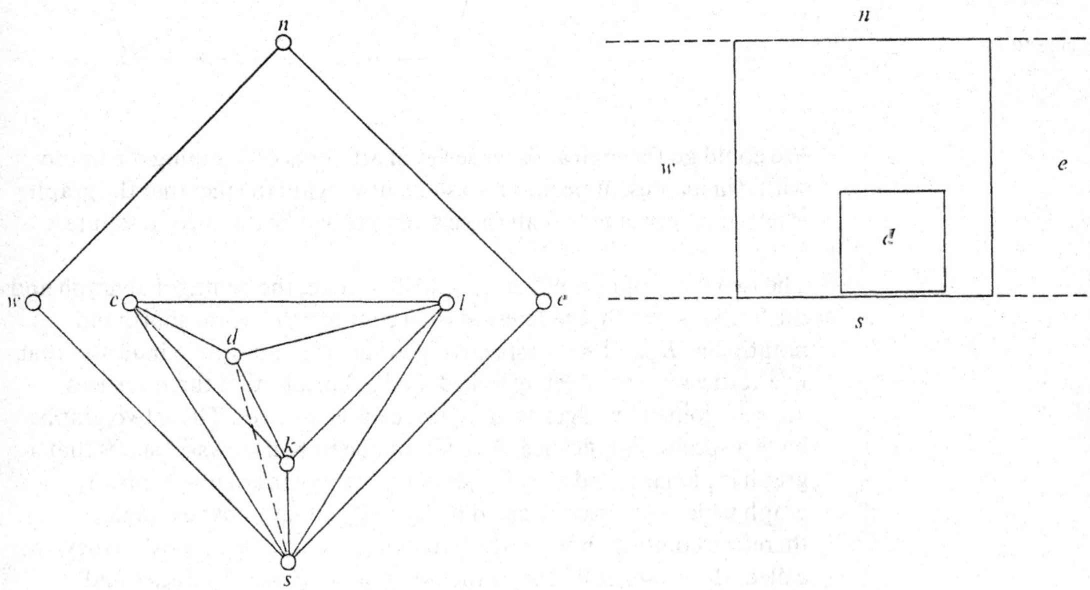

We can redraw the graph in a form which, if not familiar this time, nevertheless makes clear its structure. Ignoring for the moment the closed *cycle* of edges (*s*, *w*, *n*, *e*, *s*), the remainder of the graph may be arranged in pentagonal form. *s*, *d*, *k*, *l* and *c* are each joined by edges to every other vertex. This is the *complete graph* on five vertices.

[3] G. Kuratowski. 'Sur le problème des courbes gauches en topologie', *Fundamenta Mathematicae*, vol. 15–16, 1930, p. 271. The proof of the theorem is given in C. Berge, *The Theory of Graphs and its Applications*, London, Methuen, 1962, pp. 210–13.

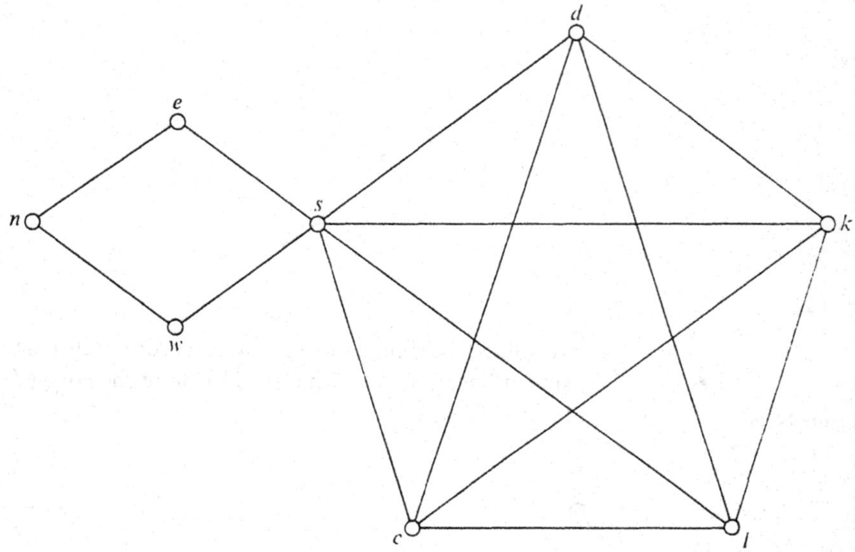

Figure 10.24

We could go through another series of attempts of planning the house with *d* in its present position, to show how again the fact that the graph is not planar means that all these attempts will be doomed to failure.

The two types of non-planar graph illustrated, the pentagonal graph and the 'utilities' graph, are referred to in a more convenient shorthand notation as K_5 and $K_{3,3}$ respectively, where the subscript 5 indicates that five vertices are completely linked, and 3,3 means that three vertices are each joined by edges to all of three other vertices. These two graphs have a special significance. A theorem due to Kuratowski[3] states that a graph is planar if and only if it does not contain as a sub-graph any graph which can be contracted to K_5 or $K_{3,3}$. Our house examples therefore comprise both of the two 'Kuratowski sub-graphs' as they are called, the presence of one or the other of which is a necessary and sufficient condition that the graph in question shall not be planar.

In the second case, of the K_5 sub-graph, the result has the general meaning, in terms of architectural planning, that there is no way whatever of arranging any five areas in plan so that every one is adjacent to all the others. Four is the largest number of areas all of which can be mutually adjacent. The complete graph on four vertices

[4] Eugène-Emanuel Viollet-le-Duc. *Discourses on Architecture*, vol. 2, Lecture XVII 'Domestic Architecture', New York, Grove Press, 1959, in particular pp. 265–76.

is planar; and the complete graph on any number of vertices greater than five cannot be planar, since we can see that it must contain K_5 as a sub-graph. This is sometimes known as the 'problem of contiguous regions' or the 'neighbouring states' theorem, and in a story attributed to Möbius takes the form of the dilemma of five sons whose father has left his land to them in his will, on the condition that it be divided so that every son is a neighbour to all the others. Of course the brothers find this impossible, and come to the reluctant conclusion that the father did not perhaps wish the estate divided at all.

The power which the 'adjacency graph' of a plan has, to reveal unsuspected implications of a series of required plan relationships which the architect may have, perhaps unconsciously, set for himself, and their conflict with other formal considerations, can be illustrated with the following example. This problem is one which dates originally from Viollet-le-Duc's *Discourses*,[4] where he engages in a long discussion of the changes in house-planning wrought by the French Revolution and the consequent democratization of society. Viollet-le-Duc points out that the social distance of former times which maintained privacy between master and servant had gone, and the need for physical barriers had arisen to replace it. In the new society 'the servant is a stranger hired by the week' and 'the life of each family must have its privacy secured' through complex and elaborate architectural planning. There must be service areas and served areas. There must be separate entrances to the house for guests and the family. 'We want an awning to shelter the carriages, but those who come and go on foot – for in a democratic society there will be such – must be able to come in the entrance hall without passing under the horses' noses.' There must be, in other words, a degree of pedestrian and vehicular segregation. There must be 'communications specially reserved' for service. All this, Viollet-le-Duc believed, meant the rejection of that symmetry which reflected earlier, less specialized, more simple domesticity. The new *individualism* in society implied isolation and the distinction of parts in the dwelling.

In 1896, Frank Lloyd Wright, who knew Viollet-le-Duc's writings very well, was presented with this same problem. Chicago society was just such a society as Viollet-le-Duc described, emphatic simultaneously about individualism and democracy. In the Aline Devin House project (Figure 10.25) Wright attempts to solve the Viollet-le-Duc problem within the constraints of symmetry. The problem, Viollet-le-Duc has

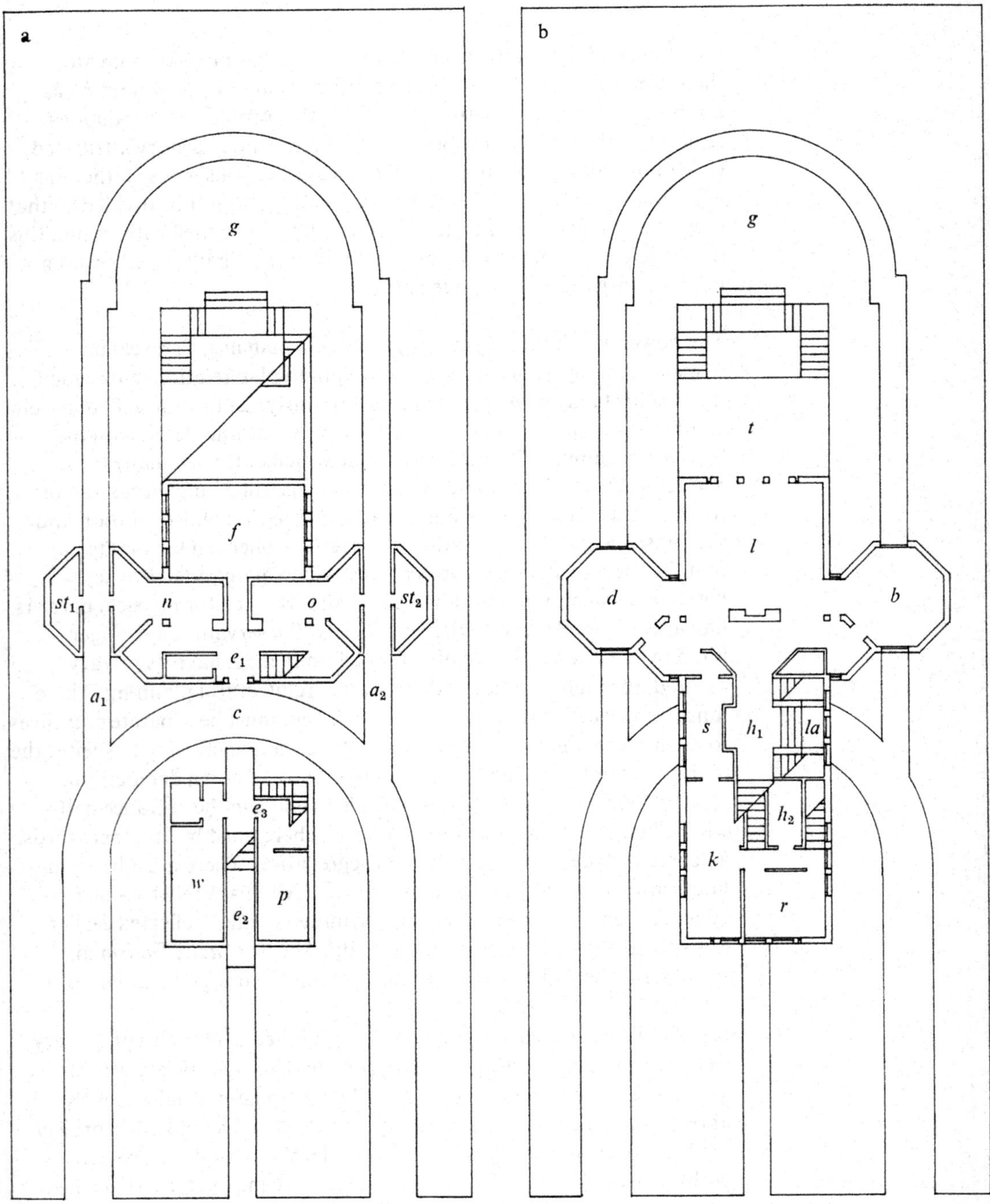
a
g
f
st_1
n
o
st_2
e_1
a_1
a_2
c
e_3
w
p
e_2
b
g
t
l
d
b
s
h_1
la
h_2
k
r

Figures 10.25 and 10.26

10.25 Frank Lloyd Wright's project (1896) for a house for Aline Devin; a, ground floor plan; b, first floor plan.

Key:

$a_{1,2}$	alleyways
b	library
c	porte-cochère
d	dining-room
$e_{1,2}$	guest and family entrances
e_3	servants' entrance
f	billiard room
g	garden
h_1	guest and family hall
h_2	servants' hall
k	kitchen
l	living-room
la	landing
m	main road
n	cellar
o	boiler room
p	pantry
r	servants' room
s	servery
$st_{1,2}$	stores
t	terrace
w	washroom

10.26 Three-dimensional projection, cut away in section down the central axis of symmetry to illustrate changes of level and the complex three-dimensional organization of the circulation system.

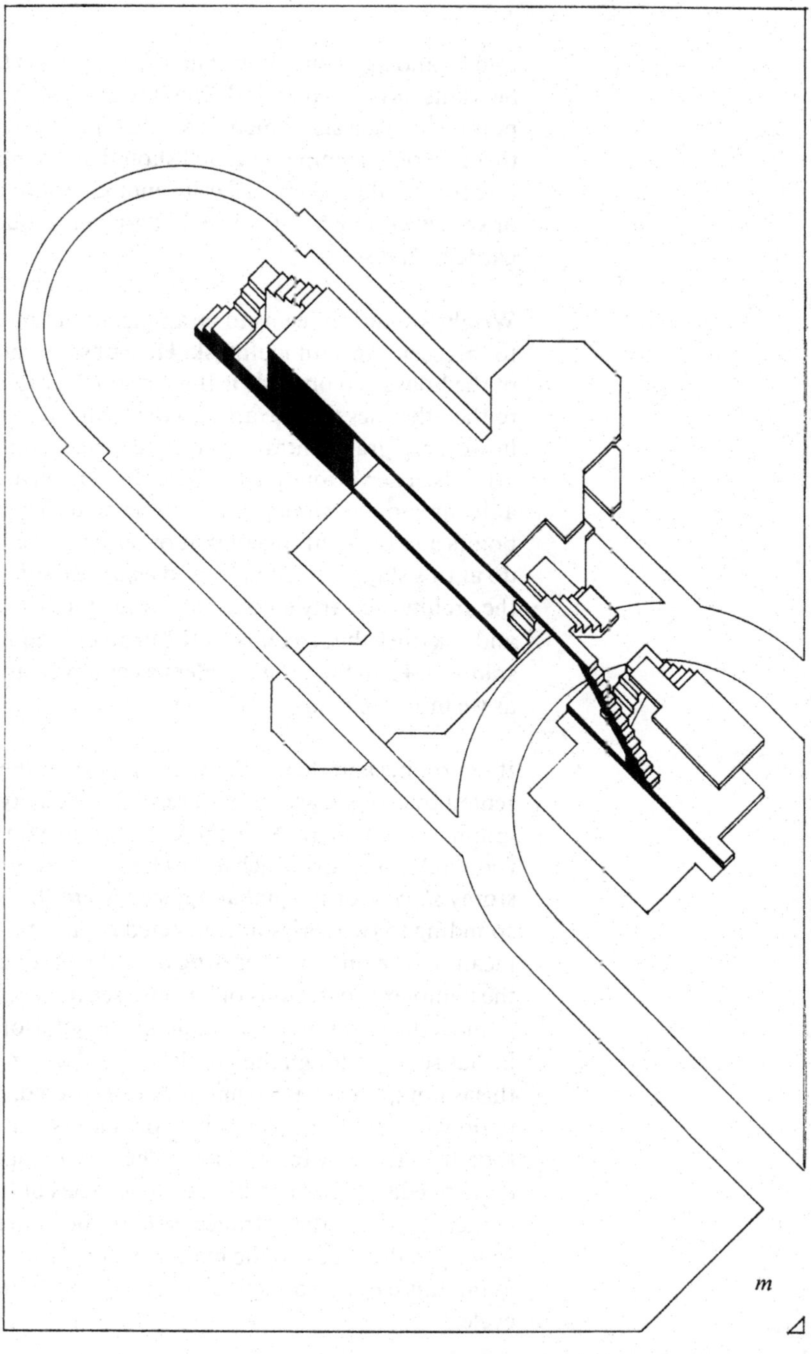

told us and as we might well imagine, appears to be a tricky one. Servants have to enter and circulate separately and yet be on hand at any point where guests or members of the family are likely to be in need of them. People coming by coach should arrive under cover, people coming on foot should be kept separate enough not to be splashed by carriages or sneezed on by horses. The living-room should communicate with the garden. And so on.

Wright's solution seems to be an ingenious and remarkable solution to this seemingly difficult task. He takes the carriageway under the belly of the house. To one side is the servants' entrance and the service area, to the other lies the entrance to the family and reception rooms. There is, however, a grand entrance fronting directly onto the street for pedestrian arrivals. The various parts of the circulation are handled with considerable three-dimensional skill in order that they be separated at some points and brought together at others (Figure 10.26). The right thing to do at this stage is to sit back and *enjoy* the solution, in the same way that the architect clearly enjoyed arriving at it. We shall do the wrong thing and ask does the *structure* of the problem require such a complicated solution. Is, in fact, the problem as complex as Viollet-le-Duc said, and as we intuitively imagine?

If the rooms and their adjacency on both of the two floors are represented by a single graph (Figure 10.27) it emerges that this 'adjacency graph' is not planar. Now there is of course no reason why a series of room relationships which are to be satisfied in a plan of more than one storey *should* form a planar adjacency graph – since the 'chains' corresponding to two series of connected rooms may cross, if necessary, by means of the one route passing over the other on a floor above. But are the complex separations of level we see in Wright's plan strictly needed to meet the requirements which, by his solution, he seems to imply that he has set? Although the graph of the plan is not planar, the reason for this is not quite what we might have expected. The problem has nothing to do with separating servants from guests and family, nor with keeping foot arrivals away from horses. The non-planarity arises because a *cycle* of edges is formed by the adjacencies of the road m to the *porte-cochère* c, of c to the entrance e_1, e_1 to the boiler-room o, o to the alleyway a_2, and a_2 back to the main road m. Inside this cycle lies the terrace t, and this clearly needs to give directly onto the garden g, outside the cycle.

Figure 10.27
'Adjacency graph' for Frank Lloyd Wright's Devin House project, illustrated in Figures 10.25 and 10.26. Adjacencies on both floors are shown in a single graph, with the staircases denoted by zig-zag lines. The graph is not planar.

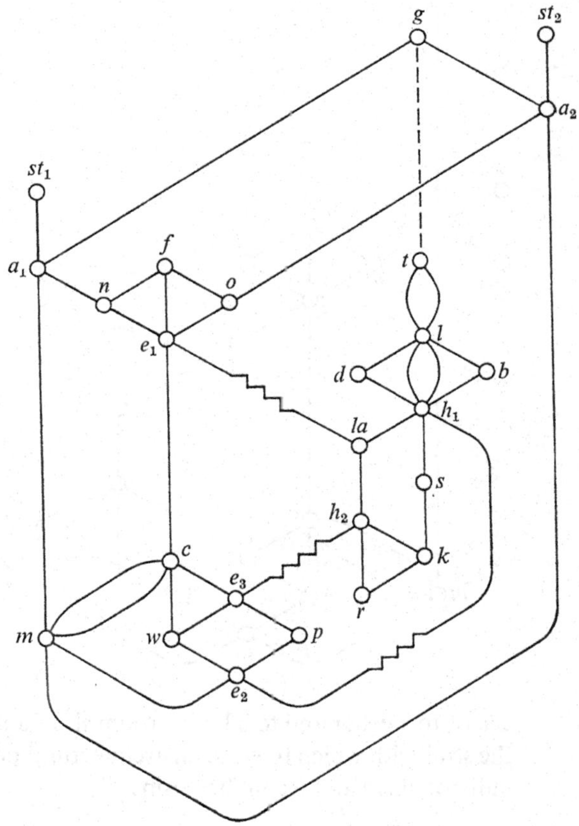

Note that for this purpose we have treated the two alleyways a_1 and a_2 as separate, a_1 giving access to the cellar *n* and a_2 to the boiler-room *o*; although they do in fact connect in the curving loop through the garden beyond the terrace. It can hardly be the intention of the plan, though, that the coalman should pass along this route, across the lake-shore panorama in full view of the terrace and garden, only to meet the wine-merchant coming around the path in the other direction. It is only the bilateral symmetry of the two pavilions which demands the doubling of this service access (just as it is symmetry which requires the pedestrian entry from the main road to enter centrally between the arms of the U-shaped carriageway; and thus necessitates the one route crossing the other in the centre of the plan at *c*). In any event the requirement that

Figure 10.28
The same adjacency graph of Figure 10.27 but with the two vertices a_1 and a_2, corresponding to the two alleyways, put together to make the single vertex $a_{1,2}$. The graph *is* now planar.

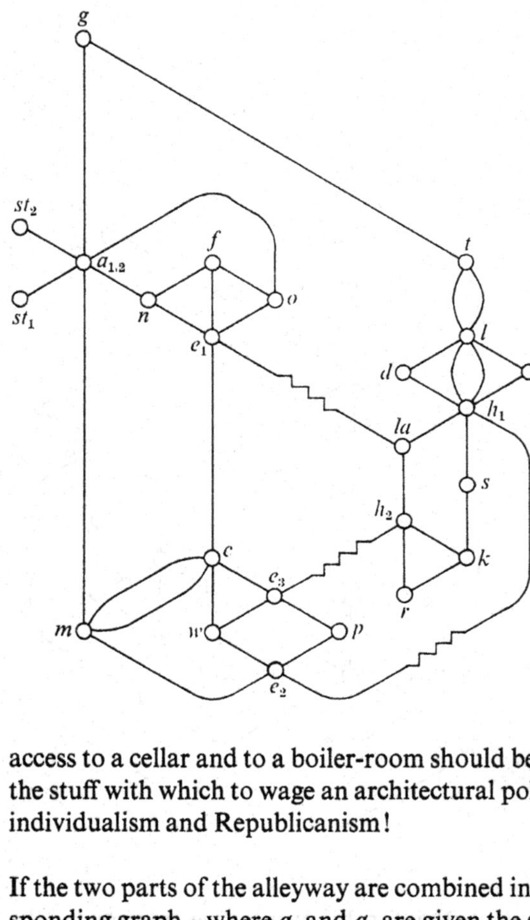

access to a cellar and to a boiler-room should be kept separate is hardly the stuff with which to wage an architectural polemic for democracy, individualism and Republicanism!

If the two parts of the alleyway are combined into one, then the corresponding graph – where a_1 and a_2 are given the single vertex $a_{1,2}$ – is planar. Figure 10.28 shows the graph, which could be realized as a one-storey plan without level changes. It probably could *not* now be interpreted symmetrically in architectural terms; although by now we should be wary enough not to say that that proves Viollet-le-Duc's point. Wright's youthful architectural dexterity is certainly fun, but it is *not* functionally necessary.

11 Electrical networks and mosaics of rectangles

In all the examples of the previous chapter the graph has been used to denote topological relationships between rooms – which room gives access to another, which room is adjacent to another. We have mentioned, though, how graphs can be used to represent not only plans and maps, but to depict all kinds of relations between groups of objects of any sort whatever – that is in general relations within *sets*. There is a close affinity between set theory and the theory of graphs.

Figure 11.1

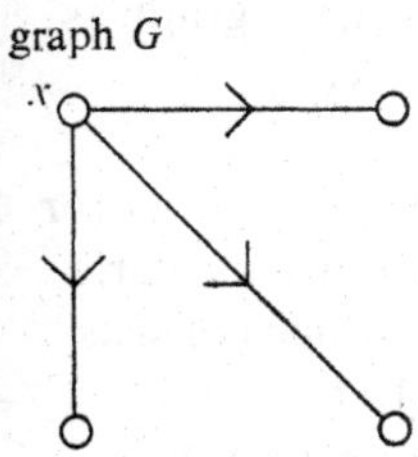

The vertices of a graph are the elements of a set. This set might be the set of rooms in a house, as above; it might be the set of bridges in the city of Königsberg, a set of people, a set of numbers. A graph G then consists of this *set* X, and in addition a *mapping* g of the set X into itself. We express this $G = (X, g)$.

Figure 11.2

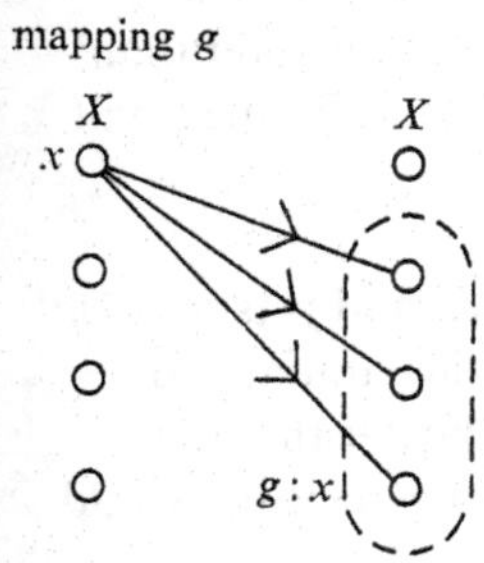

We have already met the idea of a mapping in previous chapters. The mappings dealt with in graph theory are usually one-to-many, or *multi-valued*. A multi-valued mapping g of X into X is a rule which associates to each element $x \in X$ a sub-set $g:x \subset X$ (Figure 11.1). In Figure 11.2 we illustrate this idea by arranging the elements of the set X (the vertices of the graph) in a column on the left, and listing them again in a second column on the right, and indicating the mapping by the arrows leading from x on the left to the sub-set $g:x$ which is associated to x, on the right.

Figure 11.3

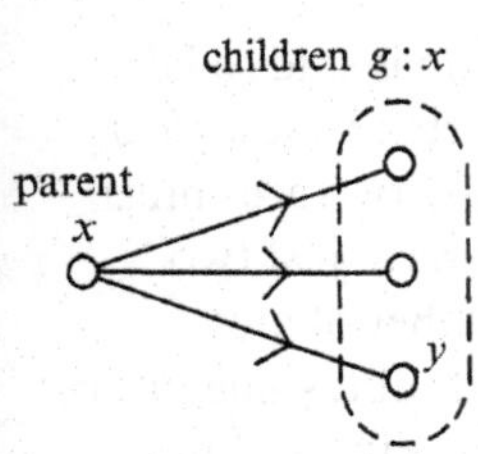

Take the example of a family tree, which is essentially a graph in which the vertices correspond to a set of people, the family (Figure 11.3). Associated to each element x of the set (each person) is the sub-set of his or her children $g:x$. This sub-set might comprise one child, several children, or possibly it might be an empty set $g:x = \emptyset$, where that person is childless.

We know already that the elements of the set are represented in a graph by points in the plane (vertices). To complete the graph we require that, where x and y are two vertices such that $y \in g:x$, they will be joined by a line. We shall ask that, unlike in the graphs of the house examples, this line carry an arrowhead pointing from x to y; this directed line is called an *arc* and it is referred to by the pair $[x, y]$.

Figure 11.4

the arc $[x, y]$: 

[1] By F. Harary, R. Z. Norman and D. Cartwright, in *Structural Models; an Introduction to the Theory of Directed Graphs*, New York, Wiley, 1965, which describes a wide range of applications of graph theory in social science and elsewhere.

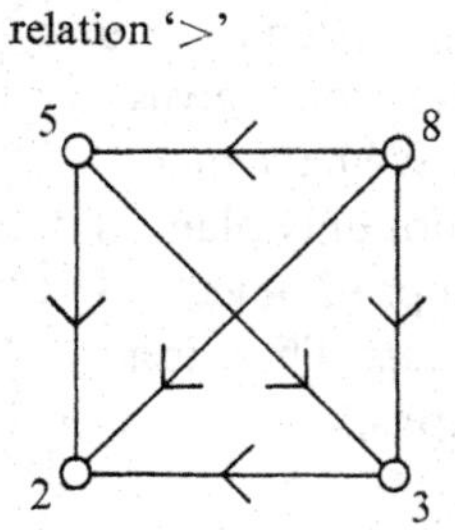

$g:8 = \{5, 3, 2\}$
$g:5 = \{3, 2\}$
$g:3 = \{2\}$
$g:2 = \emptyset$

Figure 11.5

The *arc* therefore differs from the edge in our earlier examples, in that it carries direction. We shall see what the precise relationship between the arc and the edge is shortly. Graphs composed of arcs have been called *directed graphs*, or more succinctly, *digraphs*.[1] The set of all arcs in the graph we indicate by the letter U. So instead of expressing the graph as (X, g) we can, alternatively, characterize it by the two sets (X, U).

In the family tree each arc $[x, y]$ expresses the *relation* x 'is the father (or mother) of' y. An example of a directed graph from the building world is the management technique of 'critical path analysis', which consists in drawing a graph of the set of separate jobs which go to make up a whole project or programme of building work. Each arc in that case represents the relation that job x 'must be completed before a start can be made on' y. Another relation which a graph might express in a similar way is, for a set of numbers, that the number x 'is greater than' the number y. The membership of all the sets $g: x$ is listed for the small graph of this 'greater than' relation, in Figure 11.5.

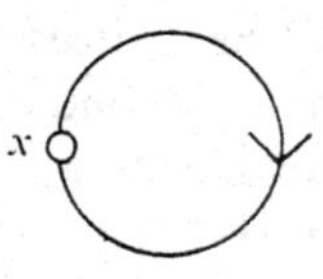

Figure 11.6

For any arc $[x, y]$ the vertex x is called its *initial vertex* and y its *terminal vertex*. It is quite permissible for an arc to have identical initial and terminal vertices, and so form a closed *loop*. This indicates that the relation expressed is *reflexive*. An example of a reflexive relation would be 'lives in the same house as'; anyone lives in the same house as himself.

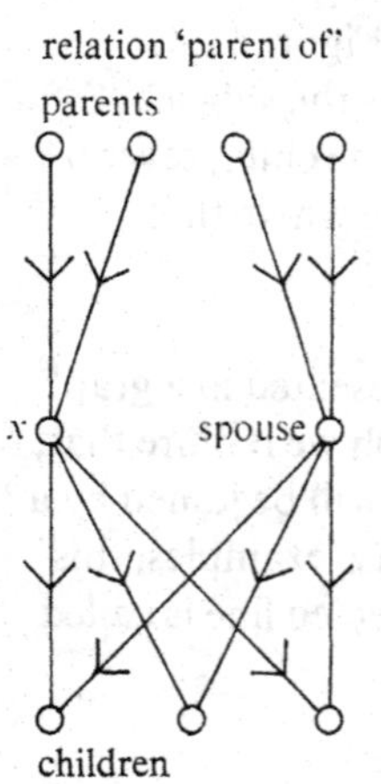

Figure 11.7

An arc whose initial vertex is x and which is not a loop is said to be *incident out from* x. Similarly an arc with terminal vertex x is *incident into* x. In a graph of a family tree we know that exactly two arcs must be incident *into* each vertex, since every person must have just two parents (Figure 11.7). There can however be any number of arcs incident *out from* a vertex in this instance, up to, no doubt, some rather indeterminate practical upper limit! Graphs of the family tree kind provide a convenient and precise form for expressing the kinship relationships which anthropologists find in primitive societies; the various incest taboos, or the complicated rules governing who is a permissible marriage partner for whom, then correspond to structural constraints on the possible forms the graph may take.

An important type of relation is the *symmetric* relation whereby if x is related to y, then necessarily y is related to x in the same way.

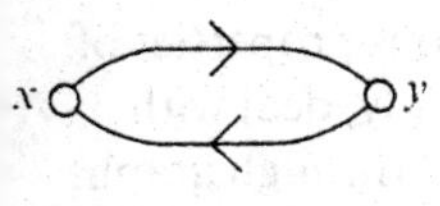

Figure 11.8

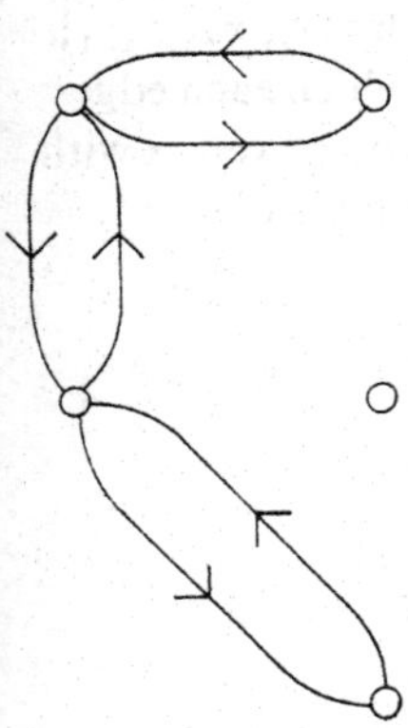

Figure 11.9

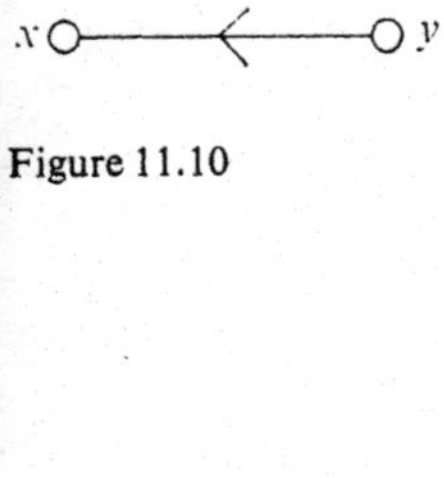

Figure 11.10

In the graph this means that the presence of an arc [x, y] implies the presence of the arc [y, x]. The graph (X, U) is thus said to be symmetric when

$$[x, y] \in U \Rightarrow [y, x] \in U.$$

An example of a symmetric relation: 'lives in the next-door house to'. If x lives next door to y, then it follows that y lives next door to x. In a symmetric graph therefore, wherever two vertices are joined, they are joined always by a pair of arcs directed in opposite senses (Figure 11.9).

If we look again at the relations which were expressed in our graphs in the earlier part of this chapter, graphs of plans and maps, then it becomes clear that these were in fact all symmetric relations. If a district x of the city of Königsberg 'is joined by a bridge to' district y, then equally y is joined to x. If the kitchen k in a house plan 'is adjacent to' the dining-room d, then d is adjacent to k; and this same consideration applied throughout the examples.

What did we with these graphs was to adopt the general convention, that each pair of oppositely directed arcs in a symmetric graph be replaced by a single line without an arrow-head. Such a line is called an *edge*. An edge is thus a more general and more inclusive term than an arc, which implies orientation. In a graph $G = (X, U)$ an edge is a pair of vertices (x, y) such that $[x, y] \in U$ *or* $[y, x] \in U$. In other words we say that x and y are joined by an edge if there is *either* an arc joining them in one or the other direction, *or* a pair of arcs joining them in both directions.

Unfortunately, for historical reasons, different conventions are observed in graph and set theory, in the use of brackets. In set theory round brackets represent an *ordered* pair, in graph theory a pair of vertices where order is not important (an edge). The reader will remember also that in earlier chapters square brackets [ ] are used to represent a closed interval while here they are used for a directed arc. The somewhat contradictory conventions might appear inconsistent and possibly confusing, but they follow more general practice in the separate literature of the two subjects, set theory and graph theory, and for that reason are maintained.

More terms which will prove useful in describing various properties of graphs, some of which have also been touched on already, deal with sequences of edges or arcs forming continuous 'routes' through graphs. In a directed graph the term *path* is used to denote a sequence of arcs $[u_1, u_2 \ldots u_k]$ of a graph (X, U) such that the terminal vertex of each arc is the initial vertex of the succeeding arc. If the path meets in turn the vertices $x_1, x_2 \ldots$ up to x_{k+1} it is convenient to denote it by $[x_1, x_2 \ldots x_{k+1}]$. The corresponding term for a sequence of k edges, in which each edge has one vertex shared with the preceding edge and the other vertex with the succeeding edge, is a *chain*, represented with the notation $(x_1, x_2 \ldots x_{k+1})$ – as in the house planning exercises.

Figures 11.11 and 11.12

path $[x_1, x_2, x_3, x_4]$ chain (x_1, x_2, x_3, x_4)

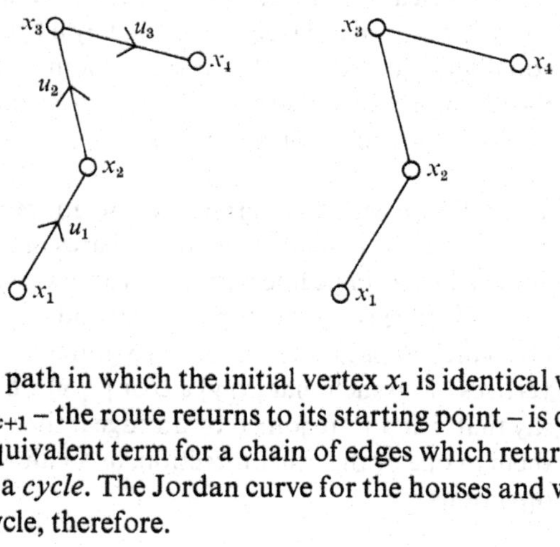

A path in which the initial vertex x_1 is identical with the terminal vertex x_{k+1} – the route returns to its starting point – is called a *circuit*. The equivalent term for a chain of edges which returns to its starting vertex is a *cycle*. The Jordan curve for the houses and wells problem was a cycle, therefore.

Figures 11.13 and 11.14

circuit $[x_1, x_2, x_3, x_1]$ cycle (x_1, x_2, x_3, x_1)

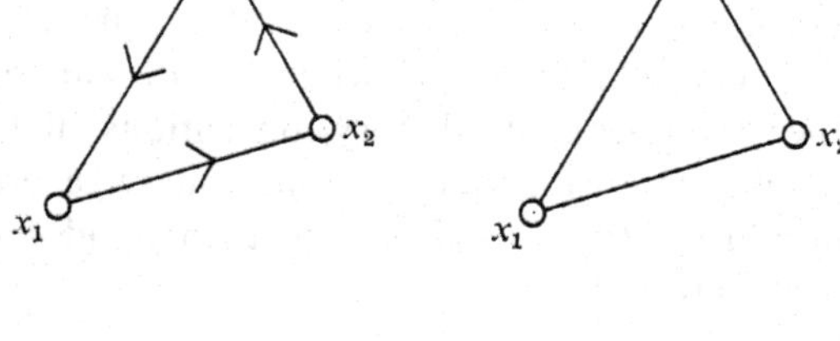

graph (X, g)

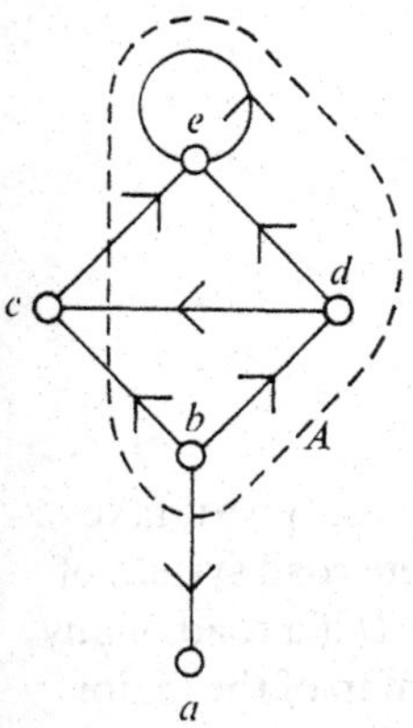

Figure 11.15

Yet another notion which we introduced earlier was that of a *sub-graph*, which bears roughly the relation to a graph that a sub-set does to a set. To be more precise, using the apparatus of formal notation which we now have introduced, a sub-graph of the graph (X, g) is defined to be a graph (A, g_A) where $A \subset X$, and in which the mapping g_A is given by

$$g_A : x = g : x \cap A.$$

Put into words, this means that a sub-graph (Figure 11.16) comprises a sub-set of the vertices of the original graph (Figure 11.15) with, for these vertices, *all* the arcs which connect them.

Figure 11.16

sub-graph (A, g_A)

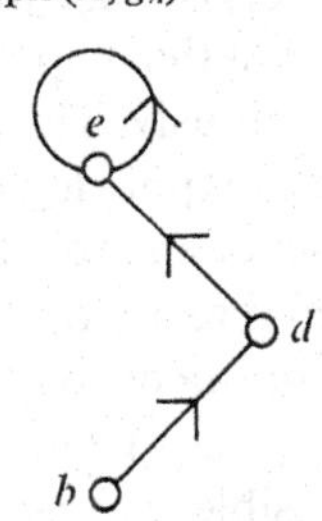

A *partial graph* of (X, g) on the other hand is defined by (X, h) where h is a new mapping such that $h : x \subset g : x$ for all elements x. A partial graph (Figure 11.17), therefore, comprises all the original vertices of the graph but a sub-set only of the arcs.

partial graph (X, h)

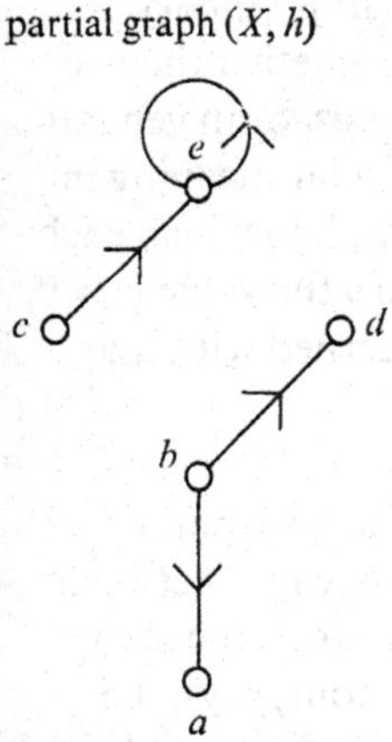

Figure 11.17

It follows that a *partial sub-graph* of (X, g), defined by (A, h_A) where $A \subset X$ and $h_A : x \subset (g : x \cap A)$, comprises a sub-set of the vertices of (X, g), with a sub-set of the arcs linking those vertices in the original graph (Figure 11.18).

partial sub-graph (A, h_A)

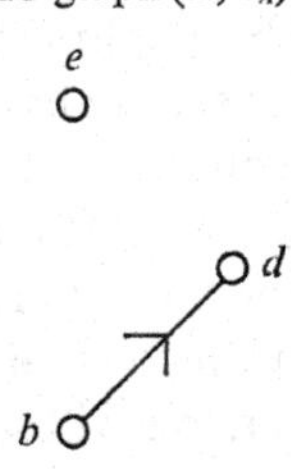

Figure 11.18

[2] Claude Berge. *The Theory of Graphs and its Applications*, London, Methuen, 1962, p. 6.

graph of roads and 'not roads'

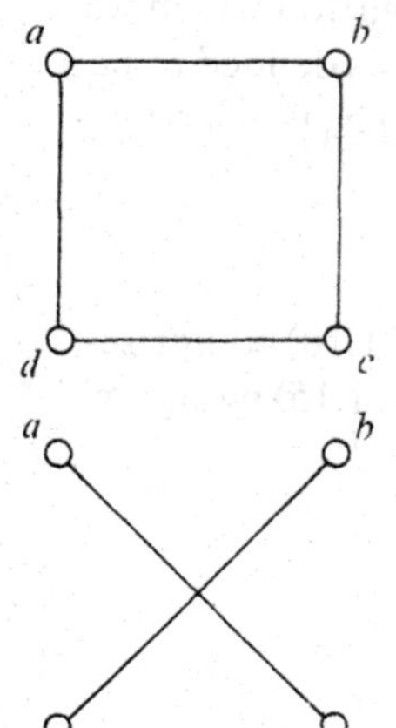

Figure 11.19

To adapt an illustration of these distinctions which Berge[2] gives; take for example the graph (X, U) representing the complete road system of London. X is the set of all road junctions, and $(x, y) \in U$ if a road of any kind joins directly the two junctions x and y. Then a map of the major roads only (but showing all junctions) is a partial graph, while a complete road plan of the West End is a sub-graph. A plan of the major roads in the West End would be a partial sub-graph.

When we talk of plans or maps as graphs, though, it is well to remember exactly what it is that each particular corresponding graph represents. In the case of a road map some edge (x, y) in the graph would usually denote the relation (as above) that junction x is joined by road to junction y. The edge, however, is in no way a symbolic picture of the road itself, as the line drawn in the map is; although it is often easy to make this confusion unconsciously. The distinction can perhaps be made clear in this way. We can quite legitimately, if somewhat perversely, draw another 'graph of a map' in which the relation expressed is that junction x 'is *not* joined directly by road to' y. Figure 11.19 shows the two graphs of a map, a graph of roads and a graph of 'not roads' (perhaps, a map and its 'anti-map'). Other yet more exotic graphs of maps would perhaps be possible, expressing different relations still.

With this point made, however, the fact remains that graph theory provides us with an ideal model for systems of routes, whether they be road maps, the patterns of circulation routes in buildings, or in general any kind of network. The word *network* has this particular meaning in the language of graph theory: it signifies a directed graph in which each arc is assigned a numerical value. We assign to the arc u the value $c \geqslant 0$, for example; this would represent some quantity concerned with that relation which the arc signifies.

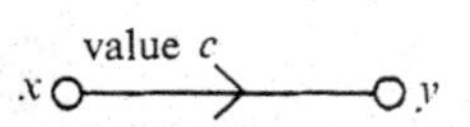

Figure 11.20

In this way the arc in a network can represent either the maximum capacity of, or else the actual volume of, flow along some link between two points in the system. This might be the flow of motor traffic along one-way streets (where c gives number of vehicles per hour, say), the flow of water along pipes (gallons per minute), the flow of commodities between manufacturer and market (tonnage per month perhaps), or, in a communication network, the flow of one-way messages like letters or telegrams – all situations in which the direction and volume of flow or traffic are measured together. We will save a fuller discussion of the subject of traffic networks for Chapter 14, in the context of circulation problems. Meanwhile, to go back to the house-planning exercises, we find a surprising application of the theory of *electrical* networks produces a graph with which we are able to represent not only the adjacencies and relative positions of rooms in a plan, but also their exact dimensions and shapes.

We illustrate the principle by taking a rather simple house plan as before: three rooms, a living-room, a dining-room and a kitchen, with a small entrance hallway and a staircase leading off it. The dimensions are given as whole numbers, which, for the sake of the example, we can assume are multiples of a basic module of 300 mm or 1 ft. The overall size of the plan rectangle is 20 modules by 21.

Figure 11.21

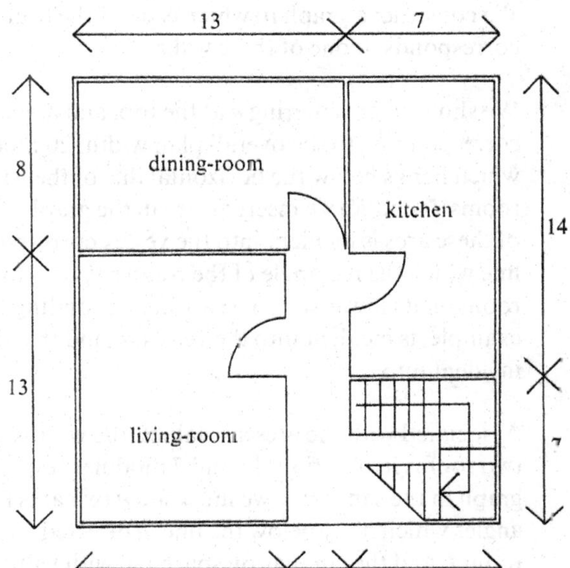

In Figure 11.22 we present the plan as a mosaic of rectangles without detail shown; and we mark those walls in the plan which lie 'horizontally' (as it is orientated on the page), by showing them as heavier lines.

We denote the four marked walls by letters: n and s for the two outside walls of the plan at the top and the bottom (on the 'north' and 'south'

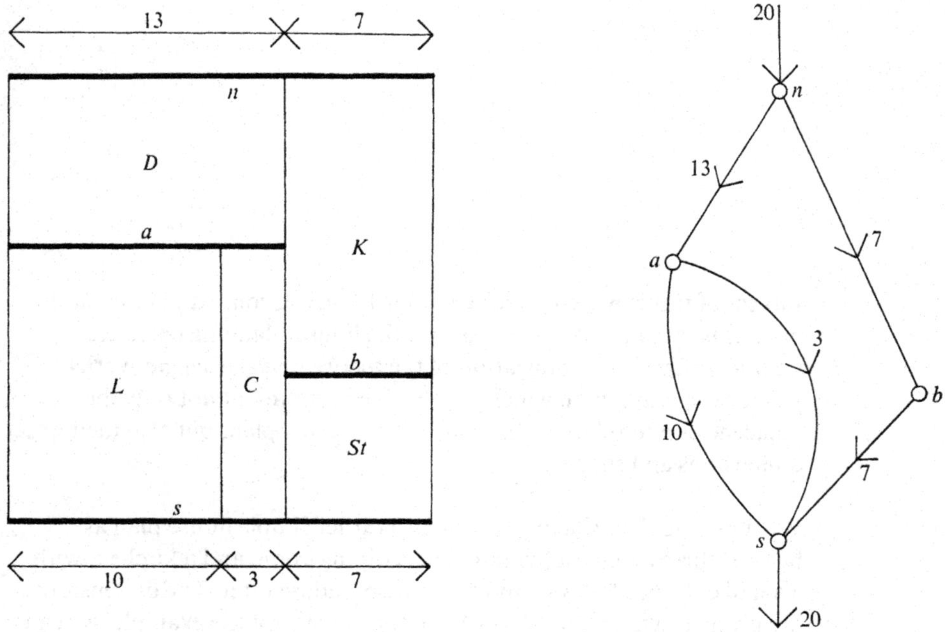

Figure 11.22

sides of the house), and *a* and *b* for the two interior walls indicated. We construct a graph in which each of the four vertices *n*, *s*, *a* and *b* corresponds to one of these walls.

We show an arc entering *n* at the top, and associate a value with this arc corresponding to the overall plan width, 20 modules. For each rectangle which hangs below the horizontal line of the wall *n* (i.e., for each of the rooms *D* and *K*) we insert an arc in the graph incident out from *n*. Each of these arcs is incident into the vertex corresponding to that horizontal line which the rectangle of the room sits on; that is, to the wall of the room on its lower side. The arc corresponding to the dining-room *D*, for example, is incident into the vertex *a*; and that corresponding to *K* is incident into *b*.

Associated with the arcs are values, the widths across the plan of the two rooms in question, 13 and 7 modules respectively. To complete the graph in the same way we must show two arcs for the two room rectangles which hang below the line *a*, directed from *a* to *s* (for the living-room *L* and the circulation space *C*), with values 10 and 3. And a single arc [*b*, *s*] (corresponding to the stairs *St*), with value 7. Finally an arc leaves *s*, with value 20, again for the total plan width.

With the special exception of the two infinite arcs value 20 (which could be imagined as entering and leaving the network to and from some external vertices at infinity), there is thus one arc for each room, with an associated value for the dimension of the room in the 'horizontal' sense. And there is one vertex for every 'horizontal' wall.

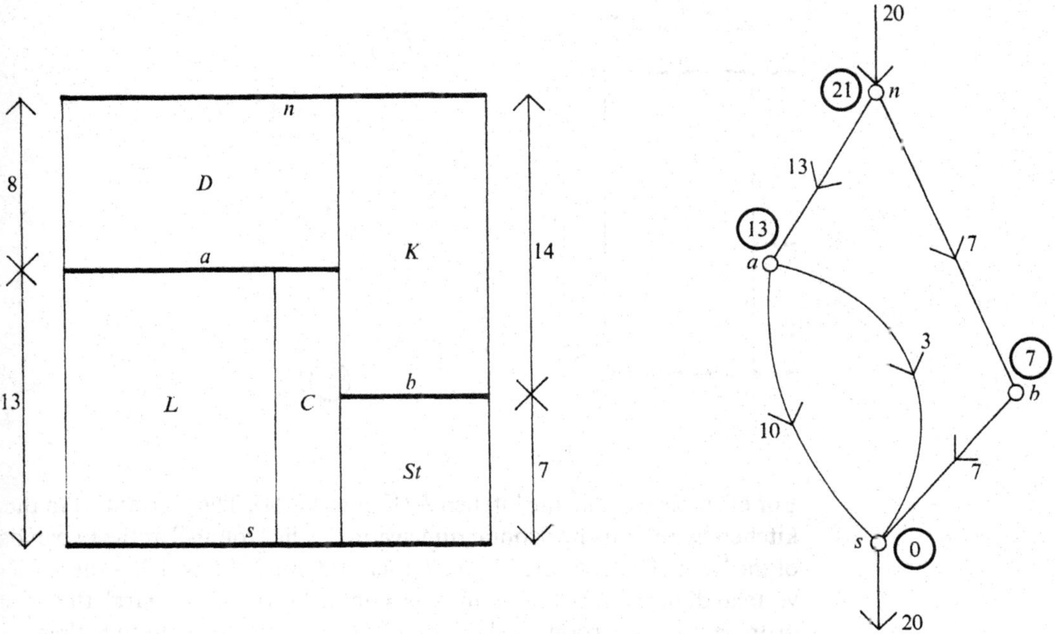

Figure 11.23

We may further associate a value with each vertex (these values are shown circled in Figure 11.23) corresponding to the 'vertical' distance of the horizontal line it represents from the bottom wall of the plan, the line *s*. The value of the vertex *s* is 0 therefore, since it corresponds to that same wall. The value of *b* is 7, of *a* 13 and of *n* 21 – the overall plan dimension from top to bottom.

The resulting graph now shows the remarkable analogy with the physics of electricity. Suppose it were an electrical circuit, with the arcs wires. The values associated with each arc are the currents in the wires, and the values of the vertices the differences in electrical potential (the voltage). This network obeys Kirchhoff's laws for electrical flow, where the *conductance* of each wire (roughly, the ease with which electricity passes) is the *proportion* of the corresponding rectangle, i.e., the ratio of its horizontal dimension to its vertical dimension.

The first of Kirchhoff's laws states that for a wire in a network with conductance C, and voltages V and V' (where $V > V'$) at its two vertices, then the current A in the wire flows from the larger voltage to the smaller and

$$A = C(V - V').$$

Transferring this back into the terms of room shapes, we see that A is the horizontal dimension of a room, while $(V - V')$ is always the vertical. Since C, as the proportion of the room, is defined as the ratio of the one to the other, the identity $A = C(V - V')$ follows.

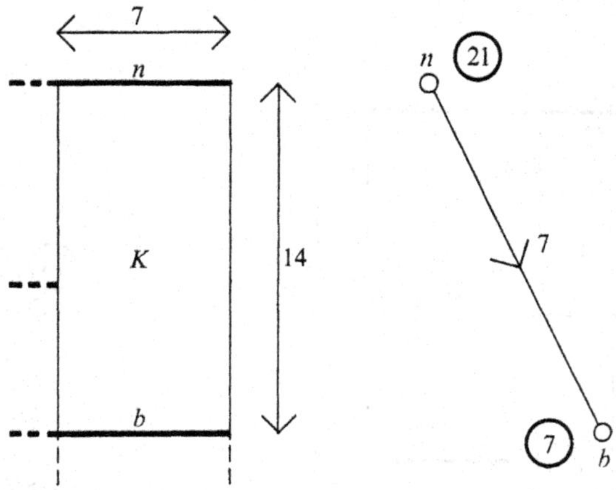

Figure 11.24

For example we take the kitchen K (Figure 11.24). The 'current' A in the kitchen 'wire' is its horizontal dimension 7. The 'voltage' at the two ends of the 'wire' V and V' are 21 and 7, their difference 14, which is the vertical dimension of the room. The 'conductance' of the 'wire', the proportion of the room, is the ratio of the one dimension to the other, $7/14 = \frac{1}{2}$. Putting these values in the equation we have

$$7 = \tfrac{1}{2}(21 - 7).$$

The second law concerns currents only, and states that the total current entering the vertex equals the total current leaving that vertex. A look at the graph shows us that in the analogy with plan dimensions this always holds true; the equivalent situation in the plan being the breaking down of some larger horizontal dimension into a number of smaller, or the adding together of smaller to give larger. A 'current' of 13 enters the vertex a, for instance, and 'currents' of 10 and 3 leave it. The direction of the arcs in the graph is the conventional direction of the flow of electricity; and the 'current' of 20 entering the network at the top (the overall width of the plan) must equal the current leaving at the bottom (that same width again).

It was by making use of this electrical analogy, in the network corresponding to a mosaic of rectangles, that the self-styled 'Important Members' of the Cambridge Trinity College Mathematical Society succeeded in 1937 in producing a solution for the old puzzle of 'squaring the square', long thought impossible. That problem consists in finding some set of squares, no two the same size, which will fit together without interstices to form a complete, larger square.

William Tutte, one of the four 'Important Members', recounts the way in which 'the key discovery of the whole research' was made by the mother of another Member, R. L. Brooks. The four friends had been, by a process of experiment, producing and cataloguing large numbers of 'squared rectangles', otherwise called 'perfect' rectangles, in the hope of

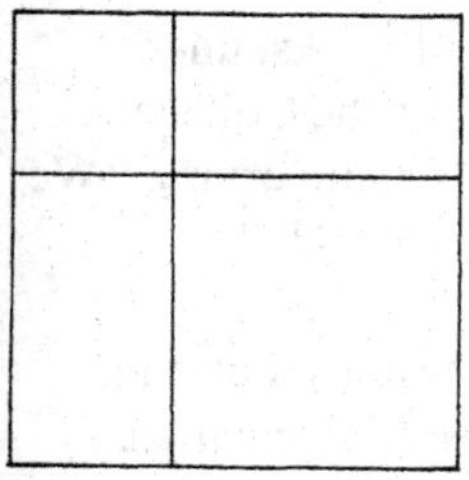

Figure 11.25

finally finding one which was a 'perfect' square. Alternatively they hoped they might find two perfect rectangles with the same overall dimensions but with no component squares in common; they would then be able to put the two rectangles together with two larger squares to form a perfect square, by the construction of Figure 11.25.

Brooks had found a rectangle with unusually small component squares (Figure 11.26) which pleased him so much that he made a jigsaw of it. Brooks's mother took the puzzle and tried to assemble the pieces. She succeeded eventually in putting them together again; but when Brooks examined her solution he found that although the rectangle had the same original overall dimensions it was not the squared rectangle he had cut up! (Figure 11.27). Here were two perfect rectangles of the same shape – but which failed 'in the worst possible way' to have no two component squares in common. Nevertheless the discovery suggested that two such rectangles might still be found. 'The Important Members met in emergency session.'

Figures 11.26 and 11.27 'Perfect' rectangles: Figure 11.26, R. L. Brooks's jigsaw, and, Figure 11.27, as re-assembled by Brooks's mother.

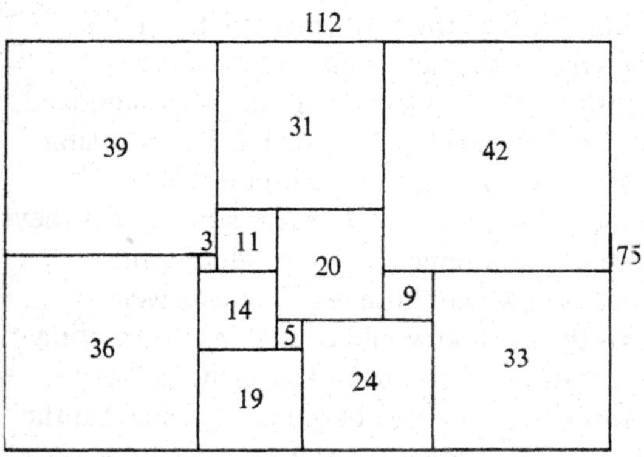

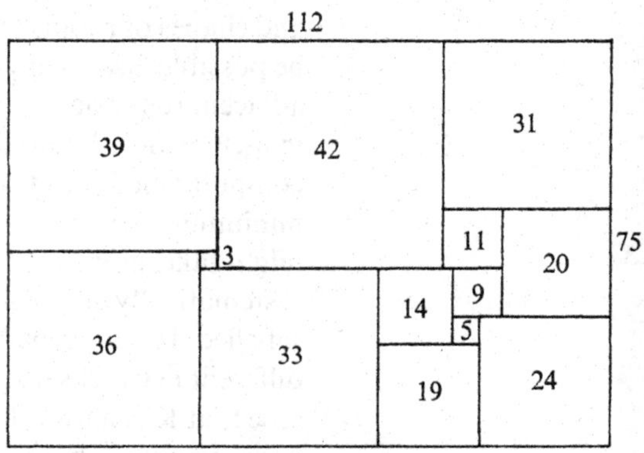

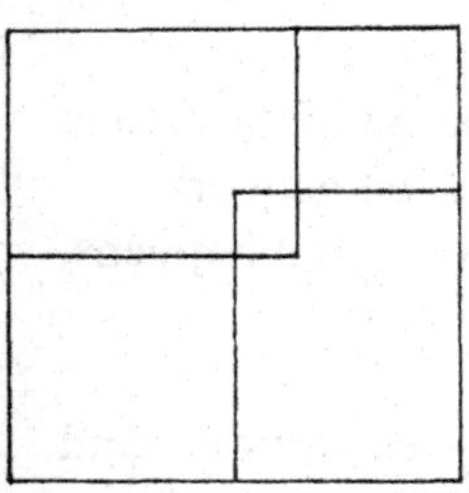

Figure 11.28

An examination of the electrical networks corresponding to the two rectangles led to the discovery of certain rotational symmetry properties of the networks, using which the Important Members could generate large numbers of pairs of rectangles of similar shape and size. They decided to experiment with a possible means of producing pairs of same-size rectangles with only one corner-component square in common, which they could then put together in the construction of Figure 11.28, again with two more large squares. 'So it came to pass that Smith and

[3] Martin Gardner. *More Mathematical Puzzles and Diversions from Scientific American*, London, Bell, 1963.

[4] Philip Steadman. *The Automatic Generation of Minimum-Standard House Plans*, Working Paper 23, University of Cambridge, Land Use and Built Form Studies, 1970.

Stone sat down to compute a complicated...pair while Brooks, unknown to them, worked on another in a different part of the College. After some hours Smith and Stone burst into Brooks's room crying "We have a perfect square!" To which Brooks replied "So have I!" '

The whole saga of the search for the perfect square is told by Tutte in Chapter 17 of Martin Gardner's delightful book *More Mathematical Puzzles and Diversions.*[3] The Important Members (R. L. Brooks, C. A. B. Smith, A. H. Stone and W. T. Tutte) published their paper on the subject of 'The Dissection of Rectangles into Squares' in the *Duke Mathematical Journal*, vol. 7, 1940, pages 312–40. Since then there has been continued interest in the subject of tiling by squares, and more recently computer methods have been used to obtain new results.

In our own architectural problems, the 'electrical network' offers the possibility for the planning of houses or other small buildings – or even of small groups of rooms within a larger building – of an economical representation of, at the same time, the relative positions, dimensions and shapes of rooms and plan. With a little further calculation it will be possible also to infer from the same graph whether one room is adjacent to the next or not, as we will show shortly. It has been suggested, in fact, by Steadman[4] that by the manipulation of such networks using computer methods it would be possible – given requirements for minimum sizes for rooms and constraints on the permissible shapes they might take, as well as 'adjacency requirements' – to produce quite systematically *all* possible plans in which those requirements were satisfied. In very broad terms the method would consist in constructing different networks in a permutational fashion, and ensuring in every case that Kirchhoff's laws for networks were obeyed; and hence that the network always bore a direct correspondence to some mosaic of rectangles – that is, to the desired plan arrangement.

To go back to the example, it is clear that the orientation of the plan in Figure 11.23 is quite arbitrary, and we can draw a second network (Figure 11.29) where the vertices are the '*vertical*' walls – as it happens, again four of them.

The two side walls on the 'east' and 'west' sides are given vertices *e* and *w*, and the two internal vertical walls labelled *c* and *d*. This second graph gives no new information on dimensions which is not contained in

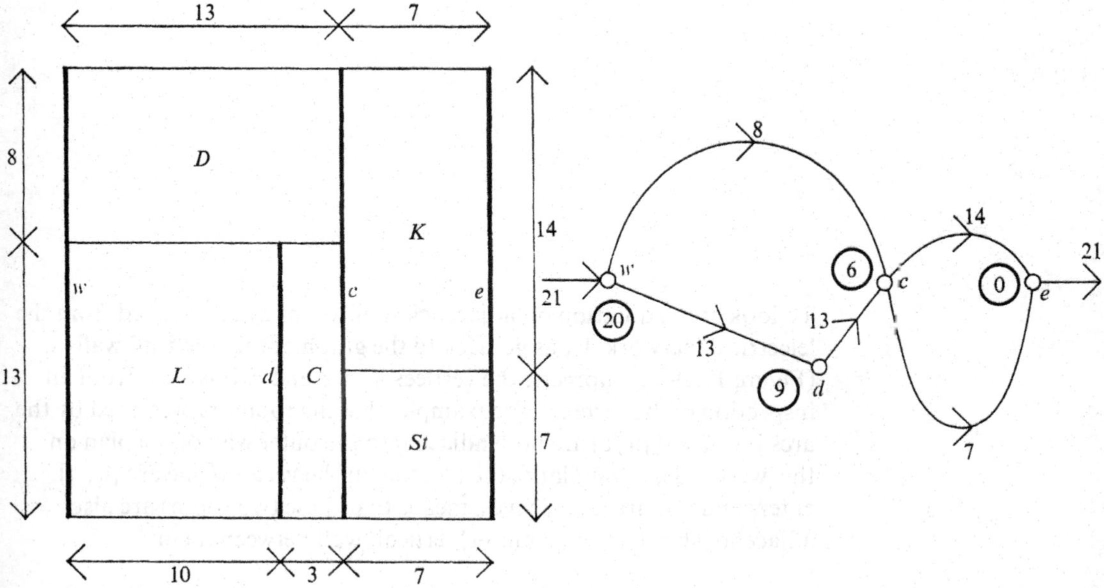

Figure 11.29

effect in the first; in fact the whole plan may be reconstructed, given just one or other of the graphs. In this case (but not, curiously, in every case) the graphs are *duals*. They contain the same number of arcs, since in both the arcs represent the rooms. And for every *face* in one graph there is a *vertex* in the other, and vice versa. Figure 11.30 shows the two graphs superimposed to illustrate this.

Figure 11.30

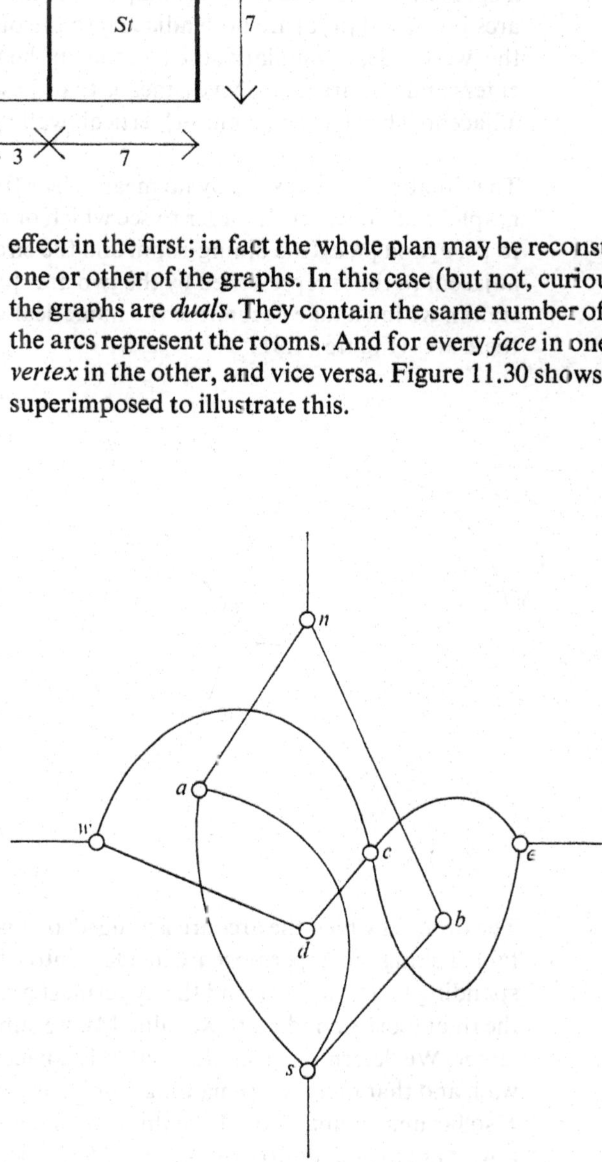

To look at the question of 'adjacencies' of rooms as determined from the 'electrical' network, let us go back to the graph of the 'vertical' walls (Figure 11.29), comprising the vertices *w*, *d*, *c* and *e*. It is clear from an inspection of the vertex *w*, for example, that the rooms represented by the arcs [*w*, *d*] and [*w*, *c*] are both adjacent to the outer wall of the plan on the 'west' side, *w*. Similarly, the fact that at the vertex *d* one arc [*w*, *d*] enters and one arc [*d*, *c*] leaves, means that these two rooms are also adjacent. They have a common 'vertical' wall between them.

The situation at vertex *c* is by no means clear from the structure of the graph itself, however. In order to see which of the rooms corresponding to [*w*, *c*] and [*d*, *c*] (the dining-room and the circulation space) are adjacent to those represented by the two arcs [*c*, *e*] (the kitchen and the stairs), we must take account of the numerical values associated with these arcs (Figure 11.31).

Figure 11.31

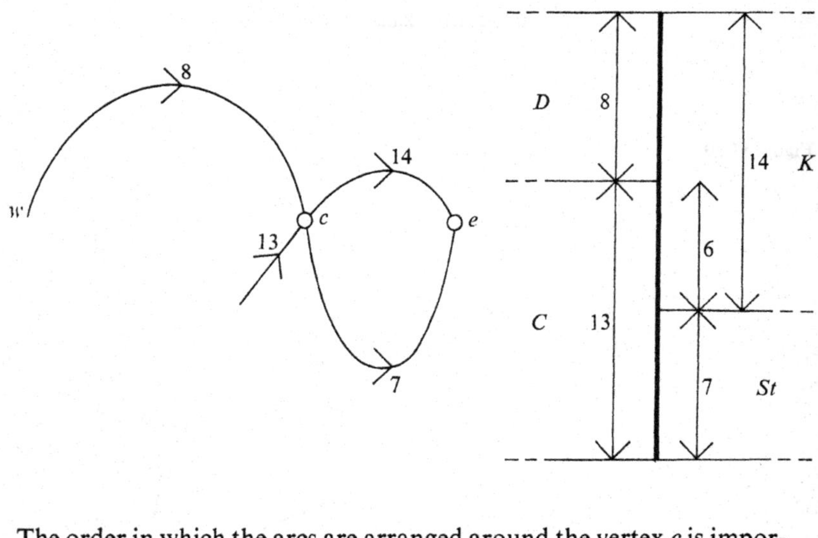

The order in which the arcs are arranged around the vertex *c* is important. Taking the uppermost arc incident into *c* from the left (corresponding to *D*, value 8), and the uppermost arc incident out from *c* on the right (corresponding to *K*, value 14), we subtract the smaller from the larger. We determine from this that *D* is adjacent to *K* along its whole wall, and that there is a remaining 6 units of 'width' along the wall of *K* so far unaccounted for. Take this remainder of 6, and take the next lowest arc incident into *c* on the left ([*d*, *c*], corresponding to *C*, value

13); again subtract the smaller from the greater. *K* is adjacent to *C* over a length of wall of 6 units, and there is now a remaining 7 units of the wall of *C* left over. We take this remainder, together with the next lowest arc incident out from *c* on the right (*St*, value 7), and the two match up.

We conclude that there are three adjacencies constituted by the arrangement, *D* with *K*, *C* with *K* and *C* with *St* (shown by dotted lines in Figure 11.32).

Figure 11.32

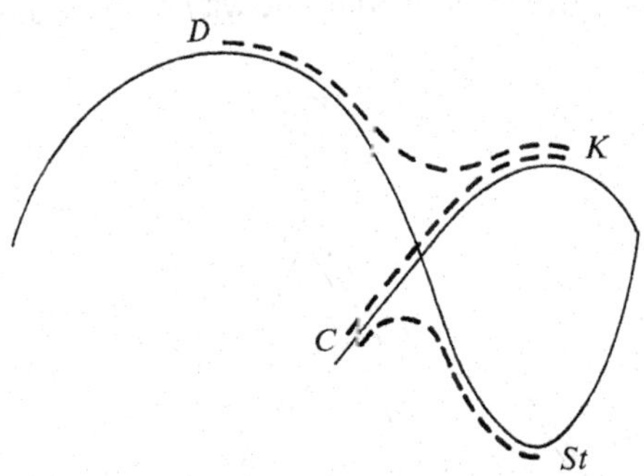

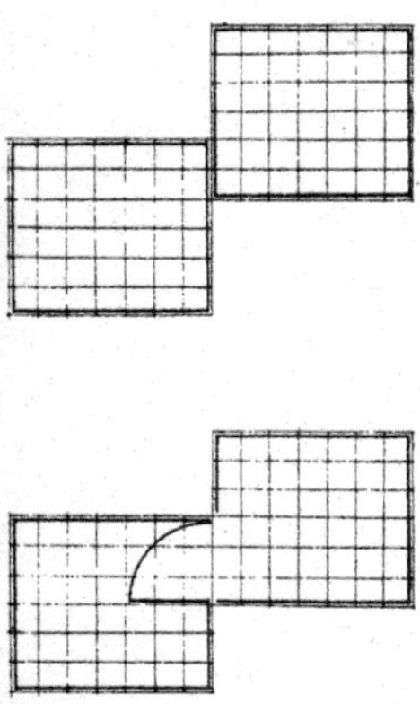

Figure 11.33

By a process of successive subtraction of the values associated with the arcs entering and leaving the vertex, in an appropriate alternating order, we can determine the adjacencies of rooms from the electrical graph in any comparable situation. What is more, with a rather peculiar arithmetic, in which any remainder in the subtraction (i.e., the overlapping of two rooms) of less than, say, 3 units were disregarded, we could make sure that only adjacencies with sufficient overlap to provide space for a door, for example, were counted. In Figure 11.33 the overlap of two modules shown would be discounted, while the width of three modules would be acceptable, assuming our module size of 300 mm, giving a door opening 900 mm across.

In the diagram made up of the two superimposed 'electrical' networks for both vertical and horizontal walls in the plan (Figure 11.30), we know that in each network there is one (finite) arc and one arc only corresponding to each room. Because the two graphs are duals, and in the nature of the way they interrelate, the edge corresponding to a room in one network *crosses* the arc corresponding to that same room in the other network. For instance, the arc [*n*, *a*] in the 'horizontal' network crosses [w, c] in the 'vertical' network, and both these arcs correspond to the dining-room D. The 'currents' in the two 'wires' are the horizontal and vertical dimensions of D respectively.

For the sake of a diagrammatic picture – and we depart here from any strict graph theoretical form of representation – let us circle the five points at which these pairs of arcs cross, and label them with letters for the rooms to which the pairs relate (Figure 11.34). We can now incorporate in this composite diagram the adjacencies of the rooms (shown by broken lines) determined in the way we have just demonstrated.

Figure 11.34

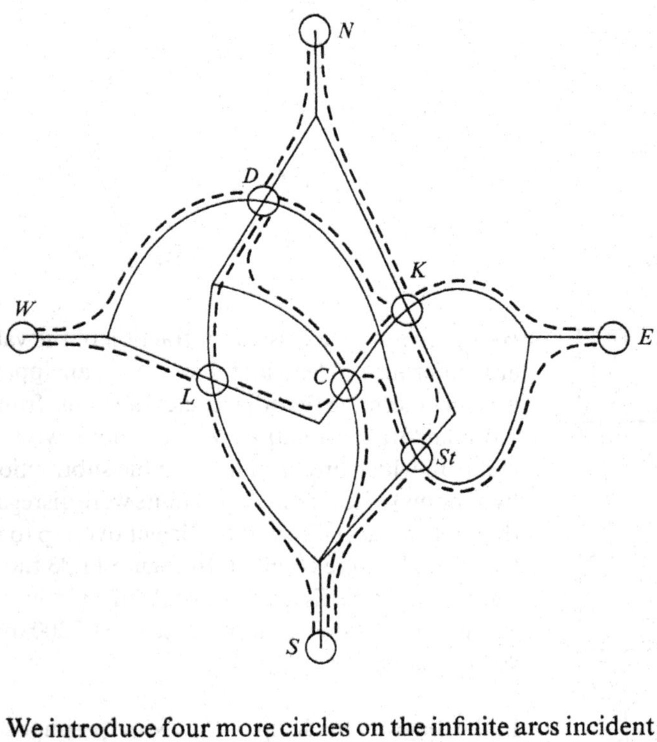

We introduce four more circles on the infinite arcs incident with n, s, w and e, to represent the four areas around the plan on the four sides, N, S, W and E, as we have done in earlier examples; and include the adjacencies of rooms to these external areas. We have now in effect produced a new graph with the circles as vertices, and the broken lines as edges, which is a complete 'adjacency graph' of the plan, of the kind we have seen before (Figure 11.35).

Figure 11.35

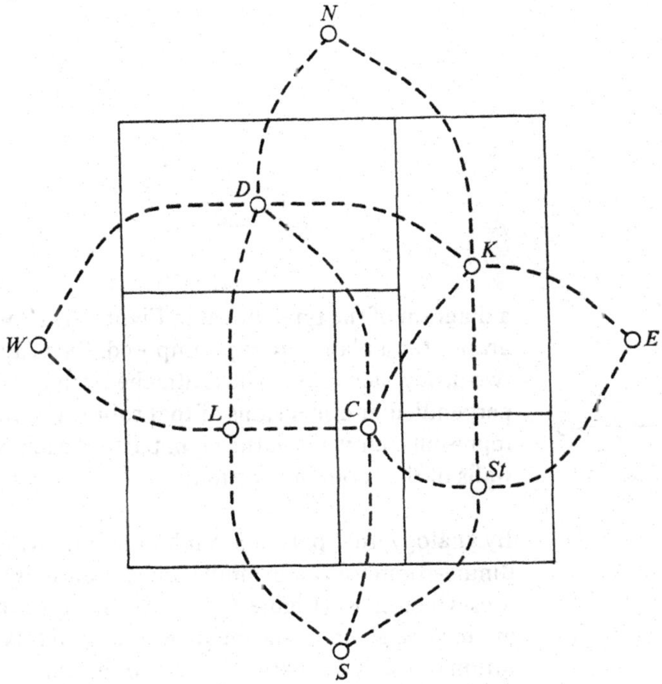

We have suggested already in Chapter 6, how in a real planning problem the architect is likely to be given not exact specifications for the dimensions and shapes of rooms required, but instead some rather loosely defined ranges which these sizes might take. His client might ask, for example, that the living-room in a house be no smaller than a certain area, and that its shape be no more elongated than a given proportion; but any increase which the architect can manage to effect on this minimum specification will be welcome. The exact final sizes will depend clearly on what can be afforded, and on the problems of fitting together the dimensions of the different rooms into the mosaic of the plan.

So far we have shown how to draw 'electrical networks' for layouts whose dimensions and adjacency relationships are already precisely worked out. In practice the architect's problem would tend, of course, to be the other way about: how, for given requirements of room size, room shape and room adjacency, can he produce a plan in which those requirements are met? If somehow the requirements could be formulated in such a way as to build up a pair of matching electrical graphs (as in Figure 11.30) in which Kirchhoff's laws were obeyed, then the problem would be solved. The corresponding plan could then just be drawn out from the diagram. But how can the imprecision of shape and size requirements, as they are likely to be set, be expressed with mathematical exactness?

Suppose it is required that a dining-room D of rectangular shape is to be not less than 8 modules in either length or breadth. We know that in

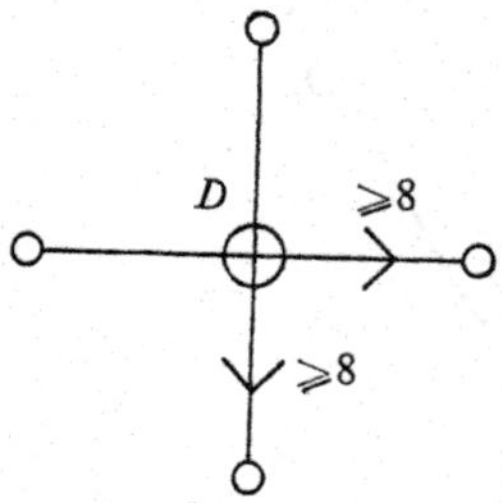

Figure 11.36

a diagram of the type shown in Figure 11.34, where the two electrical graphs for a plan were superimposed, then each room is represented by two arcs which cross, whose attached values are equal to the two perpendicular dimensions of that room. The four corresponding vertices represent four walls in the plan, parts of each of which form the four walls of the room in question.

By analogy, that part of a similar diagram which is to correspond to the dining-room for our example, must eventually take the form also of two arcs which cross (Figure 11.36). At this stage the attached value on each arc may be given by the inequality $\geqslant 8$. In this way we have a representation of the specification for the room, expression the minimum dimensional requirements. Eventually the four 'wall vertices' will be made to coincide with other vertices, corresponding either to walls of other rooms, to which the dining-room may be adjacent, or else to the outside walls of the plan as a whole.

Suppose that for the sake of the example we fix the outside shape and overall dimensions of the house plan, of which this dining-room is to form part. This plan shape is an empty rectangular 'shell', 20 modules

Figure 11.37

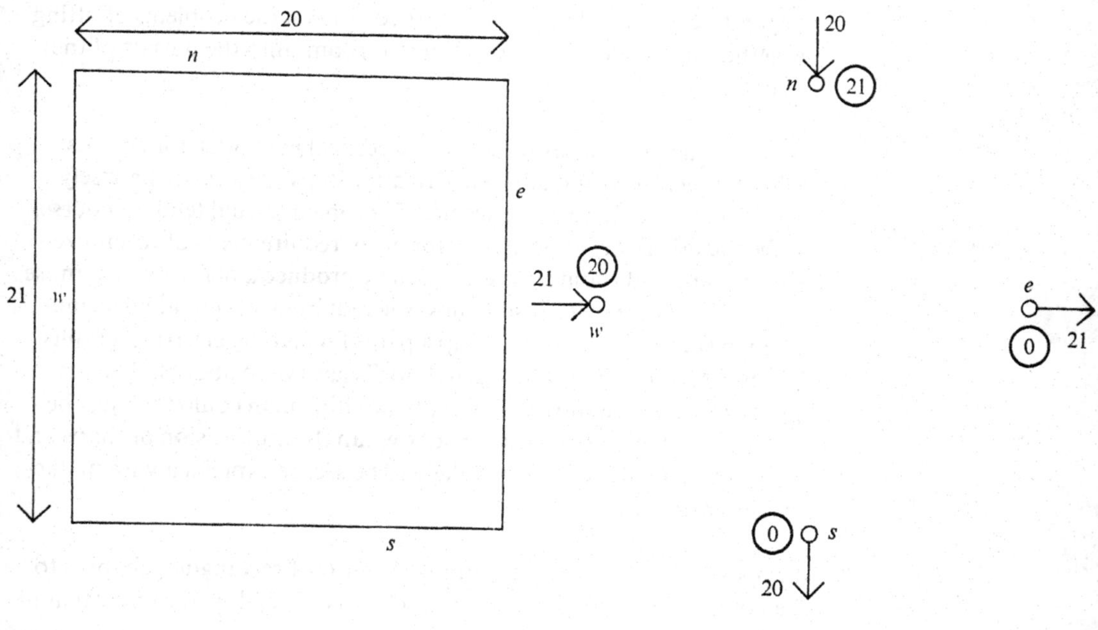

by 21, which is to be filled up with rooms (Figure 11.37). Again we know that, given this shell, whatever the exact form of the final network diagram for the layout, it will comprise at least the four vertices for the four outside plan walls *n*, *e*, *s*, and *w*. What is more, the values attached to those vertices are also fixed, as are the values attached to the infinite arcs which are incident with the vertices.

Suppose we ask that the dining-room *D* and also a kitchen *K* with minimum dimension 6 modules (and no other rooms) are to lie adjacent to the shell wall *n*. We put the 'specifications' for the rooms and the 'specification' of the plan shell together (Figure 11.38). There are two distinct permutations of position which *D* and *K* may take (without regard to their exact dimensions): where *D* is adjacent to the plan wall *w*, and *K* to *e*, or else the symmetrical reflection of that position, *K* adjacent to *w* and *D* to *e*. From these two possibilities we choose to examine the first.

Figure 11.38

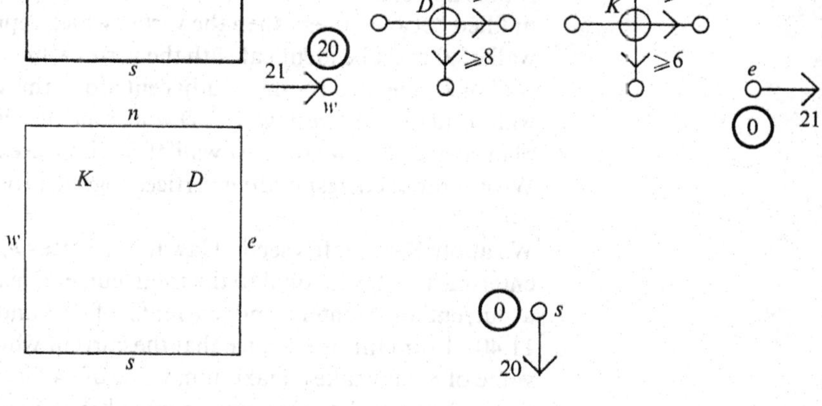

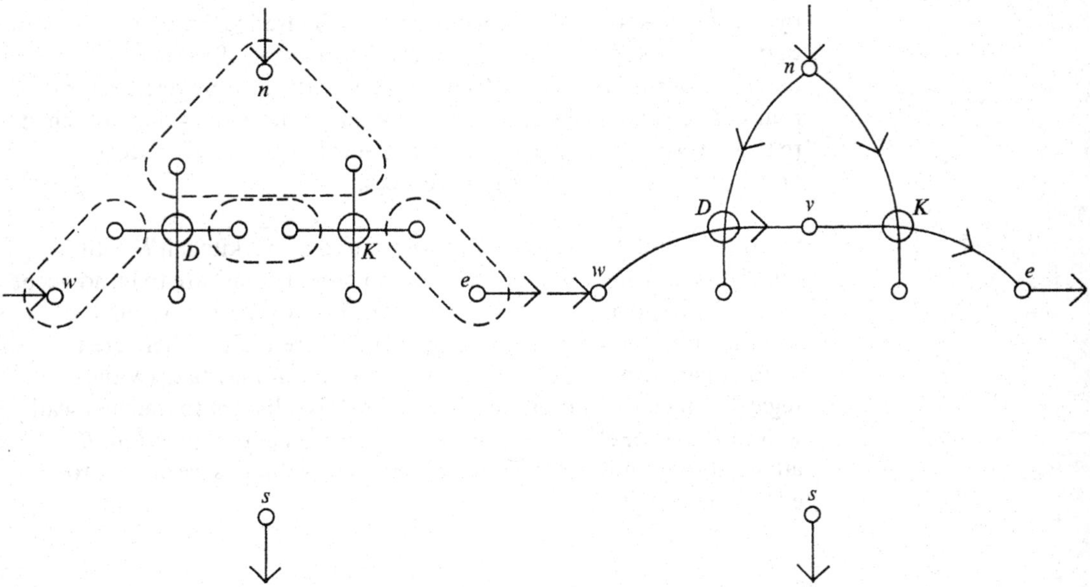

Figure 11.39

We may now infer a number of properties of the upper part of the network diagram. Both *K* and *D* are to be adjacent to *n*. This means that the 'north' walls of the two rooms become identical with the shell wall at *n*; and the three corresponding vertices in the diagram may be collapsed into a single vertex (Figure 11.39). Since *K* and *D* are the only rooms adjacent to *n*, and there must clearly be no 'dead space' left unfilled between them, then the vertex which represents the 'eastern' wall of *D* must be identical with the vertex which represents the 'western' wall of *K*. The two rooms lie adjacent along this common stretch of wall. Equally, the west wall of *D* will coincide with the west wall of the plan shell at *W*, and the east wall of *K* with the east wall of the shell at *e*. We put those corresponding vertices together too.

We apply Kirchhoff's second law to the vertex *n*. The total 'current' entering a vertex is equal to the total 'current' leaving that vertex. Here a 'current' of 20 enters, and 'currents' of $\geqslant 8$ and $\geqslant 6$ leave (Figure 11.40). From this we deduce that the current which takes a minimum value of 8, may take a maximum value of 14, i.e., ($\geqslant 8$, $\leqslant 14$); and that the current which takes a minimum value of 6 may take a maximum value of 12, i.e., ($\geqslant 6$, $\leqslant 12$).

Figure 11.40

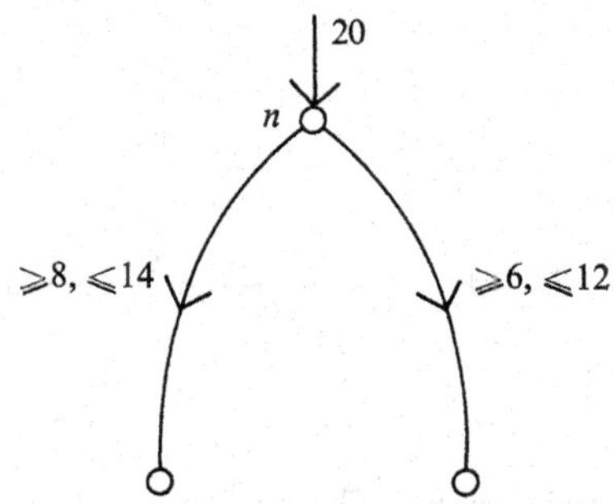

What these inequalities on the arcs mean, in terms of the plan, is that the breadth of the room *D* may permissibly vary between 8 and 14 modules and the breadth of the room *K* between 6 and 12 modules (Figure 11.41). It follows that the position of the common wall shared by *K* and *D* at the centre may permissibly vary in position between a distance of 12 modules and a distance of 6 modules from the shell at *e*. We attach this range of values (≤ 12, ≥ 6) to the *vertex v* which corresponds to that shared wall, in the same way that we attached fixed values to the vertices in the previous network diagrams.

Figure 11.41

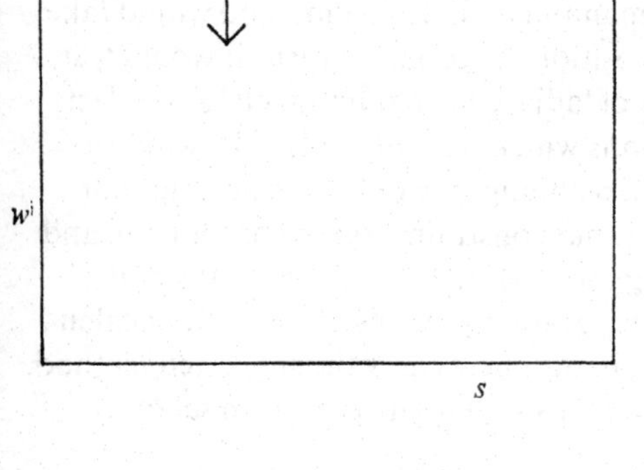

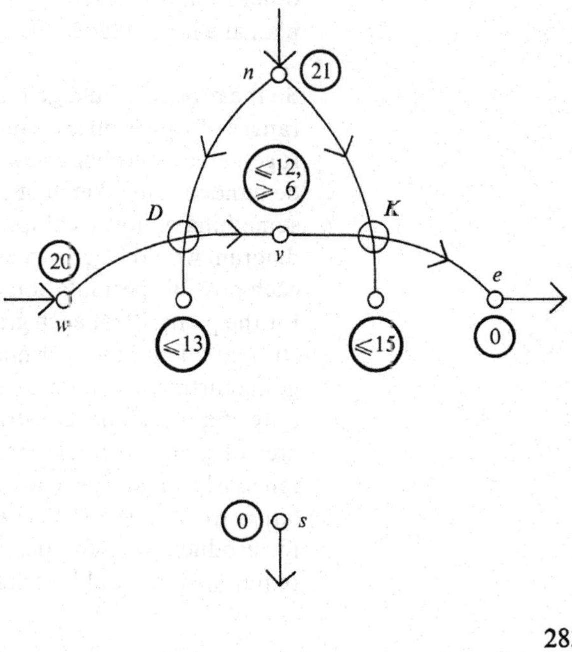

Figure 11.42

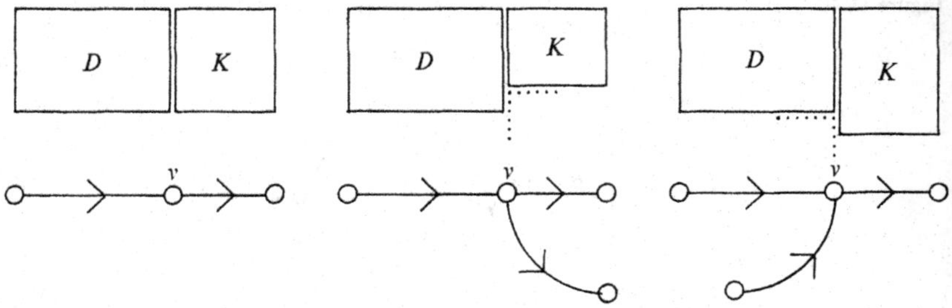

Now consider the vertex v. There is one arc $[w, v]$ incident into v, whose attached value is $\geqslant 8$; there is an arc $[v, e]$ incident out from v, with value $\geqslant 6$. There are three possibilities here (Figure 11.42). Either the two arcs ultimately will take the same value (meaning the two rooms K and D are of equal depth – that is, they stack in the north–south sense), and Kirchhoff's law will be obeyed for the vertex. Or else the value of $[w, v]$ will exceed that of $[v, e]$, or vice versa (meaning that D overlaps K by some dimension, or K overlaps D). In these latter cases at least one new arc will be needed, either incident out from v, or incident into v, respectively.

These new arcs 'grown from v' will correspond to the dimensions of other rooms, adjacent to both D and K, which are to be fitted in the plan at a later stage in the exercise.

So the process would go on, working around the plan, inferring the ranges of values on arcs and vertices, collapsing groups of vertices together, and growing new arcs as the calculations at each vertex demanded. This sketch of a procedure will perhaps serve to indicate something of how a computer representation of the electrical network diagram would be grown and manipulated. Such a method would take each possible permutation of positions of rooms in turn. It would test for the planarity of each graph of 'adjacency requirements' and reject straight off all those permutations which were inadmissible, without going further into their detailed planning. It would take account not only of dimensional constraints, but constraints on the proportion and area of rooms (through making use of Kirchhoff's first law); simultaneously 'growing' an adjacency graph by means of the kind of calculation we outlined earlier. This, in outline, is how the suggested method, for producing systematically all plans which satisfy a given set of requirements, would work.

12 Locations and associations

In the treatment of house planning problems in the last two chapters we discussed the question of the 'adjacency' of rooms; the principal, but by no means the only reason why we require rooms to be adjacent in a plan being to allow people direct access from one to the other – or possibly, in a factory or a warehouse for example, because some materials or goods are passed from one room or area to the next. Other reasons for making two rooms or areas adjacent are, as we mentioned: that if one of these areas is outside the building proper, then the 'adjacency' may allow for natural lighting or ventilation, or for a view of the outside from the room in question. And there are reasons why two rooms or areas might need to be kept separate – the opposite of an 'adjacency requirement' – for reasons of sound insulation, say, or privacy.

While the treatment of problems of plan arrangement at a small scale, as in the house, may perhaps quite realistically be put in terms of 'adjacency requirements', it becomes clear that this kind of constraint is quite inadequate for the production of workable arrangements for any larger type of plan. In a house there is no room which is any great distance from any other. But in an office block of, say, 20 storeys, and perhaps 100 m in length, the distances which the occupants may have to travel to reach one room from another, will become quite significant. We can imagine that for groups of rooms in this building – perhaps corresponding to different departments or sections of the firm occupying the building – it will not be so important that any one room is *adjacent* to another (and indeed there are limits on the numbers of rooms which may all be mutually adjacent); but it will be important for the rooms in each department to be *near* to each other, that they be grouped together – a requirement for their *proximity*.

The implication behind this 'proximity requirement' is that the occupants of the building carry out their work in some kind of regular repetitive routine; so that they more frequently make journeys between some pairs of rooms than between others. These journeys represent wasted effort and time, and so the more the architect can organize the layout of the building so as to minimize the length of these journeys from room to room, the better.

Architects, when they are planning layouts, often tend to speak of some kind of 'association' between the rooms or spaces which they are manipulating in different arrangements – and this 'association' can

[1] B. Whitehead and M. Z. Eldars. 'An Approach to the Optimum Layout of Single-Storey Buildings', *The Architects' Journal*, 17 June 1964, pp. 1373–80.

sometimes be rather nebulous and ill-defined – something to do with rooms of a like kind or classified under the same heading being grouped together. But if we are to give these problems any precision we must be more specific about the real practical reasons why two rooms should be near together – and we must separate notions about how rooms are 'associated' in some general scheme of classification from those functional and operational aspects which are important for their relative positioning in a layout. For the purposes of planning for pedestrian circulation, we will want to know who makes what journeys, of what length, and how frequently, from which rooms to which other rooms; and what 'cost' in wasted time, effort or money can be attributed to each of these journeys.

We can use a formal matrix-like notation to produce a perfectly general statement of the layout problem, where circulation is to be minimized. We introduce special meanings, for this purpose, of the terms *activity* and *location*. By 'location' we mean here some identifiable area within the building, either a work-place, a room, a 'zone' of some kind, which we may reasonably distinguish as being the origin and destination of pedestrian journeys. It may be possible to determine locations simply by dividing the floor space up into equal units with an imposed regular grid. What is important is that all locations should be roughly equivalent in size and significance.

Whether we take smaller or larger units of area for these 'locations' will depend on how reliable and detailed our information is on the frequency with which journeys are made, on how distances between locations are measured, and on the practical limits which the processes of computation involved in a solution of the problem themselves impose. We shall come back to these problems later; but let us suppose for the moment that we can define locations and represent each one by a point (perhaps its centroid), to and from which the distances separating locations are measured.

An 'activity' is more difficult to define in this context, but roughly it represents 'what goes on in a location'. In their paper on the circulation problem, Whitehead and Eldars[1] define 'activity' as 'any process which is or may be carried out at a point separate from other processes'; but this definition is a bit loose. What is important is that 'activities' are not uniquely tied to specific 'locations'. The 'location' represents in principle an empty piece of floor area or building space. Then, subsequently, that area or space is designated for a particular function, or 'activity', and the location may be delimited by partitions, possibly, or provided with appropriate furniture or equipment for the activity in question.

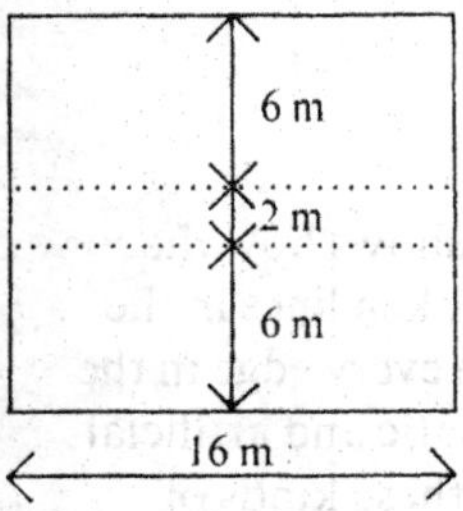

Figure 12.1

We will start with a simple example, and develop these definitions in more detail later. The floor plan of a research institute (Figure 12.1), let us imagine, is a rectangle 16 m by 14 m, and is divided by a 2-m wide corridor running centrally down the long axis of the building.

Our layout problem is to accommodate eight research workers in eight equal area offices, each of 24 m². In terms of architectural design, there is only one possible configuration of rooms, assuming that all the offices are to give on to the corridor, and all to be of the same shape, of similar character, and similarly equipped. So the layout problem is really an administrative one, of assigning a room to each worker.

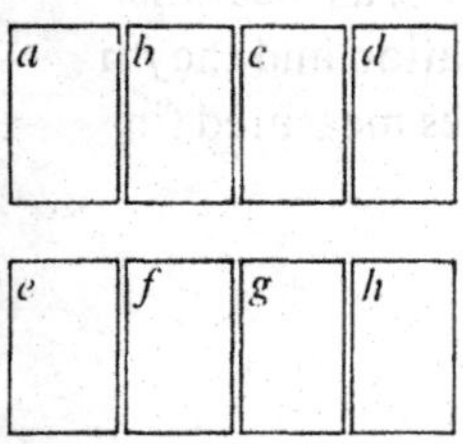

Figure 12.2

In this case the 'locations' are the rooms, and the 'activities' correspond to the individual workers. We name the rooms in the plan (Figure 12.2) by the lower-case letters *a*, *b*, *c*, *d*, *e*, *f*, *g*, *h*. And the researchers – Mrs *A*, Mr *B*, Miss *C*, Mr *D*, Dr *E*, Miss *F*, Mr *G* and Mr *H* – we will refer to by the capital initial letters of their surnames.

We can describe any particular permutation from the set of all possible allocations of rooms to workers with a Table **P**, which in the general problem represents a placing of each activity in a chosen location. Here it amounts, in effect, to a list of 'who occupies which office'. Let us choose for our first example to examine the layout where worker *A* is assigned room *a*, worker *B* room *b*, etc. The Table **P** in this case is:

Table P

Locations (offices)	*a*	*b*	*c*	*d*	*e*	*f*	*g*	*h*
Activities (workers)	*A*	*B*	*C*	*D*	*E*	*F*	*G*	*H*

Next, to describe the plan itself, we make a Table **D**, which records the distances between each pair of locations. There are a number of general problems arising from the measurement of distance for this purpose, but in the present example we shall take it that the measurement is taken from the centre of each room in question; that the door to each room is centrally placed in the corridor wall; that distances are measured to the centre-line of the corridor; and that all dimensions are taken

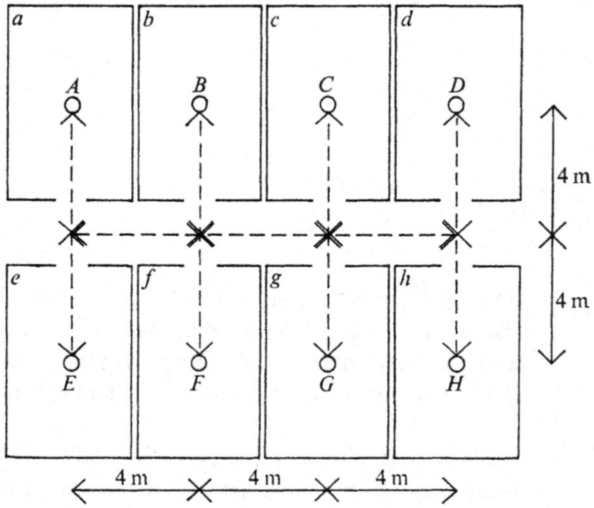

Figure 12.3

perpendicularly to the walls of the plan. Figure 12.3 shows how the assumed circulation system of the plan (shown by broken lines in the figure) thus takes the form of a rectilinear 'tree', with every edge in the tree of equal length 4 m. (This is clearly a very schematic and artificial representation: the inaccuracies involved in making these kinds of assumptions in measuring distance will be less for more extensive plans, however.)

D is a triangular table with as many columns (and rows) as locations. The entry d_{ij} records the distance between the ith location and the jth location. The table for our example, with the distances measured (in metres) along the circulation 'tree', is:

Table **D**

	a	*b*	*c*	*d*	*e*	*f*	*g*	*h*
a	·							
b	12	·						
c	16	12	·					
d	20	16	12	·				
e	8	12	16	20	·			
f	12	8	12	16	12	·		
g	16	12	8	12	16	12	·	
h	20	16	12	8	20	16	12	·

It will be clear that there can be no entries on the leading diagonal (top left to bottom right, where dots are shown), since these positions in effect record 'the distance of a location from itself'. It is also clear that only the one triangular half of the complete square table is needed, since the distance from i to j is the same as the distance from j to i, and the other triangular half would only repeat the identical information in symmetrical form. (Though this would not necessarily be true of the 'distance table' of, say, a one-way road system, where d_{ij} does not always equal d_{ji}, and the complete square table is required.)

This Table **D**, together with the Table **P**, serve to describe a particular layout solution. In order to evaluate the layout, to see how well it caters for the patterns of movement between one 'activity' – or worker in an office in this case – and another, we will need a further Table **A**, which expresses the 'association' between each pair of activities located at i and j. We have warned against loose thinking when it comes to this question of 'association'. (Sometimes the words 'affinity' or 'linkage' have been used.) Here the term is defined to mean 'the cost of separating each pair of activities, per unit distance'. But how can we measure such a 'cost'?

Here is one of the main difficulties in the circulation problem: how to determine in a particular case some set of empirically derived values for the 'association' between activities. If two research workers, in our example, are found to visit each other frequently in the course of their work, then it will be reasonable to set a high value on the association between them. If possible, we should mount a survey, to find out the frequency with which Mrs A visits Mr B, or Miss C, and so on, over a day or a week. (The assumption will be that the frequency of trips between each pair of workers remains the same, or very nearly the same, from one day to the next, or from one week to the next, or whatever time cycle is chosen; and this is an assumption which should be tested, for if the pattern of traffic is not regular and consistently repeated then none of the following analysis can properly be applied.)

A better layout, or layout of lower 'cost', will be one where the pairs of research workers who most frequently visit each other are closer together in the plan. But we cannot always simply set values in the 'association' Table **A** equal to the frequency (over some fixed period) of two-way traffic between pairs of 'activities'. For if the overall purpose is to cut down wasted time spent moving about the building, then we must accept that in many situations the value of some of the occupants' time will be greater than that of others.

In our research office, we might assume that all the researchers could be treated equally; but in a commercial firm for example, the chairman's time might be considered many times more valuable than that of his secretary – and though the secretary might make very many journeys in

[2] U. Cinar. *Facilities Planning; a Systems Analysis and Simulation Approach with Particular Reference to Hospital Design*, unpublished PhD thesis, Department of Operational Research, University of Lancaster, 1968.

the course of her day, there should be much less significance attached to these trips than to the few visits made by her employer. It has been suggested that to cope with situations of this sort, the 'association' values should consist of the basic trip frequencies 'weighted' according to the salary of the person making each trip. The cost of a layout would then represent the total cost to the firm in wages (and possibly overheads too) of time lost in travel. Crudely, if the chairman earned £10,000 a year and his secretary £1,000, then one trip made by the chairman would count for ten trips by the secretary.

This kind of measure may be appropriate in business. But in an organization such as a hospital, although a doctor's or a surgeon's time will cost more in wages than that of a nurse or a porter, perhaps it will be most important of all to reduce the length of journeys made by the patient, who receives no salary at all. For him, the importance of time saved might be literally vital.

It may be possible to deal with this difficulty by setting a subjective scale of values on the relative importance of different trips, and taking the advice of management or administrators in fixing this scale. Cinar,[2] for example, in a problem of hospital layout, establishes a series of 'weighting factors', by which the trips made by members of each grade or group in the hospital population are multiplied to produce the appropriate association values. These factors are: for medical staff 12, for nurses 3 – the ratio depending on their relative salaries – and for patients 3, their visitors 1. These last two figures are inevitably somewhat arbitrary.

In the ideal democracy of our research institute, however, we shall take it that everybody's time is valued the same. For one thing, it will simplify the arithmetic. Presented below is some tabulated data, perhaps derived from a survey, of the number of visits each worker makes per day to every other. There are no diagonal entries again of course, but the table is not symmetrical, since the number of visits worker *I* makes to worker *J* may well differ from the number *J* makes to *I*. The column and row totals show the total number of trips made to and by each individual respectively. Miss *C*, for example, works quite independently, makes no visits, and is not visited. Dr *E* by contrast has a good many contacts during the day, making 12 trips to see others and being visited 12 times himself.

Table C

		to *A*	*B*	*C*	*D*	*E*	*F*	*G*	*H*	Row totals
Visits daily by	*A*	·	0	0	0	5	2	0	1	8
	B	0	·	0	4	1	0	5	0	10
	C	0	0	·	0	0	0	0	0	0
	D	2	3	0	·	1	0	1	0	7
	E	4	1	0	1	·	2	1	3	12
	F	1	0	0	0	2	·	0	1	4
	G	0	1	0	0	0	0	·	0	1
	H	0	0	0	0	3	2	0	·	5
Column totals		7	5	0	5	12	6	7	5	

We might present the same information in graph form, as a *network* (Figure 12.4), where the vertices represent activities (research workers), the directions of the arcs indicate who visits whom, and the attached values on the arcs give the daily trip frequencies. *C* becomes an isolated vertex, and four pairs of vertices are joined only by single arcs: [*D*, *A*], [*A*, *H*], [*D*, *G*] and [*E*, *G*]. Nine other pairs of vertices are joined each by two oppositely directed arcs, representing the two-way traffic in each instance.

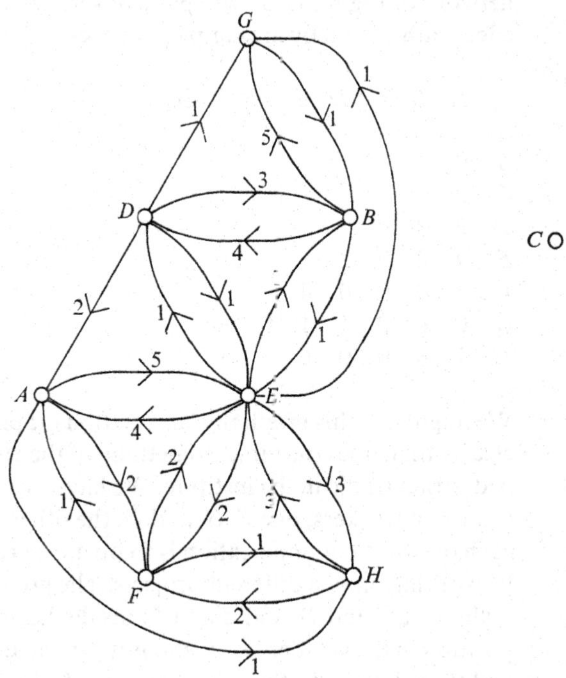

Figure 12.4
Graph showing numbers of journeys made daily by workers one to another, for the research institute example.

It is at this point in other problems, possibly, that different cost weightings would be introduced, to account for differences in the value of workers' time. The number of trips would be multiplied by the appropriate 'weighting factor' in each case, according to the salary or the importance of the person making those trips. In our example, though, we have decided to attribute exactly the same cost significance to all journeys. We therefore add together the trips made in both directions for every pair of activities (workers); and it is simply this sum of two-way traffic which, for the purposes of our example, will be the value given to the 'association' of each activity pair. Note that every single trip recorded is a *return journey*. For example, as Figure 12.4 shows, *B* visits *E* once, and *E* visits *B* once in a day. The association of *E* with *B* is therefore given the value 2. This means that *E* goes to *B and returns*, and *B* goes to *E and returns*; i.e., the distance which separates *B* from *E* will be covered four times in all.

It follows that the 'association' Table **A**, shown below, will be triangular, like the 'distance' Table **D**. Its corresponding graph is the symmetric network of Figure 12.5, with pairs of vertices shown connected by edges rather than by arcs or pairs of arcs.

Table A

	A	*B*	*C*	*D*	*E*	*F*	*G*	*H*
A	·							
B	0	·						
C	0	0	·					
D	2	7	0	·				
E	9	2	0	2	·			
F	3	0	0	0	4	·		
G	0	6	0	1	1	0	·	
H	1	0	0	0	6	3	0	·

We might call this graph the 'association graph'. It shows up, more clearly than does the table, something of the *structure* of the pattern of pedestrian traffic in the institute. We have two somewhat separate groups of workers, one *B*, *G*, and *D*, the other *A*, *E*, *F* and *H*, within each of which communication is frequent; perhaps the groups might be working on two different projects. The groups have some contact, mainly via *B* and *D*, and *A* and *E*. Inside the groups, *B* is in a central position in the smaller team, and it is principally through him that *D* and *G* are linked. In the larger *A*, *E*, *F*, *H* group it is *E* who is the main focus of communications.

These structural aspects become even more obvious if we omit progressively from the graph those edges which are of lowest value. With edges value 1 ignored, and then with value 2 omitted, the network appears as in Figures 12.6 and 12.7. By the second stage the two groups of researchers

Figure 12.5
Symmetric 'association graph' for the research institute, showing total numbers of journeys made daily in either direction between pairs of workers.

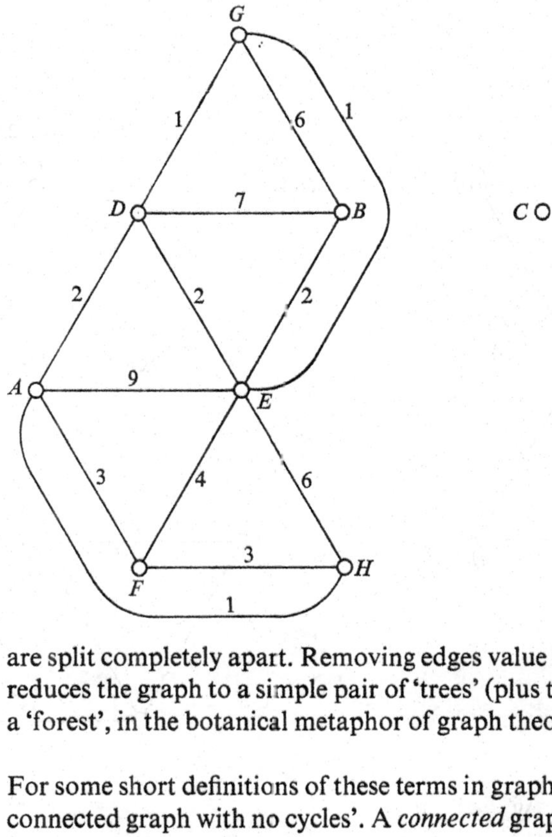

are split completely apart. Removing edges value 3, as in Figure 12.8, reduces the graph to a simple pair of 'trees' (plus the vertex *C*) – strictly, a 'forest', in the botanical metaphor of graph theory language.

For some short definitions of these terms in graph theory: a *tree* is 'a connected graph with no cycles'. A *connected* graph is a graph in which each vertex is joined to all others by chains – that is, by consecutive series of linked edges. In Figure 12.8 the edges (*H*, *F*), (*F*, *E*), (*E*, *A*) form the chain (*H*, *F*, *E*, *A*). A connected graph is thus one which consists of one part or *component* only and is not split into a number of distinct, 'disconnected' components. All our house planning graphs in the last chapter were connected, but the two graphs of Figures 12.6 and 12.7 above, are not. The graph in Figure 12.6 has the single isolated vertex at *C*; and the graph of Figure 12.7 consists of three separate components. A cycle we defined in the last chapter as a chain that returns to its starting point. Both the graph in Figures 12.6 and 12.7 contain cycles; one example in the latter is (*H*, *F*, *E*, *H*). But the last graph of the series, in Figure 12.8, does not (nor is it connected). The separate components of this final graph are thus trees; and a graph which is not connected, but consists of a number of trees, is called, naturally enough, a *forest*.

Figures 12.6, 12.7 and 12.8 'Association graph' of Figure 12.5 with edges of lowest value progressively omitted.

12.6 Edges value 1 omitted

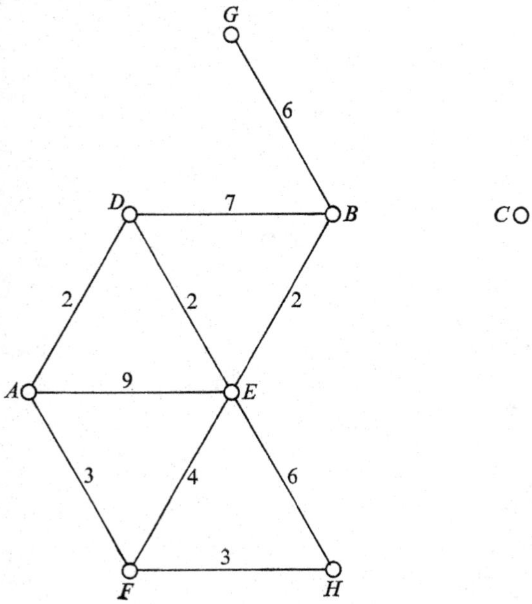

12.7 Edges value 2 omitted

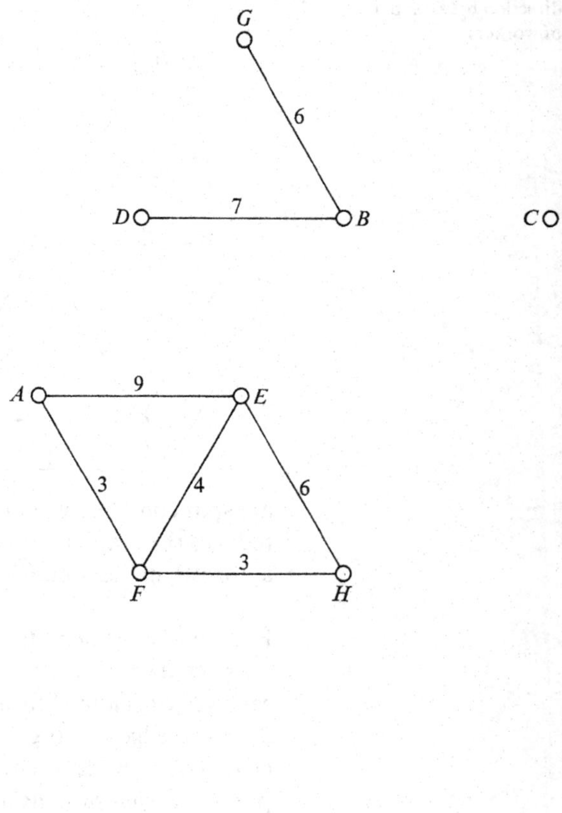

12.8 Edges value 3 omitted

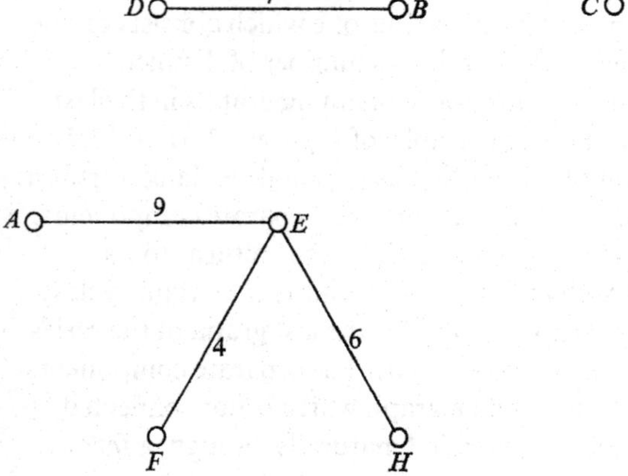

In this 'forest' graph of Figure 12.8 the importance of the positions of *B* and *E* in the two research groups emerges quite clearly. If we were to go about planning a layout intuitively, on the evidence of an inspection of this graph, it would seem sensible to divide the accommodation first into two parts, a group of three rooms together for one group of workers and a group of four rooms for the other. The eighth office goes to Miss *C*, and since she is quite independent, it does not much affect the issue which room she is given; and it will be best to place her at one of the two extreme ends of the plan.

Within the groups, a reasonable strategy would seem to be to position *E* and *B* first, and then place the other researchers around them, perhaps working in order of the relative strengths of their association with these first two. Thus in the large group we might position, after *E*, *A*, then *H*, then *F*; and in the smaller, after *B*, *D*, then *G*. Finally, we could take account of the weaker connections across the group division perhaps – the links between *A*, *E*, *D* and *B*. A plan solution worked out following this line of reasoning, and not produced by any systematic method, is shown in Figure 12.9. It can be recorded in a 'plan' Table **P**, as in our earlier convention, thus:

Table P

a	*b*	*c*	*d*	*e*	*f*	*g*	*h*
C	*B*	*E*	*F*	*G*	*D*	*A*	*H*

Figure 12.9

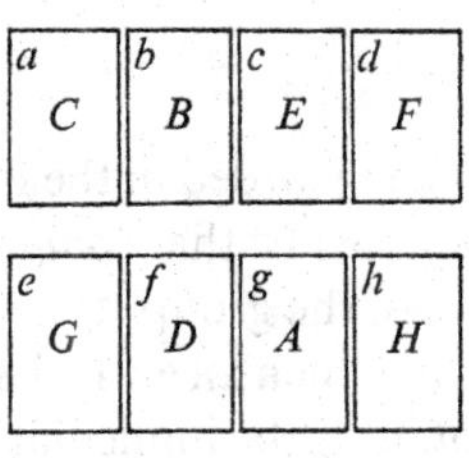

But before we embark on a discussion of the different methods for producing 'better' solutions to a circulation problem, let us return for a moment to the subject of graphs. The 'association graph' of Figure 12.5 which is essentially here a graph of *trip frequencies*, and shows the numbers of journeys made between pairs of activities, is not by any means the only type of graph we might draw of an organization. In a

[3] O. P. Tabor. *Traffic in Buildings 3; Analysis of Communication Patterns*, Working Paper 19, University of Cambridge, Land Use and Built Form Studies, 1970.

recent paper Tabor[3] has distinguished a number of possible aspects which might be represented.

Suppose that in our particular example there exists some structure of command, that some members of the research institute direct the work of others, so that the whole organization is structured in an hierarchic way, with each person directly responsible to an immediate superior. Suppose that this structure is as in Figure 12.10. The arcs in this graph signify the relationship 'is the immediate superior of'. Miss *C* is still quite independent, but each of the two separate research teams has a group leader, *E* and *B* respectively. *D* and *G* are responsible to *B*; while in the larger team there is a two-level structure of command, *E* directing *A*'s work and *A* in turn directing *F* and *H*.

Figure 12.10
Graph to show structure of command in the research institute

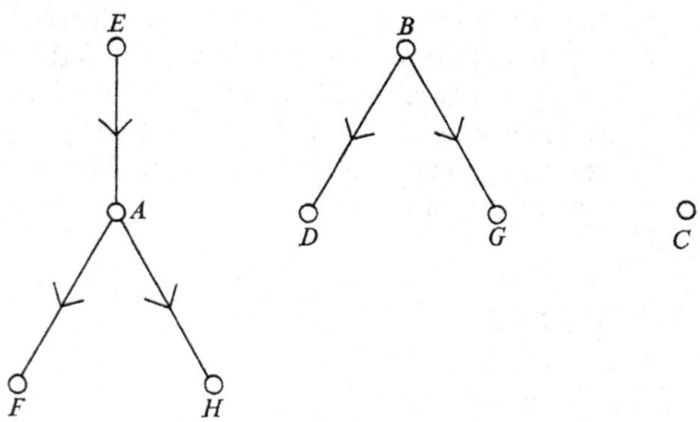

This same information can be recorded in the form of a zero-one matrix. We take for the sake of example just that *component* of the graph which corresponds to the structure of the group *A*, *E*, *F* and *H*; we can represent this as a 4×4 matrix, in which an entry 1 signifies that the worker to whom that *row* corresponds, is the immediate superior of the worker to whom the column corresponds (otherwise a 0 is entered). Thus in the first row *A* is the immediate superior of *F* and *H*.

$$
\begin{array}{c} \\ A \\ E \\ F \\ H \end{array}
\begin{array}{c} \begin{array}{cccc} A & E & F & H \end{array} \\ \begin{bmatrix} 0 & 0 & 1 & 1 \\ 1 & 0 & 0 & 0 \\ 0 & 0 & 0 & 0 \\ 0 & 0 & 0 & 0 \end{bmatrix} \end{array}
$$

An incidental but interesting property of this type of matrix is that, when multiplied by itself, the resulting product matrix shows who is who's superior *at one remove*. In this case E is two steps up the ladder of command from F and H.

$$\begin{array}{c} \\ A \\ E \\ F \\ H \end{array} \begin{array}{c} \begin{matrix} A & E & F & H \end{matrix} \\ \begin{bmatrix} 0 & 0 & 1 & 1 \\ 1 & 0 & 0 & 0 \\ 0 & 0 & 0 & 0 \\ 0 & 0 & 0 & 0 \end{bmatrix}^2 \end{array} = \begin{array}{c} \\ A \\ E \\ F \\ H \end{array} \begin{array}{c} \begin{matrix} A & E & F & H \end{matrix} \\ \begin{bmatrix} 0 & 0 & 0 & 0 \\ 0 & 0 & 1 & 1 \\ 0 & 0 & 0 & 0 \\ 0 & 0 & 0 & 0 \end{bmatrix} \end{array}$$

In a similar way the matrix of a family tree, where entries show who is who's parent, when squared will yield a matrix recording who is who's grandparent. The third power of the original matrix gives the grand-parents; and so on.

While the structure of a graph showing who is who's father must by its nature have a tree-like structure – for no person can have more than one father – graphs of the structure of command in organizations do not always necessarily share this property. It is in situations where it is most important that the responsibilities are quite unambiguous, and each person at each level must take orders from one person at the level above, and from no one else, that the organizational structure will be of strict tree form. The army is an obvious case where this applies.

We have looked so far at two types of graph, a graph of *communication* (the 'association graph'), of the frequency of journeys made by workers to each other; and a graph of *organization*, or the nominal structure of authority and responsibility. Since we would expect almost all workers to communicate with their immediate superiors by making visits to them, or vice versa, then for any particular case we would expect the two types of graph to have similarities. There might certainly be other visits made, between workers other than just those linked in the hierarchic organization graph; and so the organization graph will in all probability form a *partial graph* of the communication graph. This is true of our office example, and Figure 12.11 shows the two graphs together, the one emphasized as heavy lines drawn over the thin line edges of the other.

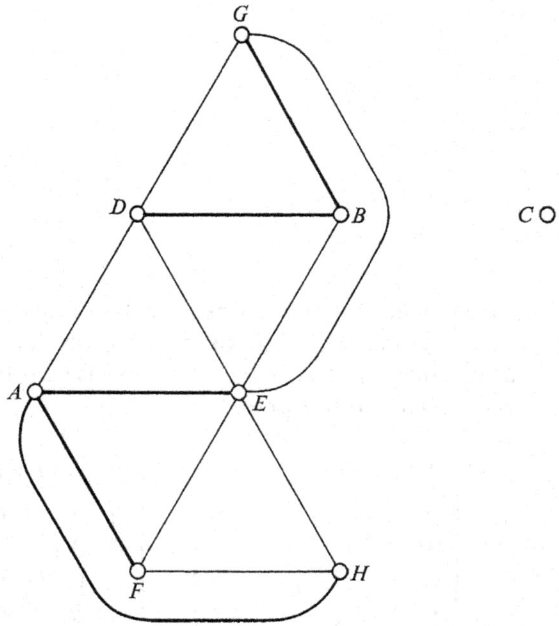

Figure 12.11
Graph of structure of command in the research institute, from Figure 12.10, as a *partial graph* of the 'association graph'

We referred to the organization graph as representing the *nominal* structure of authority – that is, how the structure is formalized and made explicit, how the organization is imagined to work by those who administer it. If the communication and organization graphs are found to differ very greatly – if the one does *not* contain the other as a partial graph, and quite a different effective hierarchy appears to exist, on the evidence of the communication graph, from that set out in the organization graph – then there are two possible explanations.

Either the institution in question does not indeed function in the way its explicit formal structure would suggest. Or else, as is perfectly likely, the communication of the organization's business is going on in other ways than by its members making personal visits (which is the only means that our communication graph above depicts) – in writing, via the telephone, or possibly by some other electronic means. An 'activity' which from a survey of pedestrian traffic might appear to be completely isolated, could in actual fact be the very nerve-centre of a traffic in papers or documents – a filing office for example, or a registry in a government ministry. One of the most difficult features of planning for pedestrian circulation is how to take account of the way in which people, when they are situated far apart in a building, may choose to use the telephone to contact colleagues who, if they were nearer, they would visit in person. Alternatively, they might save various items of business up for a single trip, where if they were closer, they would make separate journeys for each job. Or they might take a round tour of a series of offices, depending on the building's particular layout, visiting a series of colleagues in turn.

For our research office, following the example of others who have tackled circulation problems in this way, we have fixed values for the 'association' of two activities independent of their possible relative positions in a plan. We assume that the individuals concerned make journeys with particular fixed frequencies, irrespective of how far the actual distances travelled may be. We also assume that each trip is made to the destination and immediately back again, and no possible round trips are taken into account. All these assumptions are doubtful ones, and though we shall hold to them for the purposes of the analysis, we shall come back later to examine their serious implications.

A third type of graph which might be drawn for an institution or business, is a *classificatory* graph, and this might be used to record a variety of different aspects. Figure 12.12 is a classificatory graph for the research office example. We assume that *A*, *E*, *F* and *H* belong to one research team, and *B*, *D* and *G* to a second, as we suggested might be the case. The levels in the graph indicate two stages in the progressive breakdown of the whole institute into its smallest constituent parts – the individual workers. The single vertex at the topmost level represents the institute as a whole. The vertices along the bottom row are the workers. In the centre row the workers are grouped into their respective teams, and each of the two vertices represents a team.

Figure 12.12

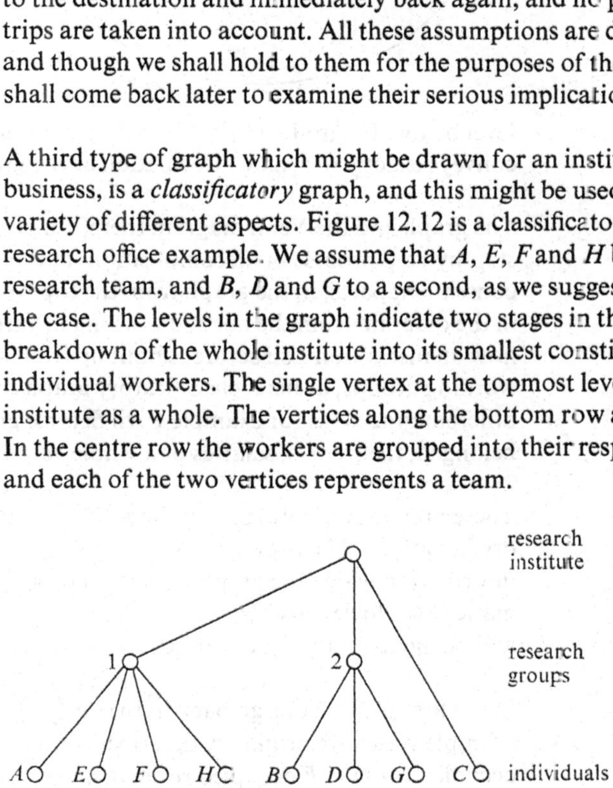

This graph-theoretical way of representing the breakdown of an organization into its constituent parts, can be directly transposed into the terms of set theory. Each vertex of the graph is in effect a set, and the topmost vertex here represents the *universal set* of all employees. At the intermediate level are two (in this case *disjoint*, or non-overlapping) sub-sets. And in the bottom row are shown all the separate one element sub-sets (employees), or *unit sets*. The arcs in the graph signify the relation that the set represented by a vertex at one level, *contains* that set represented by the vertex which the arc is incident in to at the

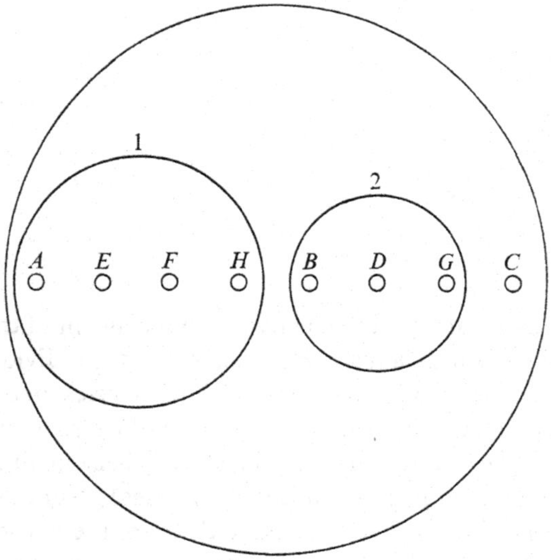

Figure 12.13

level below. Figure 12.13 is a Venn diagram which is equivalent to, and conveys exactly the same information as the graph.

The graph and the Venn diagram therefore both depict, in effect, a step-by-step process of 'decomposition' of the office structure into its constituent parts, in the graph from the top working downwards, and in the diagram from the outside working inwards. As well as showing how an institution's structure is built up from administrative units or working groups, as here, classificatory graphs could illustrate the different grades (as for example Civil Service grades) to which employees belonged, their posts, salaries or a whole variety of other characteristics.

These graphs are useful in revealing different structural properties of organizations. The problem remains as to how we can best use the information which the graphs record to produce, in some quite systematic way, 'lower cost' layouts for particular organizational structures and particular patterns of traffic.

In the first place, let us go back to the layout for the research institute example which we originally described with the Table **P** – where *A* occupied room *a*, *B* occupied room *b*, and so on. There is no obvious reason why this particular layout should be either a very good one or a very bad one, and we shall use it as a yardstick against which to measure the success of various systematic techniques for producing better solutions.

The basis for such a comparison – the 'cost' of each different layout – will be the sum, for *all* pairs of activities, of the values for the association of those activity pairs, multiplied in each case by the distance separating the two activities in the plan. To put this in formal notation, if d_{ij} is the general term in the 'distance' Table **D** – that is, the distance from the ith location to the jth location; and if a_{ij} is the general term in the 'association' Table **A** – that is, the association between the ith activity and the jth activity; then we can express the cost of the layout **P** as C_P, where

$$C_P = \Sigma_i \Sigma_j d_{ij} a_{ij}.$$

What exactly does this expression mean, and how is it derived? We take the Table **D**, and the Table **A**, which are both the same size.

Table D

	a	*b*	*c*	*d*	*e*	*f*	*g*	*h*
a	·							
b	12	·						
c	16	12	·					
d	20	16	12	·				
e	8	12	16	20	·			
f	12	8	12	16	12	·		
g	16	12	8	12	16	12	·	
h	20	16	12	8	20	16	12	·

Table A

	A	*B*	*C*	*D*	*E*	*F*	*G*	*H*
A	·							
B	0	·						
C	0	0	·					
D	2	7	0	·				
E	9	2	0	2	·			
F	3	0	0	0	4	·		
G	0	6	0	1	1	0	·	
H	1	0	0	0	6	3	0	·

Each element d_{ij} from **D** is multiplied by the corresponding element a_{ij} in **A**, and the result $d_{ij}\, a_{ij}$ is put in the same corresponding position in a new third 'cost' Table **C**.

Table C

·							
0	·						
0	0	·					
40	112	0	·				
72	24	0	40	·			
36	0	0	0	48	·		
0	72	0	12	16	0	·	
20	0	0	0	120	48	0	·

Each value in **C** will thus equal the association between a pair of activities, multiplied by the distance separating the locations in which those activities are placed.

The expression

$$\Sigma_i \Sigma_j d_{ij} a_{ij}$$

means 'the sum of all terms $d_{ij}\,a_{ij}$, for all values of i and for all values of j' – that is, the sum of all elements in Table **C**. The actual figure for this sum C_P here is 660.

For special reasons in our example here, we can put a particular interpretation on the quantity C_P. Since we have set association values equal simply to the frequency of two-way traffic between pairs of activities, then what we have done in calculating C_P is, in effect, to multiply numbers of journeys by the distances over which those journeys were made, in every case. Because we have introduced no cost weightings, C_P is therefore equal here to the total distance travelled (in metres) by all the building's occupants during one day. Or rather, as we pointed out earlier, since all journeys are *return* journeys, we should multiply C_P by two to include the return halves of the trips as well.

If we wanted then to convert to some actual monetary cost, we could divide the total distance travelled $2C_P$ by some average walking speed s (metres/hour) to give the total time spent in travel: and multiply by an hourly cost rate c (pounds/hour) assumed to be the standard cost of all workers' time, to give a total cost in pounds of £$(2 \times 660)c/s$ per day.

13 Spatial allocation procedures

Out of the many varied approaches which have been taken to solving the circulation problem, two broadly different types of method can be distinguished. Tabor[1] in his papers characterizes the two types as 'additive' and 'permutational', and other authors[2] have called them 'constructive' as against 'improvement'. To take the 'improvement' or 'permutational' methods first: the important feature of this kind of method is that a complete layout is produced – by some means or other, possibly at random – at the beginning of the process. Then the positions of activities – rooms or whatever – are permuted, swapped about in such a way as to progressively reduce the layout cost. It is in this sense that the methods consist in a process of *improvement* of some arbitrarily selected initial starting layout.

On the other hand, a typical method of the 'additive' or 'constructive' kind starts with an empty site or empty 'floor plane', and builds up a low cost layout, one activity or one room at a time. The criteria according to which each successive activity unit is positioned are made dependent on some measure of the association of that unit with all other units already placed. The one type of method starts with a not especially good layout, and attempts to improve on it. The other type attempts to build up a good layout by stages, one unit at a time.

With *permutational* methods, the most direct and unsubtle approach would be simply, of course, to go through every possible layout: to try exhaustively every different arrangement in which the activities may be assigned to locations, measure C_P in every case, and take the arrangement where its value is least. The objections to this are not theoretical, but practical.

The number of ways of arranging one activity in one location is one; of arranging two activities in two locations, two. With three activities we have three possibilities for placing the first activity. For each of these three choices there are then two empty locations left, and so two ways of placing the second activity. This gives six possibilities in all, since there is now only one location left for the third activity. By a similar argument we can see that the possibilities with four activities would be $4 \times 3 \times 2 \times 1 = 24$ in number; and in general with n activities the number of different ways these may be arranged in n locations is $n(n-1)(n-2)...1$, or *factorial* n, written $n!$ for short.

[1] O. P. Tabor. *Traffic in Buildings 1; Pedestrian Circulation*, and *Traffic in Buildings 2; Systematic Activity-Location*, Working Papers 17 and 18, University of Cambridge, Land Use and Built Form Studies, 1970.

[2] C. E. Nugent, T. E. Vollmann and J. Ruml. 'An Experimental Comparison of Techniques for the Assignment of Facilities to Locations', *Operations Research*, vol. 16, no. 1, 1968, pp. 150–73.

Values for $n!$ for successive values of n are:

n	$n!$
1	1
2	2
3	6
4	24
5	120
6	720
7	5040
8	40 320
9	362 880
10	3 628 800

So even for our very limited research institute problem the job of measuring C_P for all permutations of layout is certainly one for a computer. With ten rooms the number of possibilities is up to over 3½ million, and the addition of only one more room increases that number by a further 36 million. For a problem with any realistically large number of locations, the permutations become quite unmanageably numerous, even using computers.

Figure 13.1

In some cases the number may be somewhat reduced, by considerations of symmetry. In the research institute, since the plan is symmetrical about the axis of the corridor, then half the possible permutations of layout will be reflections of others; and the calculation of the layout cost C_P will be for all intents and purposes the same (and will give exactly the same result) for both in each pair. Equally, the layout is bilaterally symmetrical about a second axis perpendicular to the corridor; and so we can divide the effective number of possibilities to be examined in half again, to account for those layouts which are reflections one of another about this second axis.

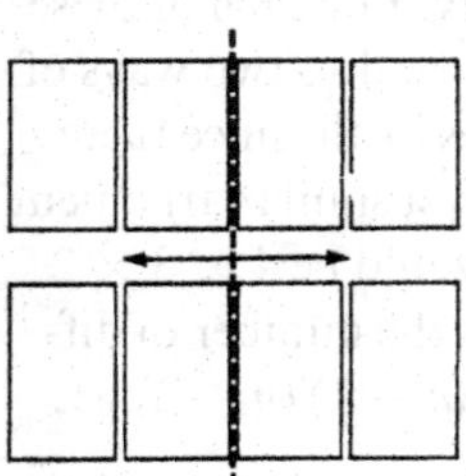

Figure 132.

But even this only brings down the number of effectively distinct arrangements for eight rooms from 40 320 to 10 080. So even symmetry considerations, where they apply, although they substantially diminish the number of alternative layouts to be evaluated, may still leave a numerical problem of gigantic proportions.

Nevertheless, it is of course just this kind of repetitive numerical task for which computers are well suited; and even if it is not possible to go

[3] *Op. cit.*

[4] G. C. Armour and E. S. Buffa. 'A Heuristic Algorithm and Simulation Approach to Relative Location of Facilities', *Management Science*, vol. 9, no. 2, 1963, pp. 294–309. Also E. S. Buffa, G. C. Armour, T. E. Vollmann, 'Allocating Facilities with CRAFT', *Harvard Business Review*, vol. 42, no. 2, 1964, pp. 136–58.

exhaustively through every single solution, it *is* feasible and economical to compare a great many. A large number of permutations of layout will differ from each other only in minute detail, and many others may well share exactly the same cost. Nugent, Vollmann and Ruml[3] have had some success with a straightforward random sampling of the set of all possible layouts for a given problem, evaluating all those chosen in the sample, and retaining the best. This technique might appear simple-minded. It may well miss the best solutions, or even miss a whole number of good solutions, depending on the relative size of the sample taken. Nevertheless, it compares well in its results for an equivalent cost in computer time with other more elaborate approaches.

No method, other than going exhaustively through all layouts, is absolutely guaranteed of finding the solution (or solutions) of lowest possible cost. All other approaches are *heuristic*: they employ strategies which will tend to lead towards better solutions, but are not bound to produce the optimum.

A technique which has been used by several authors to reduce the size of the combinatorial problem, is to examine not a complete permutation of all positions of activities, but to examine the effect of swapping only *pairs* of activities at each stage. Some particular starting layout is chosen, and *all pairs* of activities exchange their locations in turn. Any swap which effects an improvement in the layout (i.e., reduces the cost) is retained. Otherwise the two activities are returned to their former positions. The process goes on until no further improvements are possible: until a complete cycle of pair-wise swaps is tried without any change in the layout resulting.

The method described is, in broad terms, that devised by Armour and Buffa,[4] whose original paper in 1963, and their subsequent development, with other authors, of the CRAFT programs for 'facilities allocation problems' mainly in industrial plant layout, have provided the inspiration for most subsequent work on permutational methods. Armour and Buffa's basic procedure differs in detail somewhat, in that instead of making the first exchange which gives a lower cost, as soon as it is found, they measure the cost of all possible pairs of swaps in a given starting layout, before altering the plan. They then make the swap which gives the greatest cost reduction, and start the process again. The

[5] M. J. S. Beaumont. *Computer-aided Techniques for the Synthesis of Layout and Form with respect to Circulation*, unpublished PhD thesis, Department of Engineering, University of Bristol, 1967.

variation adopted here is from a suggestion of M. H. Rogers, whose work is described by Beaumont.[5]

That the number of permutational possibilities is much smaller with this kind of approach we can see by the following consideration. The number of all permutations of n objects in n positions was $n!$ The number of ways of choosing r objects from a total of n objects, irrespective of their order, is $\frac{n!}{r!(n-r)!}$. So the number of all possible pairs of activities which may be chosen for swapping with each other, from a total of n activities, is $\frac{n!}{2!(n-2)!}$. For the research office example, where $n = 8$, this number is 28. It may be necessary to go through this complete cycle of 28 swaps several times perhaps. But the total number of layouts whose cost must be measured is still enormously reduced from the 10 000 or so with which we were otherwise faced.

The course which the process of rearrangement takes, however, is dependent very much on the chosen starting layout. One failing of the method lies in the fact that from other starting positions still better layouts might be derived.

We can illustrate this with a worked example. Let us take as a starting point the layout already costed for the research institute in Chapter 12, and shown in Figure 12.2. Its circulation cost was 660. Figure 13.3 follows in diagrammatic form the various steps in the application of the method, as a tree of permutational alternatives. The order chosen for swapping activity pairs is $A \leftrightarrow B$, $A \leftrightarrow C$, $A \leftrightarrow D$, etc., to $A \leftrightarrow H$, then $B \leftrightarrow C$, $B \leftrightarrow D$...$B \leftrightarrow H$, then $C \leftrightarrow D$, $C \leftrightarrow E$..., and so on. Where any of these swaps results in a higher, or only an equal total cost, the layout is left unchanged. But where a swap results in a cost reduction, the rearrangement is immediately made, and the next swap in the sequence is then tried in the new reorganized plan.

The first swap which reduces the cost here is $B \leftrightarrow C$, from 660 to 616. The swap is represented in the figure by the edge at the extreme right of the tree at its topmost level. The next swap to result in an improvement is $B \leftrightarrow H$, bringing the cost down to 580. The successive rearrangements

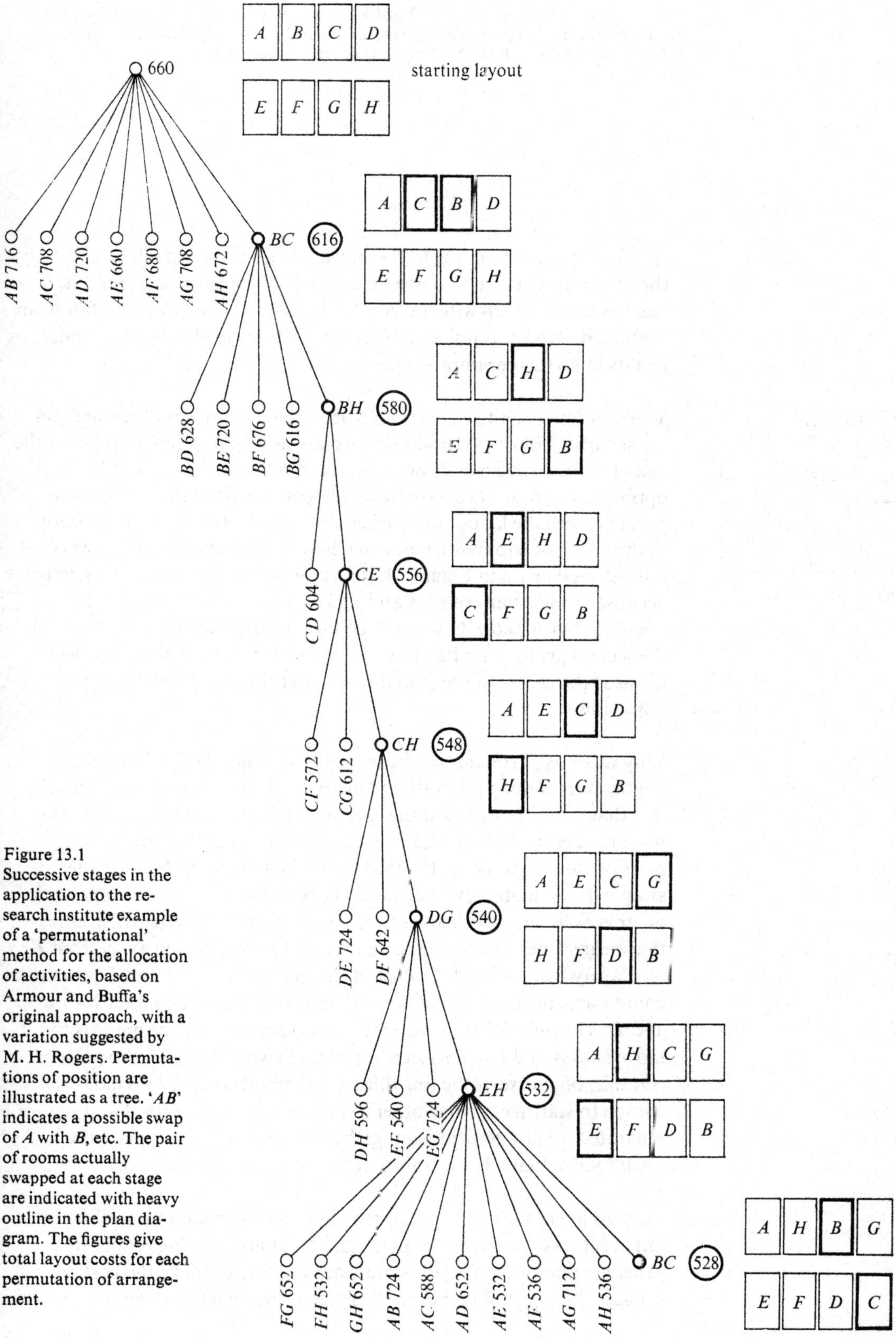

Figure 13.1
Successive stages in the application to the research institute example of a 'permutational' method for the allocation of activities, based on Armour and Buffa's original approach, with a variation suggested by M. H. Rogers. Permutations of position are illustrated as a tree. '*AB*' indicates a possible swap of *A* with *B*, etc. The pair of rooms actually swapped at each stage are indicated with heavy outline in the plan diagram. The figures give total layout costs for each permutation of arrangement.

[6] J. M. Seehof, W. O. Evans, J. W. Friedricks and J. H. Quigley. 'Automated Facilities Layout Programs' in *Proceedings of the Association for Computing Machinery, 21st National Conference, 1966*, Washington D.C., Thompson, 1966.

of the plan corresponding to these swaps are shown at each level in the diagram. The search descends ever lower down the tree, until it reaches a cost of 528 with a second exchange of $B \leftrightarrow C$, at the 36th swap evaluated. After this point another complete cycle of pair-wise exchanges results in no further improvement.

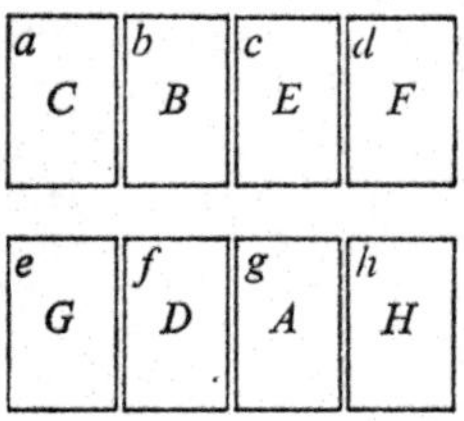

Figure 13.4

This solution is quite a reasonable one. The two groups of researchers are separated into two halves of the plan, and Miss *C* has an office at the end of the row. We can show, however, that the solution is 'sub-optimal', and that lower cost arrangements are possible, just by comparing it with the layout we worked out intuitively from an inspection of the graph of trip frequencies, in Chapter 12 (Figure 12.9). The layout is illustrated again in Figure 13.4; its cost is 492. The reason it is better is because of the positions of *A* and *E*. The group *A*, *E*, *F*, *H* remains together, but *A* and *E* have more communication with the members of the second group *B* and *D*, than do *H* and *F*, therefore to place them towards the centre of the plan instead of at the end, results in a cost reduction.

Why was this particular rearrangement not made by the systematic application of the permutational method? The reason is to do with the fact that *A* and *E* *started* at the extreme left-hand end of the plan. The two are very strongly linked to each other. The method swaps only pairs of workers at once. If either *A* or *E* is to be moved, either they swap with each other, which has no effect here; or else either one or the other is exchanged with any of the six remaining workers. In all cases this means a net cost increase, because either *A* is moved away from *E*, or *E* away from *A*; and the cost of this separation of *A* and *E* more than counteracts the lowered cost resulting from the closer contact with *B* and *D*. The only way the required improvement could be made is to move *both A* and *E at once*, and swap them with *H* and *F* respectively. But this, of course, is the one thing the method cannot do. Had we chosen to start from some other initial arrangement, with *A* and *E* separated, or else positioned together at the centre of the plan, the likelihood is that a better solution would have been reached.

Various ideas have been put forward for getting round these difficulties. Different ways of choosing an initial layout are possible: either to generate a number of layouts quite at random, and take the best, as suggested by Seehof and his colleagues.[6] This in effect combines random

[7] G. C. Armour and E. S. Buffa. 'A Heuristic Algorithm and Simulation Approach to Relative Location of Facilities', *Management Science*, vol. 9, no. 2, 1963, pp. 294–309.

[8] A suggestion made in conversation by O. P. Tabor.

sampling with the 'improvement' type of method. Or else to produce a starting layout intuitively, 'by eye', and use an improvement method to make refinements of detail, as Armour and Buffa[7] propose. Their attention in any case is directed particularly towards business management problems, of reorganizing the factory floor or warehouse plan, where some existing arrangement requires alteration.

One of the simplest expedients[8] might be just to add up the column sums in the 'association' table, and arrange them in descending order of magnitude:

E	*A*	*B*	*D*	*F*	*H*	*G*	*C*
24	15	15	12	10	10	8	0

and add up the columns sums in the 'distance' table, and arrange them in ascending order:

g	*f*	*c*	*b*	*h*	*e*	*d*	*a*
88	88	88	88	104	104	104	104

(For our particular example, because of the symmetry of the plan, the column sums of the 'distance' table give only two different values, one for the four 'inner' rooms, and one for the four 'outer'). We have now arranged activities such that the activity whose association with all others is greatest comes first, and so on; and we have arranged locations in order such that the one for which the combined sum of distances to all other locations is smallest comes first, and so on. We then simply match the two series together pair-wise,

g	*f*	*c*	*b*	*h*	*e*	*d*	*a*
E	*A*	*B*	*D*	*F*	*H*	*G*	*C*

and, if God is just, the corresponding layout should at least be better than one randomly selected, since roughly speaking the most visited activities will in this way be put in the most accessible locations. The cost of the layout achieved here (Figure 13.5) is 612, which, if not spectacularly low, is at least better than that of our previous starting layout (cost 660) for which it might usefully have been substituted.

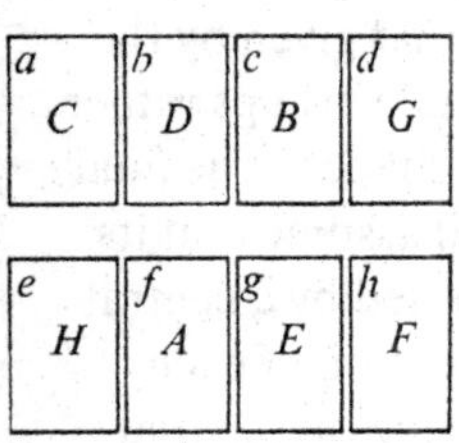

Figure 13.5

[9] M. J. S. Beaumont. *Computer-aided Techniques for the Synthesis of Layout and Form with respect to Circulation*, unpublished PhD thesis, Department of Engineering, University of Bristol, 1967.

But in whatever way the initial layout is selected, there always remains the problem that the method always follows from that point a unique sequence of moves; and from the given starting point only one final arrangement can be reached. The resulting layout, when the process of exchanging pairs of activities stops, may still be 'sub-optimal'. It has been suggested[9] that when this stage is reached, then it might prove fruitful to work back through the tree of layout permutations, making swaps which are apparently unpromising and which actually progressively increase the cost (taking them in that order); in the hope that these 'detours' might lie on the paths to still better solutions. Such a strategy however does not seem to have been much tried in practice.

The second type of systematic approach to the circulation problem we have referred to under the name of 'additive' or 'constructive' methods. The essential characteristic of these methods is that instead of making a series of changes to some initial layout, they start with one single activity in position, and build up a layout gradually by adding other activities one at a time.

Every 'additive' procedure has two essential features. The first is some kind of *spatial framework* within which the plan is assembled. Clearly, although the general overall form of the final plan is not predetermined – as with the permutational methods – there must nevertheless be some constraints on the geometry of the layout as it is built up. This might be done in the simplest way by dividing up the site or floor plane on which activities are to be arranged with a rectangular grid for example, each cell of which becomes a potential 'location'. The second requirement is for some criterion by which to decide in what order the activities should be placed, one after another, in the plan.

Another illustration using the research institute will make clear the significance of these two features. In order that the result given by an 'additive' method shall be roughly comparable with that given by the 'permutational' method previously illustrated, we shall perhaps rather artificially constrain the example here. Although the shape of the final plan is not to be completely fixed in advance, we shall assume that its general form is to consist of two rows of offices separated by a central corridor as before, with the dimensions of the corridor and of each room the same as in the earlier analysis. We can imagine the 'spatial

[10] B. Whitehead and M. Z. Eldars. 'An Approach to the Optimum Layout of Single-Storey Buildings', *The Architects' Journal*, 17 June 1964, pp. 1373–80.

framework' within which the plan is to be produced, therefore, as a corridor of indefinite length, flanked by an indefinite number of potential positions for offices either side.

Figure 13.6

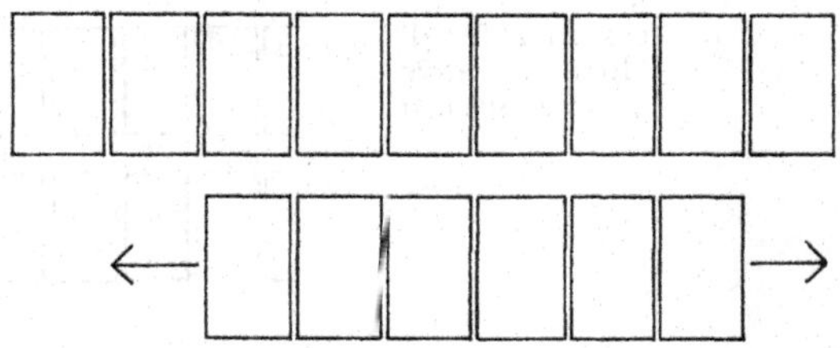

The method we shall use is based loosely on that of Whitehead and Eldars,[10] as described in their paper. It should be emphasized that in Whitehead and Eldars's own examples, however, and those of other proponents of similar methods, the form of possible layouts is not so rigidly constrained as here, either in the outline of the plan perimeter or in, for example, the position of circulation routes.

The next step is to decide an order for the placing of activities. The first activity to be positioned is that with the greatest association with all other activities. In our case, this means the research worker who is visited most *and* who makes most visits, and this is Dr *E*. Figure 13.7 will illustrate successive stages in the build-up of the plan, and the first stage shown at the top of the diagram consists in the placing of *E* in some arbitrarily chosen position in the plan framework. The next activity to be located is that which, of those remaining, has the highest association value with the activity *E* already placed – in this case *A*. The relative values for the association of all activities with *E* are shown at the top right of the figure.

Having decided to place Mrs *A* next, we must determine the exact position she is to occupy. The standard procedure for this positioning of successive activities will be to move each activity in question around all the possible positions adjacent to those activities already placed, measure the layout cost in each case, and put the activity in the position which gives the least cost. (It will not in fact be necessary to measure the *complete* layout cost for every position, only that portion of the cost dependent on the association of the activity being positioned with the activities already in place. It is only this component of the total cost which will be changed by moving the one activity around the plan.)

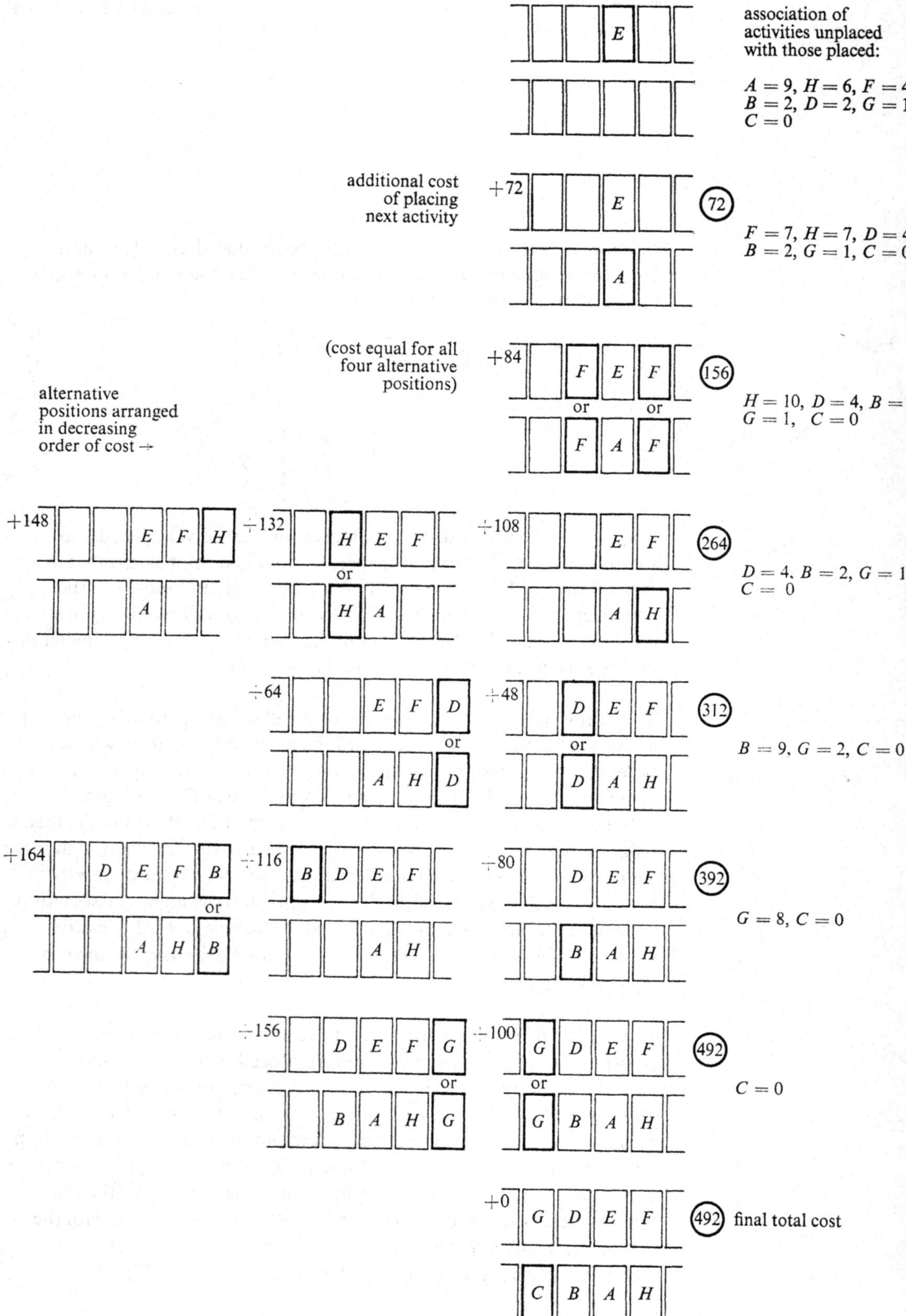
association of activities unplaced with those placed:
A = 9, H = 6, F = 4
B = 2, D = 2, G = 1
C = 0
additional cost of placing next activity
+72
E
A
72
F = 7, H = 7, D = 4
B = 2, G = 1, C = 0
(cost equal for all four alternative positions)
+84
F E F
or or
F A F
156
H = 10, D = 4, B =
G = 1, C = 0
alternative positions arranged in decreasing order of cost →
+148
E F H
A
+132
H E F
or
H A
+108
E F
A H
264
D = 4, B = 2, G = 1
C = 0
+64
E F D
or
A H D
+48
D E F
or
D A H
312
B = 9, G = 2, C = 0
+164
D E F B
or
A H B
+116
B D E F
A H
+80
D E F
B A H
392
G = 8, C = 0
+156
D E F G
or
B A H G
+100
G D E F
or
G B A H
492
C = 0
+0
G D E F
C B A H
492
final total cost

It is easy to see that in the present example the best position for *A* is in an office directly opposite *E*, since in this position the distance between the two is least. The next activity in the sequence is that which has the highest value for its combined association with *all* activities already placed. This is the rule by which successive activities are chosen in sequence for placing. The worker in question in the example is *F*. All four locations immediately adjacent to *A* and *E* in the plan are at equal distance to the positions occupied by both these activities (where distance is measured along the same kind of rectilinear 'tree' as previously); it is, therefore, a matter of indifference which of the four is chosen for *F*, since the total layout cost is the same in all cases.

Once *F* is put in, however, (arbitrarily in one of these four places), the choices for *H*, the next activity, comprise three effectively distinct positions. The cost differences for these are indicated with the diagrammatic layouts. The value given for each alternative is the combined total of the association values of *H* with *E*, *F* and *A*, multiplied by their respective distances from *H*.

This same process is gone through for each of the remaining research workers in order, and the subsequent states of the growing plan are shown in the figure, with alternative positions for each new worker at each stage illustrated and costed. The position of Miss *C*, who remains unplaced to the last, is quite optional – since she has no association with any other workers.

The final layout produced in this way is like one we have seen before. It is effectively identical with the original plan in the last chapter which

Figure 13.7 (opposite)

Successive stages in the application to the research institute example of an 'additive' method for the allocation of activities, based on the approach of Whitehead and Eldars. Each successive level in the figure corresponds to the placing of one activity, and the plan diagrams illustrate with heavy outline the alternative locations available, with their associated cost in each case, for the placing of the activity in question. The ringed figure is the cumulative total cost of that part of the layout so far completed. The values given for each activity in the left-hand column give their association with those activities already placed; the highest value at each stage determines which activity is to be located next.

we designed without the help of any systematic method. The total cost in both cases is 492.

On the showing of this one simple example, therefore, the additive method has produced a better result than the permutational method. The exact form of the solution given is due, however, to the very highly constrained 'spatial framework' we imposed on the additive example. As the number of activities handled increases, and equally if the plan form is less rigidly controlled, so the difficulties inherent in the additive type of method will loom larger.

Since the approach by its nature really implies that no prior decisions should be made on the final plan shape desired – indeed the very intention is that the application of the method itself should produce the plan form – then it is somewhat improper, on the method's own terms, to do as we have done and to fix circulation routes beforehand along which rooms are to be strung, for example, or to predetermine an envelope into which rooms are to be packed. In the example given by Whitehead and Eldars, that of the design of an operating theatre complex for a hospital, the spatial framework they establish is nothing more than a simple empty floor plane divided with a square grid which defines the 'locations', and upon which the plan may extend indefinitely, in prin-

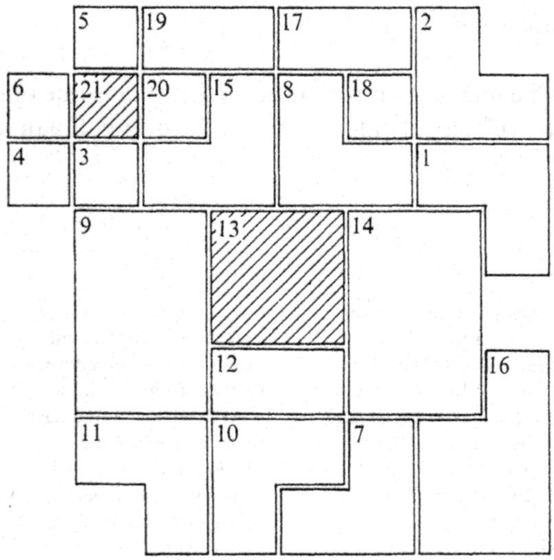

Figure 13.8
'Diagrammatic layout' achieved by Whitehead and Eldars for their hospital operating theatre example, before final modification by hand. Their method clusters the twenty-one activities into a deep plan around the principal circulation (13) and the theatres (9 and 14), in an essentially concentric arrangement. Shading denotes main circulation areas.

ciple, in any direction. In the nature of the system the plan is built up in a concentric fashion, around the first-placed activities (Figure 13.8). The criterion upon which each successive activity is given a position is that of distance to other activities, and it follows that activities tend to cluster around the centre of the plan in rings.

This might well be right for the particular kind of planning problem which Whitehead and Eldars pick upon; one where a series of 'servant spaces' are grouped around a central dominant suite of rooms, and the annular organization of the plan expresses this hierarchy. It happens too that there are no special demands for any of the central spaces, the theatres themselves, to have natural light or a view, and the plan can permissibly be a deep one with a minimum length of perimeter.

But it is only in a few types of building that this kind of situation arises. More usually we might expect to find not so pronounced a circulation hierarchy, but instead perhaps requirements for a series of loose groupings of rooms, with no very dominant links. More important, though, is the fact that for many buildings it will be desirable that the majority of rooms lie on the perimeter (so as to allow windows); and therefore with any large number of rooms the building's form must be elongated, either vertically or horizontally, and narrowed, so as to increase the perimeter area and reduce the unlit central 'core'. The basic additive method cannot effect this elongation without special constraints; since without constraint it will continue to put all activities together into an agglomerated mass.

Furthermore, in an extensive plan, we require that the circulation routes of the building, the corridors and staircases, form some coherent and economical system, and do not ramble about chaotically. But using a method which assembles a plan piece by piece and where rooms or activities are treated as relatively independent units which can be added together one by one, then it is inevitable that these overall *systematic* constraints, acting on the geometry of the building envelope and on the structure of its circulation routes, will not be satisfied. The whole must be more than the sum of the parts. The kind of plan perimeter resulting from an additive or constructive type of method in their simplest applications is often irregular and ragged; there has to be some tidying and reorganizing of the layout done by hand afterwards, before the

[11] M. J. S. Beaumont. *Computer-aided Techniques for the Synthesis of Layout and Form with respect to Circulation.*

result is acceptable as a building design. As Whitehead and Eldars put it: 'The last stage in the process is to convert the theoretical layout into practicable form.'

Brave efforts have been made, notably by Beaumont,[11] to take account of lighting and circulation considerations during the process of generating the plan itself. Beaumont's answer consists broadly in attaching to those activity units or rooms for which it is appropriate, extra, adjacent areas, one kind to act as circulation access space and another to lie adjacent to the room outside the building perimeter, and thus ensure that the room in question can be properly lit – 'a kind of floating light-well'. A series of checks made during the program's run are designed to control the positioning of activities and their associated extra areas, such that the circulation spaces should connect into continuous corridors, and that the external 'light and view' areas should not become enclosed or blocked by other units (Figure 13.9). This strategy is not, however, successful, since the attempt is still made to satisfy what are requirements governing the overall form and structure of the plan, at the level of the plan's constituent parts. Envelope and routes cannot be broken into separate 'particles' or 'atoms' and treated as such.

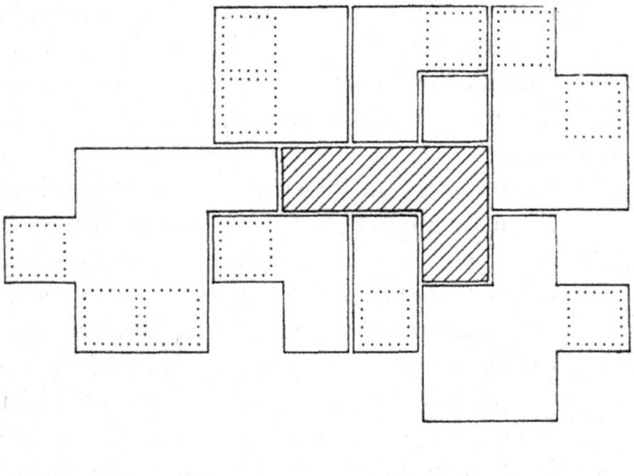

Figure 13.9
Layout produced in an exercise by Beaumont using an 'additive' method with additional constraints on the form of circulation spaces and on the plan perimeter. Shading denotes circulation; the hollow dotted squares are 'perimeter elements' which must be positioned adjacent to some external area. In this way requirements for side-light are accounted for. But the resulting plan form is ragged and the constraints do not act systematically.

What is more, natural lighting and the structure of the circulation system by no means exhaust the list of factors which affect the design of the overall building form, and whose requirements cannot be met at the detailed level in the relationship of activity units one to another. These might be considerations governing the building's structure, and the method of its construction. There might be other environmental aspects besides that of lighting to be taken account of – heating, ventilation, noise problems. And there might be limitations imposed by the size and shape of site onto which the plan is to be fitted, its orientation and relation to adjacent buildings.

So far we have been speaking mostly of additive or constructive type methods as applied to single-storey layouts. When in the next chapter we go into layouts of more than one floor we will meet new difficulties, mostly in relation to the measurement of *distance*.

14 Networks, distances and routes

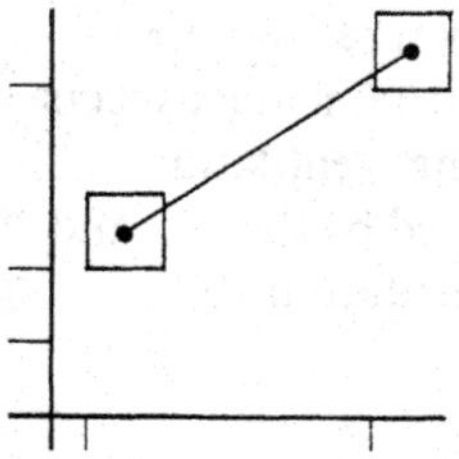

Figure 14.1

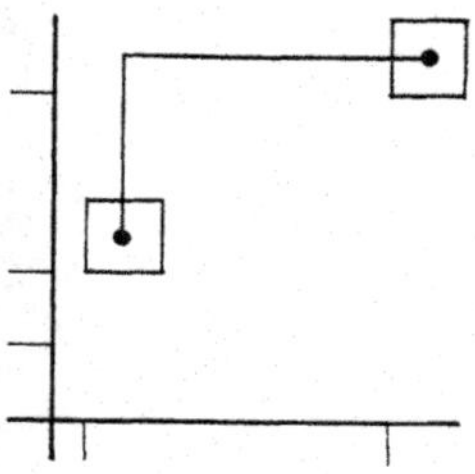

Figure 14.2

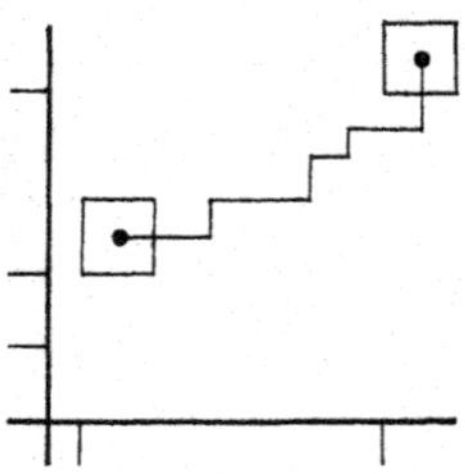

Figure 14.3

For measuring distances between 'locations' on a single floor level, there are three possibilities, representing various degrees of approximation. In the first place there is the straight line or 'airline' distance between locations (Figure 14.1) – one might use the term 'bee-line' were it not for the fact that bees frequently move in lines which are far from straight. The second is what has been called 'rectangular' distance (Figure 14.2), and the reason for using this measure is to do with the fact that in buildings planned with a rectangular geometry the circulation routes will tend to run orthogonally, following the building's main axes. To measure distance in buildings as direct straight lines is clearly very approximate, since in many cases these lines will cut diagonally across a plan, and such routes could never in reality be followed. The rectangular measure splits the distance into two components at right angles. The two types of measure, rectangular and airline, and their relationship according to Pythagoras's theorem, have been noted earlier, in Chapter 5, in the discussion of the distance between two quadrats. It is clear that the rectangular distance will be the same for journeys made in any number of 'steps', moving always orthogonally, so long as none of these steps involves a doubling back (Figure 14.3).

The third possibility is to measure, with greater or lesser accuracy, the length of 'real' routes plotted along the actual circulation system of the building in question. This is what we did with our earlier worked examples; it will be seen that even in those very simple cases several of the distances between locations were measured along routes which double back. And the more complex the plan, in general the more such detours can occur.

Figures 14.4–14.6 illustrate the three types of distance measure in the research institute plan. In one case the airline distance (Figure 14.4) between two offices is exactly equivalent to the rectangular, and in another case the rectangular distance (Figure 14.5) is equal to the real

Figures 14.4, 14.5 and 14.6

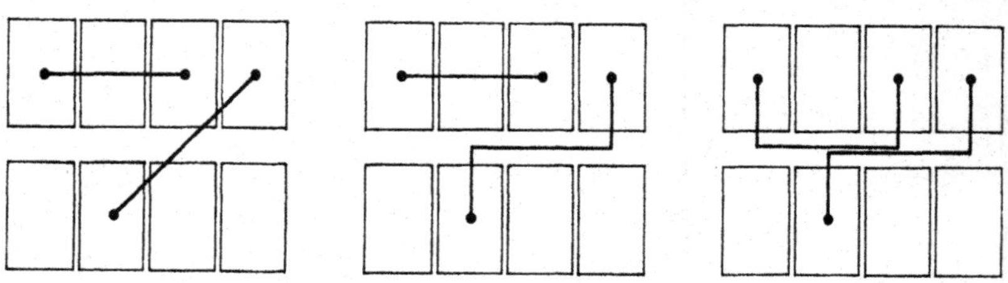

Figures 14.7 and 14.8
Diagrammatic single-storey plan layouts, each of 32 same size rooms, compared for mean journey lengths (measured as real distances) in Tabor's experiment.

[1] O. P. Tabor. *Traffic in Buildings 4; Evaluation of Routes*, Working Paper 20, University of Cambridge, Land Use and Built Form Studies, 1970, pp. 28–40.

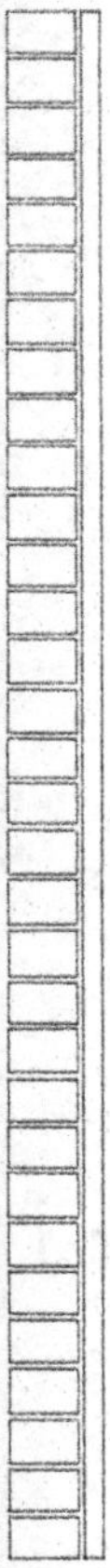

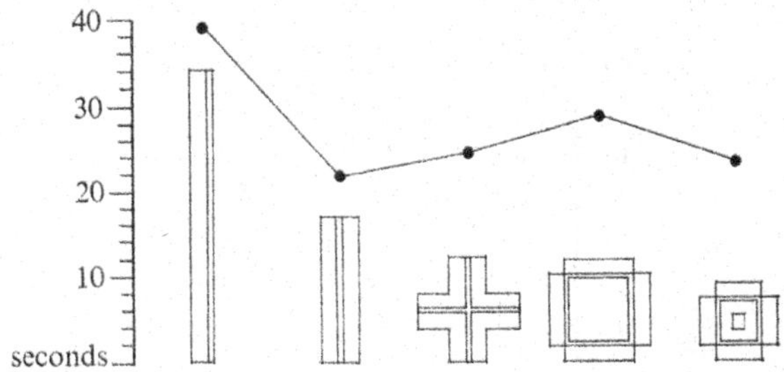

(Figure 14.6). The other examples illustrate the considerable variations possible using the different measures. The approximations involved in taking airline or rectangular distance can become very large, and the reason they are adopted at all is that in additive systems of plan generation the position and design of the circulation system is not fixed in advance, and therefore no estimate can be made of the 'real' distances which might be travelled in the final resulting layout.

Tabor has illustrated the approximations involved in relation to some hypothetical building layouts representing a range of simple geometrical forms.[1] He has planned 32 same-size rooms, on one floor, in diagrammatic buildings of five different types: a straight block with a 'single-band' layout (see Chapter 6), a similar block but with a 'two-band' plan, a four-armed cross again planned in two bands, and two square courts, one single-band, one two-band (Figure 14.7). For each form in turn he calculates all of the 496 possible journey lengths between pairs of rooms, according to the three different distance measures. The graphs of Figures 14.8 and 14.9 summarize his results, by comparing the *mean* journey lengths for the five forms. Figure 14.8 illustrates mean

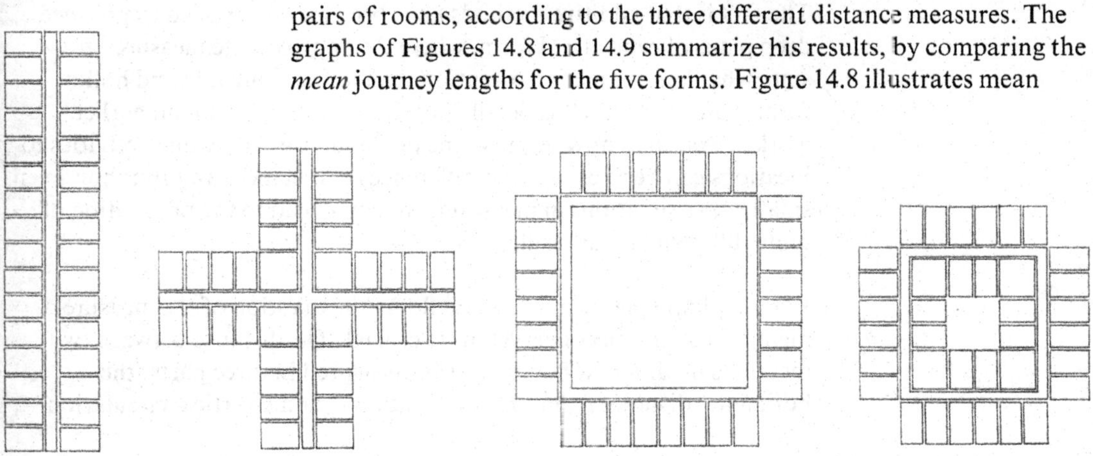

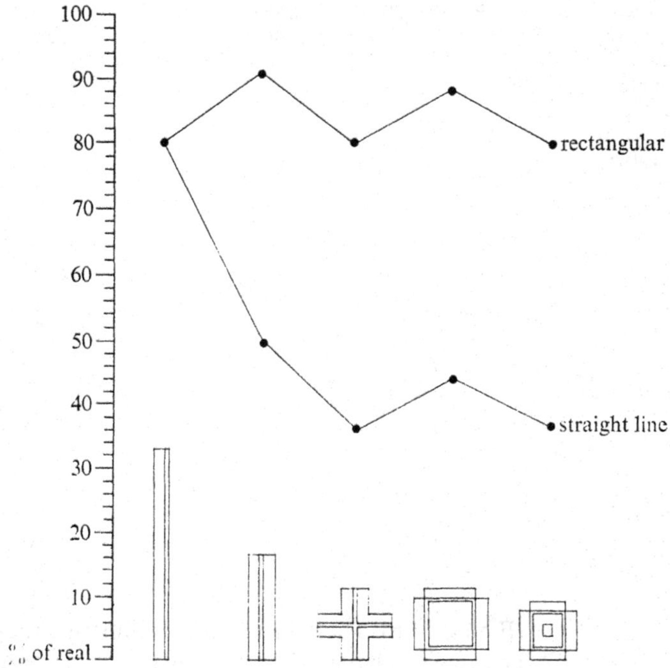

Figure 14.9
Comparison of the three types of distance measure in Tabor's diagrammatic plan forms. Mean journey lengths are calculated using straight line and rectangular measures, and plotted here as percentages of the mean values derived taking real distances.

journey lengths calculated as real distances; in Figure 14.9 the means derived from using the two more approximate measures are plotted as percentages of the real distance in each case.

The rectangular distance for all the building forms is between roughly 80% and 90% of the real figure. The showing of the straight-line measure is very much poorer. Only with the single-band slab, where very slight doubling back occurs, does the straight-line distance come near the real measure. The two-band equivalent drops to 50%; and for the court and cross forms, where it is clear that many journeys involve large detours from the straight line, the corresponding figures are between 35% and 40%, a very sorry performance.

The fact is that additive methods have been mainly applied to produce deep plans of centralized form, where the approximate measures are somewhat less of an approximation than in the elongated and hollow forms here. Nevertheless, it will have been obvious from our earlier worked examples how sensitive the procedures of allocating activities to locations can be to even slight differences of distance; and thus how even small approximations in measurement could lead to the production of quite different layout results.

For a building planned on several floors the question of the measurement of distance becomes yet more crucial. The distance between two locations on different floors must be measured in three parts: the horizontal distance from the first location to some vertical circulation

[2] See, for example, University Facilities Research Center, with Educational Facilities Laboratories Inc., *Horizontal and Vertical Circulation in University and Instructional Buildings*, New York, EFL, 1961, p. 6.

[3] Helen Parlow. 'Lift Operation and Computers', *The Architects' Journal*, 23 March 1966, p. 747.

[4] J. J. Souder, W. E. Clark, J. I. Elkind and M. B. Brown. *Planning for Hospitals; a System Approach Using Computer-aided Techniques*, Chicago, American Hospital Association, 1964.

point, a staircase or lift, the vertical distance travelled between the floors, and the horizontal distance from the lift-shaft or stair-well to the second location. Each horizontal section might be measured as a straight line, a rectangular or a real distance. To put the horizontal and vertical measures on a comparable basis it will be necessary to express each distance as a length of *time* required to travel along that part of the circulation system.

Some experimental work has been done to determine average walking speeds in buildings, both on the flat and speeds for climbing and descending stairs. The measurement of an average speed on the flat is simple enough in principle, and there is evidence to suggest that congestion is rarely bad enough in corridors and passageways to slow the walker significantly.[2] Tabor uses for his experiments an average of 1·5 metres/second, equivalent to the widely accepted figure of 3·3 miles/hour. For stairs a typical corresponding figure would be 0·3 metres/second, measured as the net speed of *vertical* movement, both for ascending and descending.

For lift travel, on the other hand, a great deal of complication may arise, both in theory and in practice. Not only may lifts travel at very different speeds, but the time taken for a lift journey will be dependent also on the waiting time, which in turn will be a function of the design and capacity of the lift installation, and of the pattern of traffic which the lifts must carry. Congestion here can have an appreciable effect on travel times, and thus the length of a lift journey, measured in terms of time, can vary even for different hours of the day.

For a given design of installation, and for a given pattern of lift use, then the arrivals of users and the movements of the lifts themselves might be simulated in detail; as for example in a computer exercise carried out by Parlow,[3] who is able by this means to make very fine calculations of the lengths of lift journeys, depending on the particular circumstances of a particular building. It is even possible, as in the COPLANNER system of Souder and his colleagues,[4] for alternative routes by stair and by lift to be compared for 'time length', and trips to be assigned to one route or the other according to the current traffic load on the lift system.

But these elaborations of treatment are only justified – and only possible – where the intention is to *evaluate* the performance, in circulation

[5] O. P. Tabor. *Traffic in Buildings 4; Evaluation of Routes*, pp. 40–58.

terms, of some layout already designed in detail. Where the purpose is to *generate* a layout, using additive methods, the problem is put the other way round.

We have discussed how with a one-storey layout, the spirit and intention of the additive type of method require that the circulation structure and the building envelope should not be fixed in advance, but instead should, in principle, be produced by the application of the method itself. Such lack of constraints on the process created its problems, just for a single-level layout. With layouts on more than one floor, the 'spatial framework' within which the plan is to be built up cannot be established at all without deciding first on a definite number of floors which the design is to comprise, and then on the particular positions which vertical circulation is to take. For without the vertical circulation fixed, then distances between locations on different floors cannot be measured to any degree of approximation at all.

A further series of experiments which Tabor[5] has made are designed to indicate just what effects these prior decisions about numbers of floors, and about the position of vertical circulation points, are likely to have on distances. He takes three of the same diagrammatic forms as before, the two-band slab, the rectangular cross and the two-band court. He takes a fixed area of accommodation – 96 equal size rooms – and plans these rooms in the different configurations. He plans a series of layouts on different numbers of floors; and since the total number of rooms remains constant throughout, so, therefore, the diagrammatic buildings become smaller in plan – though keeping the same shape – as they become taller.

A series of standard figures for dimensions and travel speeds are assumed throughout – for floor-to-floor heights, walking speed, and the vertical speed of stair travel, and standard allowances are made for waiting and door opening times for lifts. Two different speeds of lift are tried, at the rough limits of the range of speeds used in practice, 0·75 metres/second and 2·25 metres/second. For each form, and for given positions of stairs, and positions and speeds of lift, Tabor measures, in terms of travel times, all the distances between pairs of locations. He then calculates the mean travel distance for each case, as before.

The results of these experiments are complex, and the particular effects of different combinations of numbers of storeys and positions and speeds of lifts cannot be summed up simply. The general conclusions which do emerge, however, are that the increase in number of floors reduces internal distances, with the examples studied; and the *position* of vertical circulation, especially the degree to which lifts are placed eccentrically away from the middle of the plan, has at least as much effect on mean distances as does the *number* of lifts.

The implications are that fixing even just a broad spatial framework and skeleton vertical circulation system for a building on a number of floors, in itself affects in a significant way the pattern of distances between locations. This in turn puts an effective limit on the best layout cost which might be achieved inside that framework. With another framework it seems, with a different system of vertical circulation, a quite different optimal layout with a quite different layout cost could be achieved for the same given associations between activities, and using the same plan generating method.

These strictures apply yet more forcibly, of course, to the permutational kind of approach, where the exact detailed arrangement of the plan is fixed at the start, and not just its general framework; then improvements are effected within that plan. The range of solution costs possible is strictly limited by the initial choice of plan form.

With permutational methods there are no theoretical problems in predicting the form a circulation system might appropriately take. There are, by contrast, some practical problems arising here though, in the measurement of 'real' distances along some given system.

We have mentioned in Chapter 11 how the circulation routes in a building might be regarded as forming a *network* (in the special graph theoretical sense). The vertices of this network correspond to two different types of point in the plan: to the locations (or rooms) as marked by their centroids, and to the junctions in the system of horizontal and vertical routes. The edges correspond to short segments of these routes, and the values attached to the edges represent the respective lengths of the segments. Each of these 'lengths' would be appropriately calculated, for a multi-storey building, as a length of time required to move between the two points involved. The junction vertices will have at least three

[6] Described by C. Berge in *The Theory of Graphs and its Applications*, London, Methuen, 1962, p. 65.

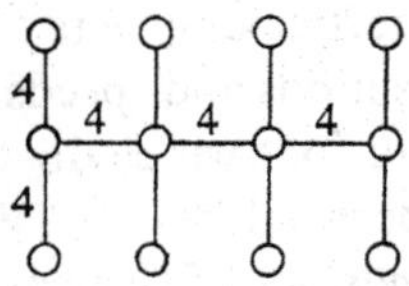

Figure 14.10

incident edges (or else they are not junctions); and the vertex marking each location will in general be a pendant or *terminal* vertex, that is, a vertex with which only one edge is incident.

The circulation network for the research centre example comprised 12 vertices and 11 edges, and it took the form of a *tree* (Figure 14.10). It happened that in that case, due to the particular dimensions of the plan, all edges were of unit length 4 m. A simple theorem states that a tree with n vertices has $(n - 1)$ edges; this may be demonstrated by the fact that the simplest possible tree has one edge and two vertices, and every time a new edge is added, if the graph is to remain a tree, then one new vertex must be added also. Between any two vertices in a tree there is a unique *chain* of edges (that is, a *simple chain*, or chain which includes no edge more than once), and this represents the shortest route between the two points. If a new edge is added to the tree but not a new vertex, then a *cycle* of edges is introduced and the network loses its tree form.

As soon as this happens, it means that alternative routes exist between some pairs of locations. Either the corridors on a single floor join in a cycle of edges, or else the cycle is formed by there being more than one vertical link between floors. We assume that, in the rational world of circulation studies, everyone travels by the shortest available route. So we must measure all possible routes, and compare them for length, to find the shortest.

In a small network (like Figure 14.10) this is an easy job to do by hand. But as the number of vertices in the network increases the process becomes more tedious. The very elegant network drawings made by students at Ulm for the plan of their own Design School give an idea of the size of problem of finding shortest routes for a medium-sized building (Figure 14.11). One might begin to look in these circumstances for methods which could be expressed as *algorithms*, and so be mechanized, using the computer.

It is worth noting in passing the suggestion of H. W. Kuhn for a *physical model* of network structure for solving the shortest route problem, made with string edges, knotted at the vertices.[6] The length of each string is proportional to the length of the edge in the network. The shortest distance between any pair of vertices is found then by picking the string mesh up at the two points in question, and pulling it taut.

Figure 14.11
Student exercises from the Ulm School of Design. Students were asked to represent the 'topological linking of the various rooms and the average traffic density' along routes, in the plan of their School's own building. Illustrated are: the basic circulation network, and networks showing information on traffic flow presented in three different schematic forms, by Karel Links, Robert Couch and Jürgen Böttcher.

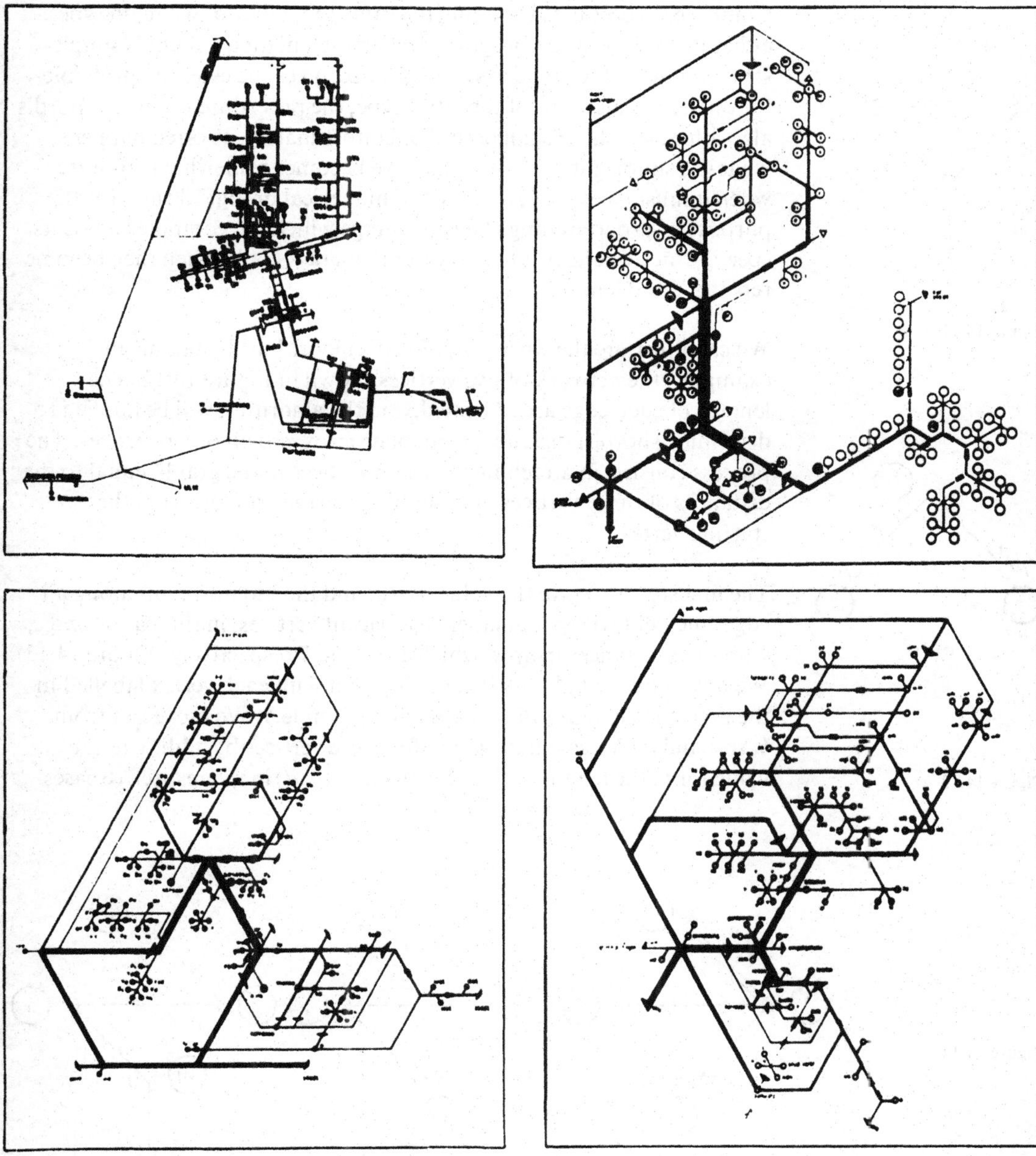

[7] L. R. Ford. *Network Flow Theory*, Santa Monica, California, RAND Corporation Paper P. 925, 1956.

Computer procedures for doing the same job fall into two main types, either matrix methods, or 'graph traversing' methods. Kuhn's simple string model, which is so easy to visualize, gives the essential principle for one of the methods of the latter type, the procedure known as Ford's algorithm, which we shall describe. Unfortunately, as often happens, what is a simple job for the human eye becomes somewhat elaborate when it must be expressed in a set of numerical or logical rules for the purpose of programming. The advantage which the computer has is its speed, which qualifies it to carry out these tasks only when they become repetitive and lengthy.

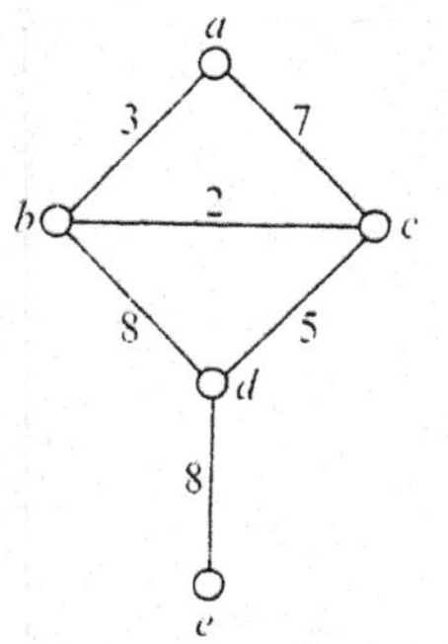

Figure 14.12

We can illustrate the working of Ford's algorithm[7] by taking an example – the network of five vertices shown in Figure 14.12. The lengths of each edge are indicated, and the algorithm will be applied to determine shortest distances from one vertex *a* to all other vertices. The process consists in attaching values to vertices in the graph, equal to the distances of those vertices, measured by various routes, from the starting vertex.

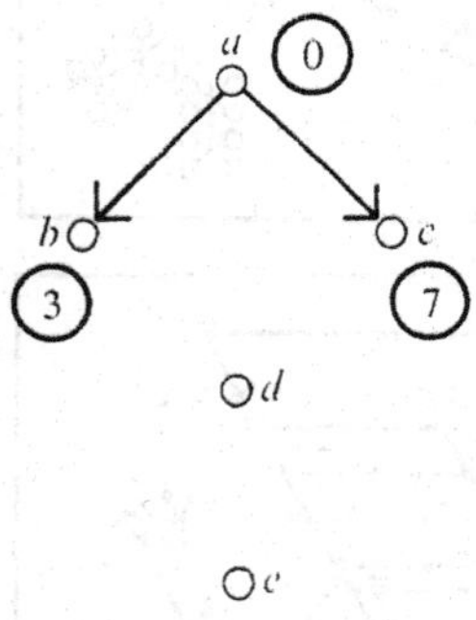

Figure 14.13

The first step is to give the value 0 to *a*, and then to move out along all edges incident with *a* and label all adjacent vertices, in this case *b* and *c*, with their distances from *a*, which are 3 and 7 respectively (Figure 14.13). We now move out along all edges incident with each vertex labelled in the first stage (Figure 14.14). Take for example *b*. We move out from *b*, to *c* and *d*. We calculate values for these vertices by adding to the value for *b*, the length of the edge (*b*, *c*) or (*b*, *d*) in the respective cases.

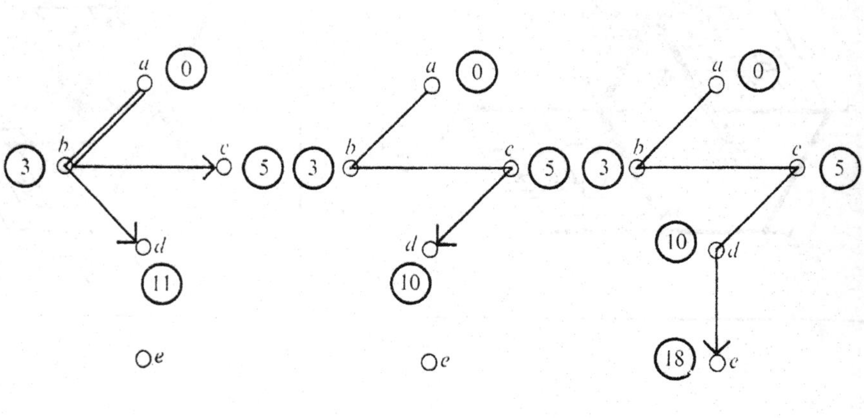

Figure 14.14

[8] Z. Prihar. 'Topological Properties of Telecommunication Networks', *Proceedings of the Institute of Radio Engineers*, vol. 44, 1956, pp. 929–33.

Two possibilities now arise. Either we find a vertex which is so far unlabelled, as for example *d*; in which case we give it the calculated value 8, which is the distance of *d* from *a*, *via b*. Or else we meet a vertex which has been already given a value previously, as with *c*. We compare the value we have for the distance via the present route with the value which the vertex already carries, and if it is less, we replace the one with the other; for the present value must equal the distance measured via a shorter route. With the vertex *c* we calculate its distance from *a* via the route (*a*, *b*, *c*) to be 5, whereas *c* has a value given in the first stage of 7 (for the route (*a*, *c*)). We therefore replace the value 7 with 5.

At every stage we move out in similar fashion from all those vertices given new values at the previous stage, changing values where required, until all vertices are labelled. The three steps needed for the example here are depicted in diagrammatic form in the figure. The value on any vertex now gives its respective shortest distance from *a*. Note how the method goes through effectively a similar process to that which is completed, instantaneously, with the tightening of the string model. The algorithm does on the other hand produce, with one application, the distances from one vertex to all the others in the network. To find *all* shortest routes, between *all* pairs of vertices, the method must be applied to every vertex in turn.

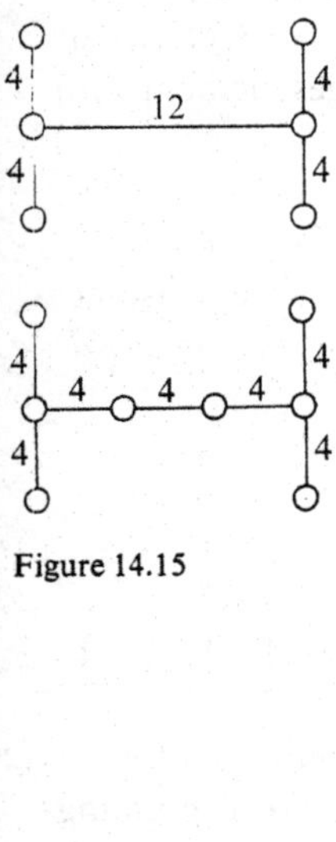

Figure 14.15

To represent the matrix type of method we shall instance an algorithm due to Prihar.[8] His method can be applied only to networks where the edges are all of equal length (like Figure 14.10), a serious limitation for this purpose, but one which can in principle be overcome by inserting extra vertices along unequal edges, so as to divide them into shorter segments of standard length (Figure 14.15).

We take a matrix **G** associated with the network. This is a zero-one matrix of the kind described earlier, in Chapter 12, in which the general element g_{ij} takes the value 1 to signify the presence of an edge in the network joining the two vertices *i* and *j*. Where no edge joins *i* and *j*, then the element takes the value 0.

The (symmetrical) matrix $\mathbf{G} = [g_{1,\,ij}]$ below corresponds to the simple (symmetrical) network of five vertices which is illustrated. (Symmetrical, in the sense that this example contains no arcs, that is no one-way links, although the method could equally cope with these.) It is clear that wherever a 1 is entered in the matrix, then this means that in the network the shortest route between the corresponding two vertices is of unit length. We start to compile a table of shortest distances, with their lengths shown as a number of edges, and we can enter the distances of length 1 straight away.

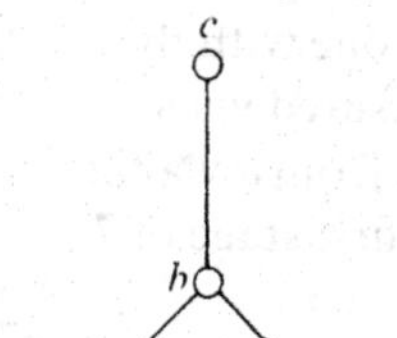

Figure 14.16

Matrix **G**

	a	*b*	*c*	*d*	*e*
a	0	1	0	1	1
b	1	0	1	0	1
c	0	1	0	0	0
d	1	0	0	0	1
e	1	1	0	1	0

Table of shortest distances

	a	*b*	*c*	*d*	*e*
a	·				
b	1	·			
c		1	·		
d	1			·	
e	1	1		1	·

The matrix is then squared. Each element in this second matrix $\mathbf{G}^2 = [g_{2,\,ij}]$ gives *the number of different chains, two edges long*, which join that pair of vertices. The element $g_{2,\,ba}$ has value 1; there is *one* two-edge chain only, (*b*, *e*, *a*). Between *e* and *a* there are two, either (*e*, *b*, *a*) or (*e*, *d*, *a*); and $g_{2,\,ea}$ takes value 2. On the leading diagonal of the matrix are the number of two-edge chains from a vertex to itself, that is, for example, (*a*, *e*, *a*), (*a*, *b*, *a*) and (*a*, *d*, *a*). The chains in this case are not *simple* chains, but *composite*, for they include the same edge more than once. The values of these diagonal elements ($g_{2,\,aa} = 3$, etc.), can be seen alternatively as representing the number of edges incident with each respective vertex.

Matrix **G**

	a	*b*	*c*	*d*	*e*
a	0	1	0	1	1
b	1	0	1	0	1
c	⓪	1	0	0	0
d	1	⓪	0	0	1
e	1	1	⓪	1	0

Matrix $\mathbf{G}^2$

	a	*b*	*c*	*d*	*e*
a	3	1	1	1	2
b	1	3	0	2	1
c	①	0	1	0	1
d	1	②	0	2	1
e	2	1	①	1	3

Table of shortest distances

	a	*b*	*c*	*d*	*e*
a	·				
b	1	·			
c	2	1	·		
d	1	2		·	
e	1	1	2	1	·

We now compare the two matrices **G** and $\mathbf{G}^2$. The element $g_{1,\,ab}$ for example, takes the value 0; but the element $g_{2,\,ab}$ in the corresponding

[9] An economic variant of the cascade method is described by J. D. Murchland in *A New Method for Finding all Elementary Paths in a Complete Directed Graph*, London School of Economics, Transport Networks Theory Group, Report 22, 1965.

position in $\mathbf{G}^2$ takes the value 2. This means that there were no one-edge links between vertices d and b, but there are 2 two-edge links ((d, a, b) and (d, e, b)). It follows that the shortest distance between d and b is two units. And this is true for any case where the element $g_{1,\,ij} = 0$, but the corresponding element $g_{2,\,ij}$ takes some value > 0. We put all these two-edge links in the shortest distance table.

We now take the cube $\mathbf{G}^3 = [g_{3,\,ij}]$ of the original matrix. The elements in this matrix give the number of *three*-edge chains between vertices. There is only one element $g_{3,\,dc}$ with a value > 0, for which the corresponding elements $g_{1,\,dc}$ and $g_{2,\,dc}$ in $\mathbf{G}$ and $\mathbf{G}^2$ were both zero. d and c are therefore the only pair of vertices which are joined neither by a one-edge nor a two-edge chain, but only by a three-edge chain. The shortest distance 3 is therefore put in the table, which is now completely filled.

Matrix $\mathbf{G}^3$

	a	*b*	*c*	*d*	*e*
a	4	6	1	5	5
b	6	2	3	2	6
c	1	3	0	2	1
d	5	2	(2)	2	5
e	5	6	1	5	4

Table of shortest distances

	a	*b*	*c*	*d*	*e*
a	·				
b	1	·			
c	2	1	·		
d	1	2	3	·	
e	1	1	2	1	·

With larger networks the process would carry on in the same way, taking ever increasing powers of the original matrix until all elements took at some stage non-zero values, and the corresponding shortest distances could be transferred to the table.

This method, of course, unlike the graph traversing technique, produces *all* shortest distances in a network at once, and not just distances from one vertex to all others. The difficulty with Prihar's particular algorithm for problems like circulation networks is the requirement that all edges be of equal length. We have chosen to describe this method because it is neat and surprising. With a complex network though, a great many extra inserted vertices may be needed, giving a matrix whose size is unmanageable even for the computer. Nevertheless there exist more economical matrix methods, notably the so-called 'cascade' algorithm which can cope with edges of unequal length.[9]

A whole range of network problems connected with measuring distance, with shortest routes and with traffic flow are described in the context of geographical and urban studies in P. Haggett and R. J. Chorley's compendious *Network Analysis in Geography* mentioned in Chapter 7, which gives a broad and comprehensive treatment of topics which it is possible only to touch on here.

We have seen how questions of the measurement of distance are important for systematic design methods based on the criterion of circulation, in particular for those of the additive type. If we should seem to have concentrated on the difficulties which surround the additive approach, then we might redress the balance perhaps by mentioning one problem which is peculiar to the permutational method. This concerns the shape and size of locations and the area required for activities.

We have made the over-simplification, in our worked examples, of assuming a standard size for locations and activities, corresponding to the area of one room in every case. This situation, of all rooms in a building being required to be the same size, is rather a rare one in practical planning, and more usually a schedule of accommodation would, of course, contain a variety of required areas. Since with these methods every activity must be capable of fitting any location, it is plain that each room cannot now be represented as a single activity. The trick is – with additive methods – simply to decide an areal module to which the size of location is made to correspond, and then to divide the area of each required room into the nearest equivalent round number of modular 'activity units'.

In Whitehead and Eldars's case, for example, the chosen size for locations and activity units is 100 ft^2 – although any convenient figure would serve – and the areas of rooms in the operating theatre group are approximated to some number of units, ranging from one unit for a lobby or store room, to six units for the theatres themselves.

All that remains is to divide the association values calculated for the room as a whole, by the number of activity units from which it is to be made up, to produce the association values for each separate unit; and to set the association between the units forming a single room to some artificially high value, so as to ensure that they are all positioned together

[10] As for example methods involving the 'hierarchical decomposition' of linkages (as used in another architectural context by C. Alexander, *Notes on the Synthesis of Form*, Cambridge, Mass., Harvard University Press, 1964); variants of the method of 'clumps' borrowed from linguistic analysis; or the taxonomic procedure of 'non-metric multidimensional scaling' – all described, and some examples of their application to office communication examples given in Tabor, *Traffic in Buildings 3; Analysis of Communication Patterns*, Working Paper 19, University of Cambridge, Land Use and Built Form Studies, 1970.

in the layout in a compact group to correspond to the area required for that room. The method then proceeds as before.

So, as we suggested when the notion of a 'location' in its special meaning was introduced, the unit need not correspond to a room always, but could be a work-space, a part of a room, or even a group of rooms treated together. Equally it is not necessary nor generally feasible, for a large problem, that an 'activity' should represent a single individual; and a great deal of study has been given in this connection to the way in which the structure of an organization might be analysed so as to find those groups of people, or groups of functions, which could most properly be treated as distinct units for the process of allocation. We have touched the fringes of this subject with the classificatory graphs of Chapter 12; some authors have borrowed the more advanced techniques of automatic classification and mathematical taxonomy for a detailed and rigorous treatment.[10] But although the principles of some of these are essentially geometrical, this is not the 'geometry of environment' but of multi-dimensional 'classification space' and their discussion goes beyond the scope of this book.

Now, although there are some small operational difficulties with additive methods in treating the shapes of rooms when they are composed of several activity units each – since despite the high association set to keep the units together, it may still be possible for the resulting shape of room to become irregular or excessively elongated, and so require subsequent tidying – nevertheless with permutational methods the question of shape and size of rooms is altogether more serious. Here in the nature of the exercise we must always exchange the position of pairs of complete activities or complete activity groups; and if their area requirements are unequal it does not help to split them into smaller modular units of area, for this simply means then that we are trying to exchange some set of units for a smaller or larger number – and one group will not fit in the free locations left by the other.

Two adjacent rooms of unequal area can be swapped possibly; but then their centroids move, and so the distance calculation will be thrown out somewhat. The CRAFT programs make use of some ingenious stratagems by which groups of rooms may be exchanged in stages, by 'carving' smaller rooms out of the area of larger ones, at their corners, and by shuffling intermediate rooms along. But this is computationally very intricate, the resulting room shapes again tend towards the irregular, and in the end the objection about the effect the process can have on the calculation of distances also still stands.

Finally, we return to the original assumptions made about the nature of journeys in a building, on the basis of which the association values were fixed, and so by extension on which the foundation of all these methods rests. Trips were assumed always to be made to their destination and directly back again, with no round trips taken account of. The association values are fixed, and no allowance is made for the quite probable eventuality that over longer distances people will tend to make fewer trips to those destinations which, if they were closer, they might visit more often. In this light it might be reasonable to suggest that the number of journeys made between a pair of activities over some fixed period be assumed to be some inverse function of the distance by which they are separated. But then the whole structure of the allocation process collapses.

A more general criticism of all the methods described is that by implication they assume (in particular the additive type) that the disposition of rooms or spaces in a plan should be worked out according to a very particular pattern of traffic, measured at one particular point in time; and then that exact form of building constructed. (Although it is true that permutational methods have been aimed rather at administrative use, more than for architectural design, for obtaining the best reorganization of functions within an existing envelope.) The building's structure may be designed to last fifty, a hundred years. But it is possible that even within a few months the structure of the organization which that building is to house will have changed, and correspondingly the whole pattern of pedestrian traffic have altered. No allowance is made for the flexibility of the plan to accept probable changes in the organization which the building accommodates, over the building's lifetime.

It is perhaps no accident then that of all building types the hospital is the one which occurs most frequently as an example for testing such methods. For here not only must many trips be made irrespective of the distances travelled; but also the pattern of hospital routine is perhaps more standardized and consistent than in other kinds of organization, and so variations in the trip pattern over a period of time may be less extreme.

[11] *Traffic in Buildings 4; Evaluation of Routes*, pp. 22–7.

Should architects be trying at all to produce plans on the basis of these types of very detailed calculation of particular movement patterns in buildings? Should their attention be given rather to trying to measure the capacity of broad *types* of building form to accommodate general *types* of traffic pattern, at a much coarser level of analysis? Some rather general questions for which it would be useful to be able to give answers with numerical meaning to, are: what typical forms of building are appropriate when the trips made by occupants are frequent, or infrequent? Are there some classes of building form whose geometry suits them better to some patterns of traffic than to others? And on the other hand, is it possible at all to characterize different organizations by the general types of traffic pattern to be found within them?

Answers to the last question lie in the study of actual patterns of travel in experimentally observed situations; which leads us off again from our main theme, into the areas of sociology and social psychology. But if we are not to look here at typical patterns of communication in organizations, seen independently of architectural forms for their accommodation, what we can do is to take the other view, to look at architectural forms independent of specific organizational patterns. This is one purpose of the experiments by Tabor which we have already described. It is quite clear from Figure 14.8 for example, how for the different type-forms chosen, the more compact plans (the two-band slab and the two-sided court) result in a greater average accessibility of one room to another – quite independently of what trips might actually be made by some set of occupants of such buildings. Hence any particular administrative layout planned within these forms (using some permutational method perhaps) would tend to start with an in-built cost advantage over those (such as the court and slab of one-band depth) where mean distances are greater.

How big an advantage this might be, would depend on the trip patterns of the particular occupant organization, and how efficiently its activities were located. If all the trips were short ones, if as Tabor puts it 'trips tend to have destinations only a few doors down the corridor', then 'distances to destinations in the rest of the building – whatever its shape – matter little'. The general effects of architectural form in relation to circulation patterns will be felt more where trips are generally longer; and to demonstrate something of these effects Tabor has designed another experiment,[11] the last we shall describe, which goes some way in

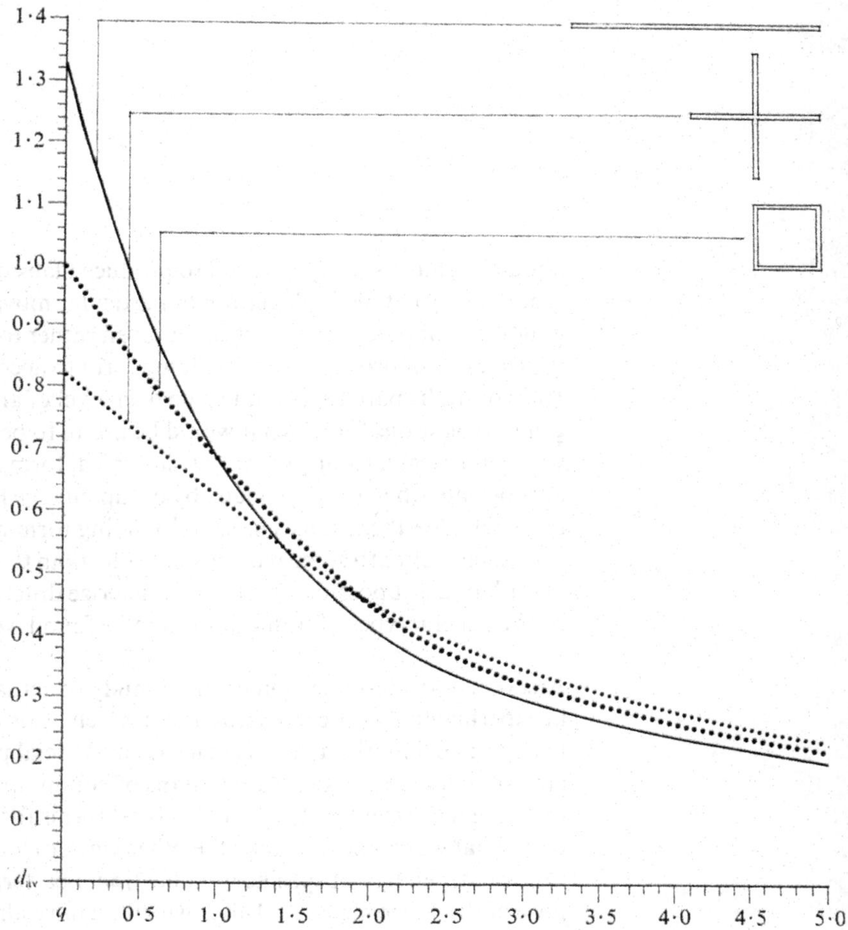

Figure 14.17
Theoretical mean journey distances d_{av} in slab, cross and court forms, for different values of q, a 'propensity' to make shorter journeys.

the suggested direction, of depicting different patterns of traffic in a broad statistical way, instead of with the detailed and peculiar descriptions demanded by automatic planning methods.

Tabor takes the three type-forms, with which we are by now familiar: the straight form, the cross and the court. There is no detailed room plan here. They are represented simply by corridor skeletons, as it were; and the total length in each form is the same. All three are single-level plans.

These are buildings, therefore, with layout and circulation systems represented at their most schematic. The patterns of traffic within the forms are specified in equally general terms. The graph (Figure 14.17) illustrates the trip lengths for each form, measured for what Tabor calls different *propensities* for those making journeys to choose neighbouring destinations. At one extreme all possible lengths of journey between all pairs of points in the plan are treated as equally probable, and the resulting mean trip lengths (d_{av}) are plotted at the extreme left of the graph where the propensity $q = 0$.

The scale along the horizontal axis represents this increasing propensity for trips to be of generally shorter length. It follows that at the extreme right, and indeed for all values of q much greater than 1, where all trips are short ones, that building form has little effect on mean journey lengths – since few journeys are diverted from their course over any distance by the building's geometry. Any ranking of the forms in order of mean journey lengths can be reversed by the effects of efficient activity location. But as q approaches 0, that is where the origin and destination of trips become effectively random, we see a ranking of the cross, court and slab forms, arranged in order of increasing journey length; which agrees with the effectively similar comparison made for the same forms, planned with room layouts of specific design, in the earlier experiment of Figure 14.8.

Guide to further reading

Throughout this book we have drawn the reader's attention to specific references which not only provide the sources of our statements but, frequently, will already have suggested ways in which the subject may be pursued further. The purpose of this section is to recommend the reading which we have found stimulating in different ways and for a variety of purposes. Many of the books we mention contain large and full bibliographies in their particular areas of study, and these may lead readers into the more specialized literature should they wish to explore the subjects in greater depth.

In our Preface we tried to say what this book does *not* do. It does not concern itself with the traditional uses of descriptive geometry in architectural drawing. There are many books on this subject, principally for engineering graphics, but R. G. Robertson's *Descriptive Geometry* (London, Pitman, 1966) will be found to be both clear and readily obtainable. Nor does our book dwell on the 'energetic-synergetic geometry' developed by R. Buckminster Fuller which has charmed recent generations of architectural students. This is adequately summarized in Robert W. Marks's *The Dymaxion World of Buckminster Fuller* (Carbondale, Southern Illinois University Press, 1966), until such time as Fuller's forthcoming *Synergetics* is published. While this 'comprehensive, mathematical, rational coordinate system' may have some relevance at the atomic level – and one or two scientists have supported this view – it is less clear why geometrical figures which exhibit 'omni-directional equilibrium of forces' should be important at the architectural scale where the uni-directional force of gravity dictates structural form. The complicated distribution of forces in lattice frames, geodesic and other dome-like structures that gravity gives rise to are well illustrated in Z. S. Makowski's book *Steel Space Structures* (London, Michael Joseph, 1965). The change of scale from atomic to engineering structures is discussed in D'Arcy Thompson's classic work *On Growth and Form* (Cambridge University Press, 1961), in a chapter 'On Magnitude'. This is certainly a work that should be looked at. As P. B. Medawar has said, D'Arcy Thompson believed 'not merely that the physical sciences and mathematics offer us the only pathway that leads to an understanding of animate nature, but also that the true beauty of nature will be revealed only when that understanding has been achieved . . . D'Arcy did away for all time with this Gothick nonsense: a clear bright light shines about the pages of *Growth and Form*, a most resolute determination to unmake mysteries'.

Unfortunately, any attempt to understand human nature and the processes of creative design may well be smothered by the Gothick mysteries which enveil the subject. This is exemplified by the literature on aesthetic measures and proportional systems, by what the Italian philosopher, Benedetto Croce, once called the 'astrology of number'. Typical examples of this mystification are Matila Ghyka's influential *Geometrical Composition and Design* (London, Tiranti, 1956), and a book by M. Borissavlievitch entitled *The Golden Number and the Scientific Aesthetics of Architecture* (London, Tiranti, 1958). An excellent refutation of this nonsense is to be found in Christopher Alexander's paper 'Perception and Modular Coordination' (*RIBA Journal*, October 1959, pp. 425–9). We have avoided overt aesthetic discussion in this book, and we hope that the chapters which do refer to proportional systems help to demystify the subject.

These then – the graphic, structural and aesthetic – are aspects of geometry and environment which our book is not about. Nor have we reviewed geometrical applications at the urban or regional scale. A stimulating series of essays, *Explorations into Urban Structure* (Philadelphia, University of Pennsylvania Press, 1964), edited by Melvin Webber, is a good start if only because Webber's own provocative essay questions the utility of geometrical concepts at this scale in an age of rapid transportation and near-instant communication. However, a recent book by Richard L. Morrill, *The Spatial Organization of Society* (Belmont, California, Wadsworth, 1970), covers the subject at the geographic scale very well indeed. Another book, Leslie J. King's *Statistical Analysis in Geography* (Englewood Cliffs, Prentice-Hall, 1969), is a first-class introduction to spatial distributions of a probabilistic nature among other related topics. A standard text on geometrical probability, with that title (*Geometrical Probability*, London, Charles Griffin, 1963), is by M. G. Kendall and P. A. P. Moran. Readers with a good background in mathematics may wish to follow developments in this area in the journal *Biometrika*.

We certainly recommend some of Christopher Alexander's essays. We have already advertised his *Notes on the Synthesis of Form* (Cambridge, Mass., Harvard University Press, 1964) as complementary to our own preoccupation with geometrical and mathematical formalism. But Alexander's later works should not be missed either: for example, 'A City is not a Tree', which appeared in *Architectural Forum* (vol. 122, no. 1, pp. 58–62; and no. 2, pp. 58–61, 1965), 'The Pattern of Streets' (*J. American Institute of Planners*, vol. 12, no. 5, 1966, pp. 273–8) and *The Atoms of Environmental Structure*, a Research and Development Paper written with B. Poyner and published by the Ministry of Public Building and Works, London, 1967.

In other subjects than architecture the 'quantitative revolution', as it has been called, is well-advanced. Richard Stone's essay 'Mathematics in the Social Sciences' (in: *Mathematics in the Social Sciences and Other Essays*, London, Chapman and Hall, 1966) is well worth reading for its uncommonly good sense. Stone is, perhaps, best known for his work in modelling the British economy: for deriving a system of mathematical expressions to simulate its ups and downs in past years. In geography, Chorley and Haggett have edited the Second Madingley Lectures (*Models in Geography*, London, Methuen, 1967) which describe the uses of mathematical models in a wide range of applications. Peter Haggett's own book, *Locational Analysis in Human Geography* (London, Methuen, 1965), provides a very catholic introduction to new techniques in this field, while *Network Analysis in Geography* (London, Edward Arnold, 1969), written with Richard Chorley, extends the subject matter more widely. In *Analytical Archaeology* (London, Methuen, 1968), David L. Clarke has reviewed comprehensively the recent impact of mathematics and computing methods on his field. It is tempting to substitute the word 'architecture' for archaeology in his opening remarks: 'Archaeology is an undisciplined empirical discipline. A discipline lacking a scheme of systematic and ordered study based upon declared and clearly defined models and rules of procedure. It further lacks a body of central theory capable of synthesizing the general regularities within its data in such a way that the unique residuals distinguishing each particular case might be quickly isolated and easily assessed. . . . Lacking an explicit theory defining entities and their relationships and transformations in a viable form, archaeology has remained an intuitive skill – an inexplicit manipulative dexterity learned by rote.'

It is now a decade ago since Edmund Leach, in *Rethinking Anthropology* (London, The Athlone Press, 1961), introduced us to the structuralist approach with this question: 'How can a modern anthropologist embark upon generalization with any hope of arriving at a satisfactory conclusion?' And his italicized answer: '*By thinking of the organiza-*

tional ideas that are present in any society as a mathematical pattern.' A stance qualified by James S. Coleman in his *Introduction to Mathematical Sociology* (New York, Free Press of Glencoe, 1964): 'This does not mean that mathematical methods developed for other sciences or as exercises in pure mathematics are not of use; to be sure, they are the only sources of mathematics that we have. It is rather to say that the necessarily difficult task of developing mathematical sociology can best be performed when our concentration remains upon the sociological problem, and the mathematical tools remain means to an end.' Which is, of course, as true of architecture as any other social science – or ought to be. On the 'uses' of pure mathematics and the nature of mathematical patterns, the reader is invited to look at G. H. Hardy's delightful *A Mathematician's Apology* (Cambridge University Press, 1967).

Apart from the areas we have mentioned, the application of mathematics in architectural design is just beginning. Our own impetus in this direction has gained momentum from parallel work by our colleagues at the centre for Land Use and Built Form Studies, Cambridge. This work is usually published in the form of Working Papers which are available from the centre. However, several substantial contributions are currently being prepared for press in a new series, *Cambridge Architectural Studies*, Cambridge University Press, which is due to start in 1971 under the editorship of Sir Leslie Martin and Lionel March. A special number of *Architectural Design*, May 1971, provides a conspectus of the centre's work on modelling, in the mathematical sense, the built environment and the activity systems which this accommodates. Three papers which make use of simple geometrical illustrations to challenge habitual assumptions concerning such things as high buildings, green belts, densities, and ribbon developments are Sir Leslie Martin's 'Architects' Approach to Architecture', May 1967, and his 'Education without Walls' of August 1968, and 'Homes beyond the Fringe' by Lionel March in August 1967. All these appeared in the *RIBA Journal* (vol. 74, no. 5, pp. 191–200; vol. 75, no. 7, pp. 356–61; vol. 74, no. 8, pp. 334–7, respectively).

Before turning to readings associated with specific chapters in this book, it may be useful to mention some introductory books on architecture for the non-architect and some elementary books on mathematics for the non-mathematician.

In general, we recommend the series of monographs published by George Braziller. The authors include Françoise Choay on *Le Corbusier*, Vincent Scully, Jr on *Frank Lloyd Wright*, Arthur Drexler on *Mies van der Rohe* and Frederick Gutheim on *Alvar Aalto* (all published in 1960). These volumes contain bibliographies which will help the reader follow up any particular interest. To Englishmen, the standard history of twentieth-century architecture is Nikolaus Pevsner's *Pioneers of Modern Design; from William Morris to Walter Gropius* (Harmondsworth, Penguin Books, 1970). An account which gives less emphasis to the English and more to the American contribution is Sigfried Giedion's *Space, Time and Architecture* (5th ed., Cambridge, Mass., Harvard University Press, 1967). Reyner Banham's *Theory and Design in the First Machine Age* (London, Architectural Press, 1960) provides a worthwhile guided tour round the Tower of Babel of architectural theories voiced in the first half of this century.

There are many books on the new mathematics now available. Our choice is bound to be arbitrary. In Britain the new mathematics has been fostered at primary school level by the Nuffield Mathematics Project which aims to change the whole attitude to the subject so that ' "Ugh, no, I didn't like maths", will be heard no more'. There is one Nuffield publication for young children which we ought to mention because of its name. The book is called *Environmental Geometry* (London, W. & R. Chambers and John Murray, 1969) and it includes such themes as shape and size constancy, and routes and directions. The numerous architectural examples are due in part to George Kasabov and a number of his first-year students at the Bartlett School of Architecture in London. It is an extremely attractive presentation which makes us feel optimistic for the long-term future of environmental understanding and design. In a growing number of primary schools, reception-class children now start their mathematical training with sets, while, in the experience of one of the authors, his daughter at six years' old found few difficulties in adding and multiplying matrices together as a game.

At the secondary level of mathematics, there are first-class text books related to the *School Mathematics Project* (Cambridge University Press) which was founded in 1961 to devise radically new mathematics courses which would reflect the up-to-date nature and usages of mathematics. In particular, Book 5 for O-level candidates, covers with admirable clarity and simplicity of presentation, together with numerous examples, many of the topics dealt with in the present book such as sets, relations, mappings, transformations, vectors and matrices, isometries and the uses of matrices to give algebraic expression to symmetry operations, as well as graphs and networks. The almost constant use of architectural or quasi-architectural problems for the sake of illustration is again quite striking. One chapter is called 'Plans and Elevations', another is in part devoted to problems of 'Heating a House', while a section on linear programming sets out the basic steps in the construction of mathematical models for a variety of practical planning problems. The four *Advanced Mathematics* books in this project continue the discussion of these topics up to A-level standard.

A very attractive introduction to the new thinking is *Mathematical Reflections* (Cambridge University Press, 1970), edited by members of the Association of Teachers of Mathematics. Of special interest to readers of the present book will be the essays on tesselations of polyominoes, the concept of mapping, transformation geometry, group calculus for geometry, enlargements, a chapter entitled 'Drawings and Representations', and a sparkling duologue by A. P. K. Cadwell called ' "I thought you were going to tell me about automorphisms." ' The latter concludes with the following statement on aesthetics and mathematics: 'We move from one space to another by making maps mapping one space into another. An artist maps from domains of ideas and emotions into ranges of experience using the patterns of his experience which he has mapped into his own mental and emotional spaces. It is his choice and use of a particular range of experience in which to express himself that is the aesthetic praxis. In the most general sense every theory is mathematical. I am trying to sketch a theory of aesthetics for you. . . . And whether you like a piece of mathematics or not is a matter of aesthetics.'

The New Mathematics by Irving Adler (New York, John Day Company, 1959), now available in paperback (London, New English Library, 1964), is addressed to the average reader who is curious about new developments in mathematics. It covers similar ground to our first four chapters and is well worth reading. A new book, *Ideas in Mathematics* by Avron Douglis (Philadelphia, W. B. Saunders, 1970) intentionally aims to convey the humane values of mathematics to non-mathematical students. It is a beautifully presented book for liberal arts students in which the earliest sections of each chapter and the beginning of almost every section have been designed to introduce their subjects gently for the benefit of 'the diffident or ill-prepared'. *An Introduction to Finite Mathematics* by J. G. Kemeny, J. L. Snell and G. L. Thompson (Englewood Cliffs, Prentice-Hall, 1957) provides a simple and clear account of matrices and vectors, among other subjects, with many interesting illustrations from the social sciences.

We would recommend the literature of 'recreational mathematics' where abstract mathematics often finds unexpected application. J. H. Cadwell in *Topics in Recreational Mathematics* (Cambridge University Press, 1966) does not only discuss the symmetry groups in a plane but he also looks at problems of stacking, nesting and space-filling, the Fibonacci sequence, the four-colour problem, procedures for dissecting plane and solid figures so that pieces may be reassembled to form other simple shapes, and some theorems related to combinatorial geometry.

Martin Gardner's regular column 'Mathematical Games' in the magazine *Scientific American* is always stimulating. Two anthologies compiled from that column are now available in paperback editions: *Mathematical Puzzles and Diversions*, London, Penguin, 1965, and *More Mathematical Puzzles and Diversions*, London, Penguin, 1965, which contain, as well as the chapter on 'Squaring the Square' by W. Tutte discussed in our Chapter 11, other sections devoted to 'Phi; the Golden Ratio' and some combinatorial problems in stacking cubelets – the so-called Soma Cube. For an absorbing example of combinatorial geometry in two and three dimensions of an architectural and modular kind, the reader may wish to glance at S. W. Golomb's *Polyominoes* (London, Allen and Unwin, 1966). 'Polyominoes' is a special coinage to denote dominoes consisting of more than two squares joined along their edges. Finally, one of the best general introductions to modern geometry is without doubt by H. S. M. Coxeter (*Introduction to Geometry*, New York, Wiley, 1961). We heartily recommend his treatment of the subject.

Chapters 1 to 3

Many of the relevant mathematical and architectural books for these chapters have already been discussed above. The most outstanding general essay on symmetry is by Hermann Weyl (*Symmetry*, Princeton University Press, 1952). This is essential reading for anyone interested in the subject. Weyl discusses a great variety of applications of the principle of symmetry in the arts, in 'inorganic' and organic nature, and the mathematical significance of symmetry. The Hungarian mathematician, Fejes Tóth (*Regular Figures*, Oxford, Pergamon Press, 1964), has produced a most comprehensive and well-illustrated account of regular figures, symmetry groups, and tesselations as well as numerous packing and covering problems. Both Weyl and Tóth reproduce plates from the great nineteenth-century collection of ornaments by the architect Owen Jones (*The Grammar of Ornament*, London, Day and Son, 1856). This classic work may be difficult to obtain, but once in the hand any effort expended in finding it will quickly be forgotten. *Egyptian Ornament* (London, Allan Wingate, 1963) contains over three hundred examples in full colour from the collection of Pavla Főrtová-Šámalová and displays the wonderful geometric imagination and inventiveness of their creators some three to four thousand years ago. An extraordinary catalogue, by Daniel Sheets Dye (*A Grammar of Chinese Lattice*, Cambridge, Mass., Harvard University Press, 1949) of windows and grilles constructed by Chinese workmen between 1000 BC and AD 1900 presents many other variations on the frieze and wallpaper groups. Many of these Chinese examples are reminiscent of the complicated lattice designs found in original Tudor lead glazing and ribbed ceilings sampled in *The Domestic Architecture of England during the Tudor Period* (London, Batsford, 1929) by Thomas Garner and Arthur Stratton.

The general notion of mapping and transformation is, of course, fundamental to geography. We have found applications and discussions in that field to be stimulating in our own work. Mention must be made of William Bunge's *Theoretical Geography* (Lund, C. W. K. Gleerup, 1966) and particularly his chapters on metacartography, on distance, nearness and geometry, and on the meaning of spatial relationships. J. P. Cole and C. A. M. King's *Quantitative Geography* (London, Wiley, 1968) is less philosophical and more practical, but its great merit lies in its numerous worked examples, and the large number of maps and diagrams which will commend the book to those who find visual presentation easy to follow. The chapter 'Mathematics' provides a quick summary of the relevance of new mathematics in geography. It covers much of the same material as in the present book, but naturally takes most of its illustrations from applications to spatial organization at a larger scale than the architectural. Their chapter concludes with a good bibliographical section on books about the new mathematics.

Chapters 4 to 7

Two paperbacks, A. E. Coulson's *An Introduction to Vectors* and *An Introduction to Matrices* (London, Longman, 1969) provide very lucid accounts of their topics and are well worth obtaining. Our approach to the description of shape, particularly rectangular configurations, was suggested by exercises in James F. Gray's *Sets, Relations and Functions* (New York, Holt, Rinehart and Winston, 1962). The notation in Chapter 6 is based on a paper by Lionel March (*A Set-Theoretic Description of Built Forms*, Working Paper 1, University of Cambridge, Land Use and Built Form Studies, 1970) and is meant to be suggestive only, and will not be found elsewhere. But, the concepts of 'interval' and 'vector' are well established in mathematics, and the idea of combining them would seem natural enough. What is not yet clear is whether useful and powerful theorems can be derived using this particular formalism. However, we hope that the potential elegance and economy of a set-theoretic approach to the mathematical description of rectangular spaces will be appreciated from our examples. The chapter on irregular polygons is based on a further paper by March (*The Description and Measurement of Irregular Polygons*, Working Paper 8, University of Cambridge, Land Use and Built Form Studies, 1970). The work was prompted originally by investigations into the shapes of plots of land and building sites in urban areas. The $2 \times n$ matrix representation of an n-gon was suggested by Michael Trace who, with March, derived the u, v formula for its area.

L. A. Lyusternik's *Convex Figures and Polyhedra* (New York, Dover, 1963) is available in paperback and is a fascinating introduction to the subject of convexity which, as the author remarks, 'offers gratifying material for the popularization of mathematics among younger students'. Another Soviet classic in this field, and available in English, is *Convex Figures* by I. M. Yaglom and V. G. Boltyanski (New York, Holt, Rinehart and Winston, 1961). We also recommend *Euclidean Geometry and Convexity* by Russell V. Benson (New York, McGraw-Hill, 1966). On the whole these three books are at undergraduate level.

Chapter 8

Bruce Martin's book *The Co-ordination of Dimensions for Building* (London, Royal Institute of British Architects, 1965) gives a short history of the subject of modular coordination, a brief and clear account of the theory, and an exhaustive bibliography, historically arranged, including articles and books published between 1936 and 1965. The first entry in this bibliography is, appropriately, the third volume of A. F. Bemis's *The Evolving House*, subtitled 'Rational Design' (Cambridge, Mass., Technology Press, M.I.T., 1936). Of the three volumes it is this one which contains the most interesting and important sections, on modular coordination and rationalization of housing design.

Some recent books and publications which cover the more theoretical and mathematical, as opposed to practical and technical aspects of the subject, include Ezra Ehrenkrantz's *The Modular Number Pattern* (London, Tiranti, 1956), which describes work carried out at the Building Research Station (a kit for the construction of a three-dimensional clear plastic model of the 'number pattern' of the title is included with the book): the reports of the European Productivity Agency Project (First and second reports, Paris OEEC, 1956 and 1961, respectively); and the 'Modular Primer' by E. Corker and A. Diprose which was published as a supplement to the *Modular Quarterly* (no. 1, 1963), the journal of the Modular Society, devoted to the discussion and promotion of modular design.

As an introduction to the principal ideas of number theory we could not do better than to recommend Tobias Dantzig's classic *Number; the Language of Science* (4th edition, London, Allen and Unwin, 1968) which is a fascinating and entertaining account of the historical development of the concept of number, which Dantzig describes as 'A Critical Survey Written for the Cultured Non-Mathematician'.

The main work on the application of the theory of numbers to problems of combinations of dimensions in building is a very different kind of book. This is P. H. Dunstone's single-minded, almost obsessional working out and presentation in tabular form of many hundreds of combinatorial possibilities for groups of two or three sizes (*Combinations of Numbers in Building*, London, Estates Gazette, 1965). Like *The Modular Number Pattern* there is a 'do-it-yourself' kit provided with the book, consisting of tables and 'combigraphs'. The basic theory of these effectively Diophantine problems is outlined by Dunstone, while formulations of some expressions relating critical numbers and numbers of combinations are given by J. H. Clarke in a paper 'Linear Diophantine Equations Applied to Modular Co-ordination' (*Australian Journal of Applied Science*, vol. 15, no. 4, 1964, pp. 345–8).

Chapter 9

The literature of architectural systems of proportion is enormous, and we do not intend to do any more here than to select one or two of the most important contributions to the subject. We have warned earlier of the dangers of numerological brain fever, with which some authors might infect the unwary reader, and prescribed Christopher Alexander's refreshing paper ('Perception and Modular Coordination', *RIBA Journal*, October 1959, pp. 425–9) as an antidote. Rudolf Wittkower's 'The Changing Concept of Proportion' (*Daedalus*, 1960, pp. 199–215) is another such tonic; and certainly Wittkower's masterly and influential study of Renaissance systems of proportion and their significance in relation to musical theory and philosophical thought, *Architectural Principles in the Age of Humanism* (London, Tiranti, 1962) should not be missed out.

Perhaps even more influential on a modern generation of architects has been Le Corbusier's *The Modulor* (London, Faber & Faber, 1954) and its sequel *Modulor 2* (Boulogne sur Seine, Editions de L'Architecture D'Aujourd'hui, 1956). These volumes tell, in Le Corbusier's inimitable style, the story of how the Modulor system was invented and applied, what the world thought of it, how its dimensions cropped up in unexpected places, and much more. The most comprehensive and level-headed treatment of the whole history of proportions in architecture, from Classical and Renaissance to nineteenth-century and modern examples is P. H. Scholfield, *The Theory of Proportion in Architecture* (Cambridge University Press, 1958). This has a mathematical appendix and a full bibliography.

Chapters 10 and 11

'The number of books on graph theory is very small,' says Oystein Ore, whose own little book *Graphs and Their Uses* (New York, Random House, 1963) is the best short introduction to the subject. Since it is said that the beginnings of graph theory came from Euler's discussion of the Königsberg bridge problem, it is an odd coincidence that two of the major books on graph theory should be by the German author D. König and the Frenchman Claude Berge. Of these two, Berge's book *Théorie des Graphes et ses Applications*, is available in English translation (*The*

Theory of Graphs and its Applications, London, Methuen, 1962) and covers the whole subject very fully. The reader is warned that there is some variation in terminology between authors, and in particular that Ore uses 'arc' to mean a special form of what is in Berge's terms a 'chain'. The terms 'circuit' and 'path' are also used somewhat differently by Ore (who provides a glossary of definitions). And other authors use 'node' and 'link' for 'vertex' and 'edge'. We have chosen to follow Berge for terminology throughout.

Two other books which are strongly recommended, since besides giving theoretical treatment they both illustrate a large variety of applications of graph theory, are Harary, Norman and Cartwright's *Structural Models* (New York, Wiley, 1965)—which the authors describe as an introduction to the use of mathematical models in the social sciences—and R. G. Busacker and T. L. Saaty, *Finite Graphs and Networks; an Introduction with Applications* (New York, McGraw-Hill, 1965). The latter covers applications in human science, computing science, operations research, various combinatorial problems, and in the solution and study of puzzles and games.

Three papers cover specifically the use of planar graphs in problems of architectural plan layout. These are P. H. Levin's 'The Use of Graphs to Decide the Optimum Layout of Buildings' (*The Architects' Journal*, 7 October, 1964, pp. 809–15) which describes a 'pencil and paper' method; 'Computer-Aided Plant Layout' by M. Krejcirik (*Computer-Aided Design*, Autumn 1969, pp. 7–19); and an article in the Canadian magazine *Habitat* (vol. 12, no. 2, 1969, pp. 13–18) by Jean Cousin entitled 'Architecture et Topologie' which has been republished in English translation in *Architectural Design* ('Topological Organization of Architectural Space', pp. 491–3) in October 1970.

Chapter 11 follows closely the argument of a recent paper by Steadman (*The Automatic Generation of Minimum-Standard House Plans*, Working Paper 23, University of Cambridge, Land Use and Built Form Studies, 1970). Since the main text of this chapter was written our attention has been drawn to the work at Carnegie-Mellon University in Pittsburgh of John Grason, who has taken an extraordinarily similar approach but has pursued it much further and has produced a series of computer programs which achieve substantially the aims for an automatic layout system which are only tentatively outlined here. Grason's work has so far only been published in the USA in working papers and in a doctoral thesis 'Methods for the Computer-Implemented Solution of a Class of "Floor Plan" Design Problems' (submitted 1970) which describes his method at considerable length (xerox copies available on demand from University Microfilms, Ann Arbor, Michigan, USA). But his work in this area predates the publication of Steadman's paper, and the thesis offers a complete demonstration and confirmation of the potential of the approach, some aspects only of which we have proposed independently here. The account of the discoveries made by the 'Important Members' in the process of their successful attempt at 'squaring the square', given by W. Tutte in Martin Gardner's second puzzle book, has already been mentioned.

The constraints imposed on the rectangular plan perimeter and on the minimum dimensions and proportions of rooms in the hypothetical house planning examples of Chapter 11, may be compared with the recommendations made in actual practice for dimensional standards in public authority housing in a number of official British government and government agency publications. These include the report of the Parker Morris Committee on housing, made in 1961 (Ministry of Housing and Local Government, *Homes for Today and Tomorrow*, London, HMSO, 1961), the National Building Agency's *Generic Plans* and *Metric House Shells* (London, National Building Agency, 1969 and 1968, respectively), and the former Ministry of Housing and Local Government's *Space in the Home* (London, HMSO, 1969).

Chapters 12 to 14

Much of the substance of our last three chapters, 12 to 14, – in particular the account given of various experiments in Chapter 14 – is drawn from the work of our colleague, Philip Tabor, at the centre for Land Use and Built Form Studies, Cambridge. His series of papers published by the centre under the general title of 'Traffic in Buildings', and referred to repeatedly in our main text already, cover 'Systematic Activity-Location', 'Analysis of Communication Patterns' and 'Evaluation of Routes' (Working Papers 18–20, University of Cambridge, Land Use and Built Form Studies, 1970). A review paper 'Pedestrian Circulation' (Working Paper 17, 1970) which covers the whole subject at a more general level, is in turn based partly on our Chapters 12 and 13 here, in what must almost by now amount to a case of bibliographical incest. The material contained in all four of Tabor's papers is also to be brought together in one of the volumes – scheduled for publication in 1971 – of the *Cambridge Architectural Studies* series already mentioned.

Apart from this, the literature of systematic design procedures based on circulation criteria consists mainly of technical papers scattered through a wide variety of specialized journals, or else is to be found in doctoral theses; and has not otherwise been collected together in more accessible form. We have given references to a number of the most important papers in the main text, but we would recommend again two papers specially, as representing the first and most basic statements of the 'permutational' and 'additive' methods respectively. These are Armour and Buffa's original paper published in *Management Science* (vol. 9, no. 2, 1963, pp. 294–309), and the paper of Whitehead and Eldars in *The Architects' Journal* (17 June 1964, pp. 1373–80).

One exception to these general remarks about research papers is Souder, Clark, Elkind and Brown's substantial book on hospital planning (*Planning for Hospitals*, Chicago, American Hospital Association, 1964) which, though perhaps it has not had a wide circulation outside the USA, does describe a very complete planning system with elaborate computer aids, including the use of computer generated graphic displays. An interesting simulation model of the frequency of pedestrian trips is proposed, plans may be laid out on the cathode ray screen using a light-pen and trips routed through these plans, and the operation of lift systems is simulated, as we have mentioned in Chapter 14. More generally, Britton Harris has used some of the systematic 'activity-location' procedures described in Chapter 13 to illustrate some broad points about the design and planning processes, in a stimulating essay published in *Architectural Design* (June 1970, pp. 315–6: 'One case of computer optimization related to design method').

P. Haggett and R. J. Chorley's *Network Analysis in Geography*, referred to above, is an excellent review of many topological and geometric network problems, and covers in detail the questions of shortest paths and network structure which we touch on in Chapter 14. This book, too, contains a lengthy bibliography which will lead the interested reader on into the related geographical and operations research literature. Especially recommended in the geographical context are William Garrison's paper on the 'Connectivity of the Interstate Highway System' (*Papers and Proceedings of the Regional Science Association*, vol. 6, 1960, pp. 121–37) and K. J. Kansky's book *Structure of Transportation Networks* (University of Chicago, Department of Geography Research Paper no. 84, 1963), both of which use graph theoretical measures to express various properties and characteristics of actual transport network systems. The basic graph theory textbooks mentioned previously in relation to Chapters 10 and 11 cover network and shortest path problems. The concept of 'distance' at the geographical scale is the subject of an excellent review and bibliography by Gunnar Olsson (*Distance and Human Interaction, a Review and Bibliography*, Philadelphia, Regional Science Association, 1965).

Mathematical notation

General symbols

$x \equiv y$	x is identical to y
$x = y$	x is equal to y
$x \neq y$	x is not equal to y
$x \doteq y$	x is approximately equal to y
$x \overset{\text{e}}{-} y$	x is equivalent to y
$x \to y$	x tends to y (also x maps onto y, see below)
$x < y$	x is less than y
$x \leqslant y$	x is less than or equal to y
$x \dagger y$	general composition of x and y
$x + y$	addition of x and y
$x.y,\ xy$	product of x and y
x/y	division of x by y
$x \Rightarrow y$	statement x implies statement y
$x \Leftrightarrow y$	statement x is equivalent to statement y
$+\sqrt{x}$	positive square root of x
$\lvert x \rvert$	absolute value, or positive value, of x
$\ln x$	natural logarithm of x
$\sin x$, $(\cos x)$	(co)sine of angle x
arc $\sin x$, (arc $\cos x$)	the angle whose (co)sine is x
$\Sigma_i\, x_i$	sum of all terms x_i for relevant values of i
$n!$	factorial n, $n(n-1)(n-2) \dots 2.1$, for integral n
$\binom{n}{r}$	number of r-combinations of n things, $n!/r!(n-r)!$
$n \bmod r$	residue modulo r

Page numbers indicate where the symbol is first used or defined.

Sets

		Page
$\{x \mid x \text{ has properties } y\}$	set of all elements x for which the statement 'x has properties y' is true	16
$\{x, y, \dots\}$	set with listed elements (no order implied)	15
$(x, y, \dots)$	set with ordered elements	148
$x \to y$	element x maps onto element y	13
$M: X \to Y$	mapping of set X onto Y	16
N	set of all natural numbers, 1, 2, 3, ...	17

Z	set of all integers, 0, ± 1, ± 2, ...	17
Z_+	set of all positive integers and zero, 0, 1, 2, ...	17
Z_{+a}	set of all positive integers a and above, a, $a + 1$, $a + 2$, ...	209
Q	set of rational numbers, m/n such that $m, n \in Z$	17
Q'	set of all irrational numbers, $x \in R$ but $x \notin Q$	17
R	set of all real numbers	16
R_+	set of all positive real numbers	206
R	set of points on a line in one dimension	16
R^2	set of points in a plane in two dimensions	150
R^3	set of points in three-dimensional space	150
A, B	typical sets	124
$a \in A$	element a exists in set A	15
$a \notin A$	element a does not exist in set A	15
I	universal set	125
$\emptyset$	null set	125
A'	complement of set A	125
$A \subset B$	set A contained in, is a sub-set of, B	125
$A \not\subset B$	set A is not contained in B	147
$A \cup B$	union of sets A and B (cup)	124
$A \cap B$	intersection of sets A and B (cap)	124
$\bigcup_{i=1}^{n} A_i$	multiple union of sets A_1, A_2, ... A_n	127
AB	cartesian product of sets A and B	148
$P(A)$	power set of A	156
$N(A, r)$	circular neighbourhood of a set of points A	164

Transformations and symmetry groups

$\mathbf{I}$	identity transformation	44
$\mathbf{T}$	translation	43
$\mathbf{T}_2 . \mathbf{T}_1$	product of two translations	43
$\mathbf{T}^n$	n repeated translations	50
$\mathbf{T}^{-1}$	inverse translation	44
$\mathbf{S}$	general rotation	46
$\mathbf{S}_\theta$	rotation through θ	113
$\mathbf{S}(O)$	rotation about O	46
$\mathbf{S}_2\mathbf{S}_1$	product of two rotations	46
$\mathbf{S}^n$	n repeated rotations	52
$\mathbf{H}$	halfturn, rotation through π	51
$\mathbf{R}$	reflection	48
$\mathbf{R}_2\mathbf{R}_1$	product of two reflections	48
$\mathbf{R}^{-1}$	inverse reflection	49
$\mathbf{G}$	glide reflection	52
C_n	cyclic group of order n (Figure 3.1)	52, 56
D_n	dihedral group of order n (Figure 3.1)	50, 56
F_1^1, etc.	frieze groups (Figure 3.3)	62
W_1^1, etc.	wallpaper groups (Figures 3.8 to 3.12)	64 ff.

Matrices and vectors

$\mathbf{a} = \begin{bmatrix} a_1 \\ a_2 \\ \vdots \\ a_n \end{bmatrix} = [a_i]$	n-dimensional column vector	90
$\mathbf{a}^{-T} = [a_1\, a_2 \ldots a_n]$	transpose of **a**, n-dimensional row vector	91
$\mathbf{a}.\mathbf{b}$	inner product of vectors **a** and **b**	94
i, **j**	orthogonal unit vectors	108
$\mathbf{e}_j$	elementary vector with jth element equal to 1 and all others zero	117
$\lVert\mathbf{a}\rVert_1$	taxicab norm, $a_1 + a_2$	188
$\lVert\mathbf{a}\rVert_2$	euclidean norm, $+\sqrt{(a_1^2 + a_2^2)}$, the length of vector **a**	109
$\overline{AA'}$	geometrical vector from point A to A'	42, 105
$\mathbf{A} = [a_{ij}]$	matrix with element a_{ij} in the ith row and jth column	101
$\mathbf{A}_{mn}$	matrix with m rows and n columns	101
det **A**	determinant of a square matrix	104
I	identity matrix	103
O	null matrix	150
$\mathbf{E}_{ij}$	2×2 zero-one matrix	150
$\mathbf{P}j_1 j_2 j_3 \ldots j_n$	permutation matrix	117
R	reflection matrix	186
S	spin, or rotation, matrix	184

Components, panels and blocks

$\langle i \rangle$	modular component	124
$\langle i; j \rangle$	modular panel	128
$\langle i; j; k \rangle$	modular block	129
$[a_1, a_2]$	component A	145
$[a_1, a_2; b_1, b_2]$	panel AB	151
$[a_1, a_2; b_1, b_2; c_1, c_2]$	block ABC	168
d A	dimension, or length, of component A	127
cm A	centre of mass, or centroid, of component A	127

Graphs

X	set of vertices	263
x	typical vertex	263
$G(X, g)$	graph which is a mapping, g, of a set of vertices X into itself	263
$g{:}x$	subset of vertices X associated with vertex x	263
(x, y)	edge	265
$[x, y]$	arc	263
$(x, y, \ldots z)$	chain	266
$[x, y, \ldots z]$	path	266
$(x, y, \ldots z, x)$	cycle	266
$[x, y, \ldots z, x]$	circuit	266
K_5	Kuratowski sub-graph with 5 vertices	256
$K_{3,3}$	Kuratowski sub-graph with 6 vertices	256

Index

Numbers in *italic* indicate pages on which figures fall.